*Introduction to Soil Physics*

# Introduction to Soil Physics

## DANIEL HILLEL

*Department of Plant and Soil Sciences*
*University of Massachusetts*
*Amherst, Massachusetts*

Academic Press
San Diego  New York  Boston
London  Sydney  Tokyo  Toronto

Academic Press, Inc.
*A Division of Harcourt Brace & Company*
525 B Street, Suite 1900, San Diego, California 92101-4495

*United Kingdom Edition published by*
ACADEMIC PRESS LIMITED
24-28 Oval Road, London NW1 7DX

Library of Congress Cataloging in Publication Data

Hillel, Daniel.
    Introduction to soil physics.

    Based on the author's Applications of soil
physics and Fundamentals of soil physics.
    Bibliography: p.
    Includes index.
    1. Soil physics.   I. Title
S592.3.H55        631.4'3        81-10848
ISBN  0-12-348520-7              AACR2

PRINTED IN THE UNITED STATES OF AMERICA
    96  97  QW  9  8  7

*This book is only a small clearing at the edge of the woods*
*where students might observe a few of the trees*
*as they prepare to set out independently*
*to explore the great forest which yet lies beyond.*

# Contents

## Part IV   THE GASEOUS PHASE

## 8.  Soil Air and Aeration

## Part V   COMPOSITE PROPERTIES AND BEHAVIOR

## 9.  Soil Temperature and Heat Flow

## 10.  Soil Compaction and Consolidation

## *Part VI   THE FIELD–WATER CYCLE AND ITS MANAGEMENT*


## 12.   Infiltration and Surface Runoff

## 13.   Internal Drainage and Redistribution Following Infiltration

## 14.   Groundwater Drainage

## 15.  Evaporation from Bare-Surface Soils

## 16.  Uptake of Soil Moisture by Plants

## 17.  Water Balance and Energy Balance in the Field

## Bibliography

# Preface

This book is a unified, condensed, and simplified version of the recently issued twin volumes, "Fundamentals of Soil Physics" and "Applications of Soil Physics." It is meant to serve as a textbook for undergraduate students in the agronomic, horticultural, silvicultural, environmental, and engineering sciences. Nonessential topics and complexities have been deleted, and little prior knowledge of the subject is assumed. An effort has been made to provide an elementary, readable, and self-sustaining description of the soil's physical properties and of the manner in which these properties govern the processes taking place in the field. Consideration is given to the ways in which the soil's processes can be influenced, for better or for worse, by man. Sample problems are provided in an attempt to illustrate how the abstract principles embodied in mathematical equations can be applied in practice. The author hopes that the present version will be still more accessible to students than its precursors and that it might serve to arouse their interest in the vital science of soil physics.

*Daniel Hillel*
*Amherst, Massachusetts*

# Part I    BASIC RELATIONSHIPS

# 1    *The Task of Soil Physics*

The soil beneath our feet is the basic substrate of all terrestrial life. The intricate and fertile mix composing the soil, with its special life-giving attributes, is a most intriguing field of study. The soil serves not only as a medium for plant growth and for microbiological activity per se but also as a sink and recycling factory for numerous waste products which might otherwise accumulate to poison our environment. Moreover, the soil supports our buildings and provides material for the construction of earthen structures such as dams and roadbeds.

The attempt to understand what constitutes the soil and how it operates within the overall biosphere, which is the essential task of soil science, derives both from the fundamental curiosity of man, which is his main creative impulse, and from urgent necessity. Soil and water are, after all, the two fundamental resources of our agriculture, as well as of our natural environment. The increasing pressure of population has made these resources scarce or has led to their abuse in many parts of the world. Indeed, the necessity to manage these resources efficiently on a sustained basis is one of the most vital tasks of our age.

That knowledge of the soil is imperative to ensure the future of civilization has been proven repeatedly in the past, at times disastrously. In many regions we find shocking examples of once-thriving agricultural fields reduced to desolation by man-induced erosion or salinization resulting from injudicious management of the soil–water system. Add to that the shortsighted depletion of unreplenished water resources as well as the dumping of poisonous wastes—and indeed we see a consistent pattern of mismanagement. In view of the population–environment–food crisis facing the world, we can ill afford to continue squandering and abusing such precious resources.

*3*

The soil itself is of the utmost complexity. It consists of numerous solid components (mineral and organic) irregularly fragmented and variously associated and arranged in an intricate geometric pattern that is almost indefinably complicated. Some of the solid material consists of crystalline particles, while some consists of amorphous gels which may coat the crystals and modify their behavior. The adhering amorphous material may be iron oxide or a complex of organic compounds which attaches itself to soil particles and binds them together. The solid phase further interacts with the fluids, water and air, which permeate soil pores. The whole system is hardly ever in a state of equilibrium, as it alternately wets and dries, swells and shrinks, disperses and flocculates, compacts and cracks, exchanges ions, precipitates and redissolves salts, and occasionally freezes and thaws.

To serve as a favorable medium for plant growth, the soil must store and supply water and nutrients and be free of excessive concentrations of toxic factors. The soil–water–plant system is further complicated by the facts that plant roots must respire constantly and that most terrestrial plants cannot transfer oxygen from their aerial parts to their roots at a rate sufficient to provide for root respiration. Hence the soil itself must be well aerated, by the continuous exchange of oxygen and carbon dioxide between the air-filled pores and the external atmosphere. An excessively wet soil will stifle roots just as surely as an excessively dry soil will desiccate them.

These are but a few of the issues confronting the relatively new science of soil physics, a field of study which has really come into its own only in the last generation. Definable as the study of the state and transport of all forms of matter and energy in the soil, soil physics is an inherently complex subject, a fact which may account for its rather late development.

Our present-day knowledge of the soil physical system is still rather fragmentary. Hence, we continue to search and re-search for answers to the numerous newly arising questions. The business, and fun, and occasional agony of science is the continuing endeavor to achieve a coordinated understanding and explanation of observable phenomena without ever resting on yesterday's conclusions. Consequently a valid book on soil physics should reflect the complexity of the system even while attempting to present a coordinated and logical description of what is admittedly only a partial knowledge of it.

# 2    *General Physical Characteristics of Soil*

## A. Introduction

The term *soil* refers to the weathered and fragmented outer layer of the earth's terrestrial surface. It is formed initially through disintegration and decomposition of rocks by physical and chemical processes, and is influenced by the activity and accumulated residues of numerous species of microscopic and macroscopic plants and animals. The physical weathering processes which bring about the disintegration of rocks into small fragments include expansion and contraction caused by alternating heating and cooling, stresses resulting from freezing and thawing of water and the penetration of roots, and scouring or grinding by abrasive particles carried by moving ice or water and by wind. The chemical processes tending to decompose the original minerals in the parent rocks include hydration, oxidation and reduction, solution and dissociation, immobilization by precipitation or removal of components by volatilization or leaching, and various physico-chemical exchange reactions. The loose products of these weathering processes are often transported by running water, glaciers, or wind, and deposited elsewhere.

Soil formation processes continue beyond the initial weathering of rocks and minerals. In the course of soil development, the original character of the material is further modified by the formation of secondary minerals (e.g., clay minerals) and the growth of organisms which contribute organic matter and bring about a series of ongoing physicochemical and biochemical reactions in addition to those experienced by the original mineral material.

The process of soil development culminates in the formation of a characteristic *soil profile*, to be described later in this chapter.


## B. Soil Physics Defined

Throughout this book we shall be considering the soil from the viewpoint of soil physics, which can be described as the branch of soil science dealing with the physical properties of the soil, as well as with the measurement, prediction, and control of the physical processes taking place in and through the soil. As physics deals with the forms and interrelations of matter and energy, so soil physics deals with the state and movement of matter and with the fluxes and transformations of energy in the soil.

On the one hand, the fundamental study of soil physics aims at achieving a basic understanding of the mechanisms governing the behavior of the soil and its role in the biosphere, including such interrelated processes as the terrestrial energy exchange and the cycles of water and transportable materials in the field. On the other hand, the practice of soil physics aims at the proper management of the soil by means of irrigation, drainage, soil and water conservation, tillage, aeration and the regulation of soil heat, as well as the use of soil material for engineering purposes. Soil physics is thus seen to be both a basic and an applied science with a very wide range of interests, many of which are shared by other branches of soil science and by other interrelated sciences including terrestrial ecology, hydrology, microclimatology, geology, sedimentology, botany, and agronomy. Soil physics is likewise closely related to the engineering profession of soil mechanics, which deals with the soil mainly as a building and support material.


## C. Soil as a Disperse Three-Phase System

Natural systems can consist of one or more substances and of one or more phases. A system comprised of a single substance is also monophasic if its physical properties are uniform throughout. An example of such a system is a body of water consisting entirely of uniform ice. Such a system is called homogeneous. A system comprised of a single chemical compound can also be heterogeneous if that substance exhibits different properties in different regions of the system. A region inside a system which is internally uniform physically is called a phase. A mixture of ice and water, for instance, is chemically uniform but physically heterogeneous, as it includes two phases. The three ordinary phases in nature are the solid, liquid, and gaseous phases. A system containing several substances can also be monophasic. For example, a solution of salt and water is a homogeneous liquid. A system of

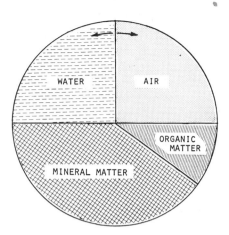

**Fig. 2.1.** Schematic composition (by volume) of a medium-textured soil at a condition considered optimal for plant growth. Note that the solid matter constitutes 50% and the pore space 50% of the soil volume, with the latter divided equally between water and air. The arrows indicate that these components can vary widely, and in particular that water and air are negatively related so that an increase in one is associated with a decrease of the other.

several substances can obviously also be heterogeneous. In a heterogeneous system the properties differ not only between one phase and another, but also among the internal parts of each phase and the boundary between the phase and its neighboring phase or phases. Interfaces between phases exhibit specific phenomena resulting from the interaction of the phases. The importance of these phenomena, which include adsorption, surface tension, and friction, depends on the magnitude of the interfacial area per unit volume of the system. Systems in which at least one of the phases is subdivided into numerous minute particles, which together exhibit a very large interfacial area per unit volume, are called *disperse systems*. Colloidal sols, gels, emulsions, and aerosols are examples of disperse systems.

The soil is a heterogeneous, polyphasic, particulate, disperse, and porous system, in which the interfacial area per unit volume can be very large. The disperse nature of the soil and its consequent interfacial activity give rise to such phenomena as adsorption of water and chemicals, ion exchange, adhesion, swelling and shrinking, dispersion and flocculation, and capillarity.

The three phases of ordinary nature are represented in the soil as follows: the solid phase constitutes the *soil matrix*; the liquid phase consists of soil water, which always contains dissolved substances so that it should properly be called the *soil solution*; and the gaseous phase is the *soil atmosphere*. The solid matrix of the soil includes particles which vary in chemical and mineralogical composition as well as in size, shape, and orientation. It also contains amorphous substances, particularly organic matter which is attached to the mineral grains and often binds them together to form aggregates. The organization of the solid components of the soil determines the geometric characteristics of the pore spaces in which water and air are transmitted and retained. Finally, soil water and air vary in composition, both in time and in space.

The relative proportions of the three phases in the soil vary continuously, and depend upon such variables as weather, vegetation, and management. To give the reader some general idea of these proportions, we offer the rather simplistic scheme of Fig. 2.1, which represents the volume composition of a medium-textured soil at a condition considered to be approximately optimal for plant growth.

## D. Volume and Mass Relationships of Soil Constituents

Let us now consider the volume and mass relationships among the three phases, and define some basic parameters which have been found useful in characterizing the physical condition of a soil.

Figure 2.2 is a schematic representation of a hypothetical soil showing the volumes and masses of the three phases in a representative sample. The masses of the phases are indicated on the right-hand side: the mass of air $M_a$, which is negligible compared to the masses of solids and water; the mass of water $M_w$; the mass of solids $M_s$; and the total mass $M_t$. These masses can also be represented by their weights (the product of the mass and the gravitational acceleration). The volumes of the same components are

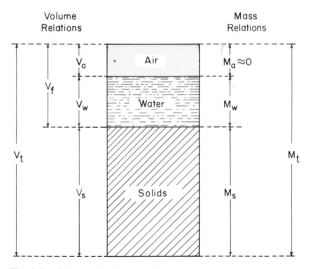

**Fig. 2.2.**  Schematic diagram of the soil as a three-phase system.

indicated on the left-hand side of the diagram: volume of air $V_a$, volume of water $V_w$, volume of pores $V_f = V_a + V_w$, volume of solids $V_s$, and the total volume of the representative soil body $V_t$.

On the basis of this diagram, we can now define terms which are generally used to express the quantitative interrelations of the three primary soil constituents.

### 1. DENSITY OF SOLIDS (MEAN PARTICLE DENSITY) $\rho_s$

$$\rho_s = M_s/V_s \qquad (2.1)$$

In most mineral soils, the mean density of the particles is about 2.6–2.7 $gm/cm^3$, and is thus close to the density of quartz, which is often prevalent in sandy soils. Aluminosilicate clay minerals have a similar density. The presence of iron oxides, and of various heavy minerals, increases the average value of $\rho_s$, whereas the presence of organic matter lowers it. Sometimes the density is expressed in terms of the *specific gravity*, being the ratio of the density of the material to that of water at 4°C and at atmospheric pressure. In the metric system, since the density of water at standard temperature is assigned the value of unity, the specific gravity is numerically (though not dimensionally) equal to the density.

### 2. DRY BULK DENSITY $\rho_b$

$$\rho_b = M_s/V_t = M_s/(V_s + V_a + V_w) \qquad (2.2)$$

The dry bulk density expresses the ratio of the mass of dried soil to its total volume (solids and pores together). Obviously, $\rho_b$ is always smaller than $\rho_s$, and if the pores constitute half the volume, $\rho_b$ is half of $\rho_s$, namely 1.3–1.35 $gm/cm^3$. In sandy soils, $\rho_b$ can be as high as 1.6, whereas in aggregated loams and in clay soils, it can be as low as 1.1 $gm/cm^3$. The bulk density is affected by the structure of the soil, i.e., its looseness or degree of compaction, as well as by its swelling and shrinkage characteristics, which are dependent upon clay content and wetness. Even in extremely compacted soil, however, the bulk density remains appreciably lower than the particle density, since the particles can never interlock perfectly and the soil remains a porous body, never completely impervious.

### 3. POROSITY $f$

$$f = V_f/V_t = (V_a + V_w)/(V_s + V_a + V_w) \qquad (2.3)$$

The porosity is an index of the relative pore volume in the soil. Its value generally lies in the range 0.3–0.6 (30–60%). Coarse-textured soils tend to be less porous than fine-textured soils, though the mean size of individual pores is greater in the former than in the latter. In clayey soils, the porosity is highly variable as the soil alternately swells, shrinks, aggregates, disperses, compacts, and cracks. As generally defined, the term porosity refers to the volume fraction of pores, but this value should be equal, on the average, to the areal porosity (the fraction of pores in a representative cross-sectional area) as well as to the average lineal porosity (being the fractional length of pores along a straight line passing through the soil in any direction). The total porosity, in any case, reveals nothing about the *pore size distribution*, which is itself an important property to be discussed in a later section.

### 4. VOID RATIO *e*

$$e = (V_a + V_w)/V_s = V_f/(V_t - V_f) \qquad (2.4)$$

The void ratio is also an index of the fractional volume of soil pores, but it relates that volume to the volume of solids rather than to the total volume of soil. The advantage of this index over the previous one ($f$) is that a change in pore volume changes the numerator alone, whereas a change of pore volume in terms of the porosity will change both the numerator and denominator of the defining equation. Void ratio is the generally preferred index in soil engineering and mechanics, whereas porosity is the more frequently used index in agricultural soil physics. Generally, *e* varies between 0.3 and 2.0.

### 5. SOIL WETNESS

The wetness, or relative water content, of the soil can be expressed in various ways: relative to the mass of solids, relative to the total mass, relative to the volume of solids, relative to the total volume, and relative to the volume of pores. The various indexes are defined as follows (the most commonly used are the first two).

a. *Mass Wetness w*

$$w = M_w/M_s \qquad (2.5)$$

This is the mass of water relative to the mass of dry soil particles, often referred to as the *gravimetric water content*. The term *dry soil* is generally defined as a soil dried to equilibrium in an oven at 105°C, though clay will often retain appreciable quantities of water at that state of dryness. Mass wetness is sometimes expressed as a decimal fraction but more often as a percentage. Soil dried in "ordinary" air will generally contain several per cent more water than oven-dry soil, a phenomenon due to vapor adsorption and often referred to as soil *hygroscopicity*. In a mineral soil that is saturated, *w* can range between 25 and 60% depending on the bulk density. The saturation water content is generally higher in clayey than in sandy soils. In the case of organic soils, such as peat or muck, the saturation water content on the mass basis may exceed 100%.

b. *Volume Wetness* $\theta$

$$\theta = V_w/V_t = V_w/(V_s + V_f) \qquad (2.6)$$

The volume wetness (often termed volumetric water content or volume fraction of soil water) is generally computed as a percentage of the total volume of the soil rather than on the basis of the volume of particles alone. In sandy soils, the value of $\theta$ at saturation is on the order of 40–50%; in medium-textured soils, it is approximately 50%; and in clayey soils, it can approach 60%. In the latter, the relative volume of water at saturation can exceed the porosity of the dry soil, since clayey soils swell upon wetting. The use of $\theta$ rather than of $w$ to express water content is often more convenient because it is more directly adaptable to the computation of fluxes and water quantities added to soil by irrigation or rain and to quantities subtracted from the soil by evapotranspiration or drainage. Also, $\theta$ represents the depth ratio of soil water, i.e., the depth of water per unit depth of soil.

c. *Degree of Saturation* $s$

$$s = V_w/V_f = V_w/(V_a + V_w) \qquad (2.7)$$

This index expresses the volume of water present in the soil relative to the volume of pores. The index $s$ ranges from zero in dry soil to unity (or 100%) in a completely saturated soil. However, complete saturation is seldom attained, since some air is nearly always present and may become trapped in a very wet soil.

### 6. Air-Filled Porosity (Fractional Air Content) $f_a$

$$f_a = V_a/V_t = V_a/(V_s + V_a + V_w) \qquad (2.8)$$

This is a measure of the relative air content of the soil, and as such is an important criterion of soil aeration. The index is related negatively to the degree of saturation $s$ (i.e., $f_a = f - s$).

### 7. Additional Interrelations

From the basic definitions given, it is possible to derive the relation of the various parameters to one another. The following are some of the most useful interrelations.

(1) Relation between porosity and void ratio:

$$e = f/(1 - f) \qquad (2.9)$$

$$f = e/(1 + e) \qquad (2.10)$$

(2) Relation between volume wetness and degree of saturation:

$$\theta = sf \tag{2.11}$$

$$s = \theta/f \tag{2.12}$$

(3) Relation between porosity and bulk density:

$$f = (\rho_s - \rho_b)/\rho_s = 1 - \rho_b/\rho_s \tag{2.13}$$

$$\rho_b = (1 - f)\rho_s \tag{2.14}$$

(4) Relation between mass wetness and volume wetness:

$$\theta = w\rho_b/\rho_w \tag{2.15}$$

$$w = \theta\rho_w/\rho_b \tag{2.16}$$

Here $\rho_w$ is the density of water ($M_w/V_w$), approximately equal to 1 gm/cm$^3$. Since the bulk density $\rho_b$ is generally greater than water density $\rho_w$, it follows that volume wetness exceeds mass wetness (the more so in compact soils of higher bulk density).

(5) Relation between volume wetness, fractional air content, and degree of saturation:

$$f_a = f - \theta = f(1 - s) \tag{2.17}$$

$$\theta = f - f_a \tag{2.18}$$

A number of these relationships are derived or proven at the end of this chapter, and the derivation or proof of the others is left as a useful exercise for students. Of the various parameters defined, the most commonly used in characterizing soil physical properties are the porosity $f$, bulk density $\rho_b$, volume wetness $\theta$, and mass wetness $w$.

## E. The Soil Profile

Having defined the soil's components and their proportions, let us now consider a composite soil body as it appears in nature. The most obvious, and very important, part of the soil is its surface zone. An examination of that zone will reveal much about processes taking place through the surface, but will not necessarily reveal the character of the soil as a whole. To get at the latter, we must examine the soil in depth, and we can do this, for instance, by digging a trench and sectioning the soil from the surface downward. The vertical cross section of the soil is called the *soil profile*.

The soil profile is seldom uniform in depth, and typically consists of a succession of more-or-less distinct layers, or strata. Such layers may result from the pattern of deposition, or sedimentation, as can be observed in wind-deposited (aeolian) soils and particularly in water-deposited (alluvial) soils. If, however, the layers form in place by internal soil-forming (pedogenic) processes, they are called *horizons*. The top layer, or *A horizon*, is the zone of major biological activity and is therefore generally enriched with organic matter and often darker in color than the underlying soil. Next comes the *B horizon*, where some of the materials migrating from the A horizon (such as clay or carbonates) tend to accumulate. Under the B horizon lies the *C horizon*, which is the soil's parent material. In the case of a residual soil formed in place from the bedrock, the C horizon consists of the weathered and fragmented rock material. In other cases, the C horizon may consist of alluvial, aeolian, or glacial sediments.

The A, B, C sequence of horizons is clearly recognizable in some cases, as for example in a typical zonal soil such as a *podzol*. In other cases, no clearly developed B horizon may be discernible, and the soil is then characterized by an A, C profile. In still other cases, as in the case of very recent alluvium, hardly any profile differentiation is apparent. The character of the profile depends primarily on the climate, and secondarily on the parent material, the vegetation, the topography, and time.

The typical development of a soil and its profile, called *pedogenesis*, can be summarized as follows: The process begins with the physical disintegration or "weathering" of the exposed rock formation, which thus forms the soil's parent material. Gradual accumulation of organic residues near the surface brings about the development of a discernible A horizon, which may acquire a granular structure stabilized to a greater or lesser degree by organic matter cementation. (This process is retarded in desert regions.) Continued chemical weathering (e.g., hydration, oxidation, and reduction), dissolution, and reprecipitation may bring about the formation of clay. Some of the clay thus formed tends to migrate, along with other transportable materials (such as soluble salts) downward from the A horizon and to accumulate in an intermediate zone (namely, the B horizon) between the A horizon and the deeper parent material of the so-called C horizon. Important aspects of soil formation and profile development are the twin processes of *eluviation* and *illuviation* (washing out and washing in, respectively) wherein clay and other substances emigrate from the overlying *eluvial* A horizon and accumulate in the underlying *illuvial* B horizon, which therefore differs from the A horizon in composition and structure. Throughout these processes, the profile as a whole deepens as the upper part of the C horizon is gradually transformed, until eventually a quasi-stable condition is approached in which the counter processes of soil formation and of soil erosion are more or less in balance. In arid regions, salts such as calcium sulfate and calcium carbonate, dissolved from the upper part of the soil, may precipitate at some depth to form a cemented "pan." Numerous variations of these processes are possible, depending on local conditions. The characteristic depth of the soil, for instance, varies from location to location. Valley soils are typically deeper

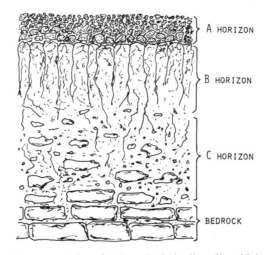

**Fig. 2.3.** Schematic representation of an hypothetical soil profile, with its underlying parent rock. The A horizon is shown with an aggregated crumblike structure, and the B horizon with columnar structure.

than mountain soils, and the depth of the latter depends on slope steepness. In some cases, the depth of the soil is a moot question, as the soil blends into its parent material without any distinct boundary. However, the biological activity zone seldom extends below 2–3 m, and in some cases is shallower than 1 m.

A hypothetical soil profile is presented in Fig. 2.3. This is not a typical soil, for in the myriad of greatly differing soil types recognized by pedologists it is well nigh impossible to define a single typical soil. Our illustration is only meant to suggest the sort of differences in appearance and structure likely to be encountered in a soil profile between different depth strata. Pedologists classify soils by their genesis and recognizable characteristics (see Fig. 2.4). However, pedological profile characterization is still somewhat qualitative and not sufficiently based on exact measurements of pertinent physical properties such as hydraulic, mechanical, and thermal characteristics.

**Sample Problems**

**1.** Prove the following relation between porosity, particle density, and bulk density:

$$f \overset{?}{=} (\rho_s - \rho_b)/\rho_s = 1 - \rho_b/\rho_s$$

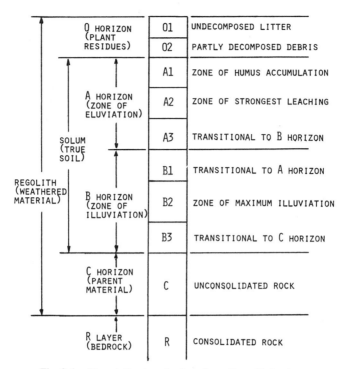

| | | |
|---|---|---|
| O HORIZON (PLANT RESIDUES) | O1 | UNDECOMPOSED LITTER |
| | O2 | PARTLY DECOMPOSED DEBRIS |
| A HORIZON (ZONE OF ELUVIATION) | A1 | ZONE OF HUMUS ACCUMULATION |
| | A2 | ZONE OF STRONGEST LEACHING |
| | A3 | TRANSITIONAL TO B HORIZON |
| B HORIZON (ZONE OF ILLUVIATION) | B1 | TRANSITIONAL TO A HORIZON |
| | B2 | ZONE OF MAXIMUM ILLUVIATION |
| | B3 | TRANSITIONAL TO C HORIZON |
| C HORIZON (PARENT MATERIAL) | C | UNCONSOLIDATED ROCK |
| R LAYER (BEDROCK) | R | CONSOLIDATED ROCK |

**Fig. 2.4.** Descriptive terminology for soil profile horizons.

Substituting the respective definitions of $f$, $\rho_s$, and $\rho_b$, we can rewrite the equation as

$$V_f/V_t \overset{?}{=} 1 - (M_s/V_t)/(M_s/V_s)$$

Simplifying the right-hand side, we obtain

$$V_f/V_t \overset{?}{=} 1 - (V_s/V_t) = (V_t - V_s)/V_t$$

But since $V_t - V_s = V_f$, we have

$$V_f/V_t = V_f/V_t$$

Q.E.D.

**2.** Prove the following relation between volume wetness, mass wetness, bulk density, and water density ($\rho_w = M_w/V_w$):

$$\theta \overset{?}{=} w\rho_b/\rho_w$$

Again, we start by substituting the respective definitions of $\theta$, $w$, $\rho_b$, and $\rho_w$:

$$V_w/V_t \overset{?}{=} [(M_w/M_s)(M_s/V_t)]/(M_w/V_w)$$

Rearranging the right-hand side,

$$\frac{V_w}{V_t} = \frac{V_w}{M_w}\frac{M_w}{M_s}\frac{M_s}{V_t} = \frac{V_w}{V_t}$$

Q.E.D.

**3.**  A sample of moist soil having a wet mass of 1000 gm and a volume of 640 cm³ was dried in the oven and found to have a dry mass of 800 gm. Assuming the typical value of particle density for a mineral soil, calculate the bulk density $\rho_b$, porosity $f$, void ratio $e$, mass wetness $w$, volume wetness $\theta$, water volume ratio $v_w$, degree of saturation $s$, and air-filled porosity $f_a$.

Bulk density:

$$\rho_b = \frac{M_s}{V} = \frac{800 \text{ gm}}{640 \text{ cm}^3} = 1.25 \text{ gm/cm}^3$$

Porosity:

$$f = 1 - \frac{\rho_b}{\rho_s} = 1 - \frac{1.25 \text{ gm/cm}^3}{2.65 \text{ gm/cm}^3} = 1 - 0.472 = 0.528$$

Alternatively,

$$f = V_f/V_t = (V_t - V_s)/V_t$$

Since

$$V_s = \frac{M_s}{\rho_s} = \frac{800 \text{ gm}}{2.65 \text{ gm/cm}^3} = 301.9 \text{ cm}^3$$

Hence

$$f = \frac{640 \text{ cm}^3 - 301.9 \text{ cm}^3}{640 \text{ cm}^3} = 0.528 = 52.8\%$$

Void ratio:

$$e = \frac{V_f}{V_s} = \frac{V_t - V_s}{V_s} = \frac{640 \text{ cm}^3 - 301.9 \text{ cm}^3}{301.9 \text{ cm}^3} = 1.12$$

Mass wetness:

$$w = \frac{M_w}{M_s} = \frac{M_t - M_s}{M_s} = \frac{1000 \text{ gm} - 800 \text{ gm}}{800 \text{ gm}} = 0.25 = 25\%$$

Volume wetness:

$$\theta = \frac{V_w}{V_t} = \frac{200 \text{ cm}^3}{640 \text{ cm}^3} = 0.3125 = 31.25\%$$

(*Note:* $V_w = M_w/\rho_w$, wherein $\rho_w$, the density of water, equals approximately 1 gm/cm$^3$.) Alternatively,

$$\theta = w\frac{\rho_b}{\rho_w} = 0.25\frac{1.25 \text{ gm/cm}^3}{1 \text{ gm/cm}^3} = 0.3125 = 31.25\%$$

Water volume ratio:

$$v_w = \frac{V_w}{V_s} = \frac{200 \text{ cm}^3}{301.9 \text{ cm}^3} = 0.662$$

Degree of saturation:

$$s = \frac{V_w}{V_t - V_s} = \frac{200 \text{ cm}^3}{640 \text{ cm}^3 - 301.9 \text{ cm}^3} = 0.592 = 59.2\%$$

Air-filled porosity ($f_a$):

$$f_a = \frac{V_a}{V_t} = \frac{640 \text{ cm}^3 - 200 \text{ cm}^3 - 301.9 \text{ cm}^3}{640 \text{ cm}^3} = 0.216 = 21.6\%$$

4. How many centimeters (equivalent depth) of water are contained in a soil profile 1 m deep if the mass wetness of the upper 40 cm is 15% and that of the lower 60 cm is 25%? The bulk density is 1.2 gm/cm$^3$ in the upper layer and 1.4 in the deeper layer. How much water does the soil contain in cubic meters per hectare of land?

Recall that $\theta = w(\rho_b/\rho_w)$(where $\rho_w = 1$).

Volume wetness in upper layer: $\theta_1 = 0.15 \times 1.2 = 0.18$.

Equivalent depth of water in upper 40 cm = $0.18 \times 40 = 7.2$ cm.

Volume wetness in lower layer: $\theta_2 = 0.25 \times 1.4 = 0.35$.

Equivalent depth of water in lower 60 cm = $0.35 \times 60 = 21.0$ cm.

Total equivalent depth of water in 100 cm profile = $7.2 + 21.0 = 28.2$ cm.

Area of hectare = 10,000 m$^2$; volume of soil (1 m deep) per hectare = 10,000 m$^3$.

Volume of water contained in 1 m deep soil per hectare = $10,000 \times 0.282 = 2820$ m$^3$.

# Part II    THE SOLID PHASE

# 3    *Texture, Particle Size Distribution, and Specific Surface*

## A. Introduction

Having introduced the concept of the soil as a three-phase system, let us now take a closer look at the solid phase, which is, to begin with, the permanent component of the soil and the one which gives substance to the whole. Conceivably, one could have soil without air, or without water, and in a vacuum without both (as is the case with the "soil" found on the moon), but it would be difficult to imagine a soil in any circumstances without the solid phase. The material of which the soil solid phase is composed includes discrete mineral particles of various sizes, as well as amorphous compounds, with the latter generally attached to, and sometimes coating, the particles. As the content of the amorphous material, such as hydrated iron oxides and humus, is generally (though not invariably) small, we can in most cases represent the solid phase as consisting by and large of distinct particles, the largest among which are visible to the naked eye and the smallest of which are colloidal and can only be observed by means of an electron microscope.

In general, it is possible to separate soil particles into groups and to characterize the soil in terms of the relative proportions of its particle-size groups, which may differ from one another in mineral composition as well as in particle size. It is these attributes of the soil solid phase, particle size and mineral composition, which largely determine the behavior of the soil: its interactions with fluids and solutes, as well as its compressibility, strength, and thermal regime.

To characterize the soil material physically and quantitatively, we must define its pertinent and measurable properties. A distinction should be made in this context between static and dynamic soil properties. *Static properties* are intrinsic to the material itself and are unaffected by any external variables. *Dynamic properties*, on the other hand, are manifested in the response of the body to externally imposed effects, such as mechanical stresses tending to cause deformation and failure or the entry of water. In this chapter we shall concentrate our attention upon static properties of the soil solid phase, such as texture, particle size distribution, and specific surface, insofar as these are permanent and immutable attributes of the soil material capable of being measured objectively and having a bearing upon soil behavior.

## B. Soil Texture

The term *soil texture* refers to the size range of particles in the soil, i.e., whether the particles of which a particular soil is composed are mainly large, small, or of some intermediate size or range of sizes. As such, the term carries both qualitative and quantitative connotations. Qualitatively, it represents the "feel" of the soil material, whether coarse and gritty or fine and smooth. An experienced soil classifier can tell, by kneading or rubbing the moistened soil with his fingers, whether it is coarse textured or fine textured and can also assess in a semiquantitative way to which of the several intermediate textural "classes" the particular soil might belong. In a more rigorously quantitative sense, however, the term soil texture denotes the measured distribution of particle sizes or the proportions of the various size ranges of particles which occur in a given soil. As such, soil texture is a permanent, natural attribute of the soil and the one most often used to characterize its physical makeup.

The traditional method of characterizing particle sizes in soils is to divide the array of possible particle sizes into three conveniently separable size ranges known as *textural fractions* or *separates*, namely, *sand*, *silt*, and *clay*. The actual procedure of separating out these fractions and of measuring their proportions is called *mechanical analysis*, for which standard techniques have been devised. The results of this analysis yield the *mechanical composition* of the soil, a term which is often used interchangeably with soil texture.

Unfortunately, there is as yet no universally accepted scheme for classification of particle sizes, and the various criteria used by different workers are rather arbitrary. For instance, the classification standardized in America by the U.S. Department of Agriculture differs from that of the International Soil Science Society (ISSS), as well as from those promulgated by the American Society for Testing Materials (ASTM), the Massachusetts Institute of Technology (MIT), and various national institutes abroad. The classification followed by soil engineers often differs from that of agricultural soil scientists.

The same terms are used to designate differing size ranges, an inconsistency which can result in considerable confusion.

A number of the often used particle size classification schemes are compared in Fig. 3.1.

In the first place, the problem is what should be the upper limit of particle size which can properly be included in the definition of soil material. Some soils contain large rocks which obviously do not behave like soil although, if numerous, might affect the behavior of the soil in bulk. It is more or less conventional to define *soil material* as particles smaller than 2 mm in diameter. Larger particles are generally referred to as *gravel*, and still larger rock fragments, several centimeters in diameter, are variously called *stones*, *cobbles*, or if very large, *boulders*. Where gravel and stones occupy enough of the soil's volume to influence soil physical processes significantly, their volume fraction and size range should be reported along with the specification of the finer soil material.

The largest group of particles generally recognized as soil material is *sand*, which is defined as particles ranging in diameter from 2000 $\mu$m (2 mm) down to 50 $\mu$m (USDA classification) or to 20 $\mu$m (ISSS classification). The sand fraction is often further subdivided into subfractions such as coarse, medium, and fine sand. Sand grains usually consist of quartz, but may also be frag-

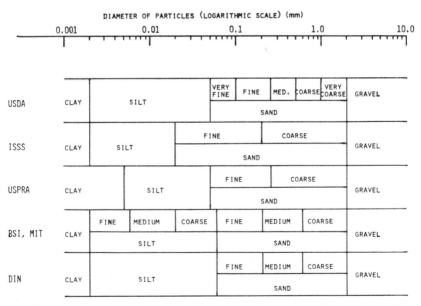

**Fig. 3.1.** Several conventional schemes for the classification of soil fractions according to particle diameter ranges: U.S. Department of Agriculture (USDA); International Soil Science Society (ISSS); U.S. Public Roads Administration (USPRA); German Standards (DIN); British Standards Institute (BSI); Massachusetts Institute of Technology (MIT).

ments of feldspar, mica, and occasionally heavy minerals such as zircon, tourmaline, and hornblende, though the latter are rather rare. In most cases, sand grains have more or less uniform dimensions and can be represented as spherical, though they are not necessarily smooth and may in fact have quite jagged surfaces (Fig. 3.2), which, together with their hardness[1], account for their abrasiveness.

The next fraction is *silt*, which consists of particles intermediate in size between sand and *clay*, which, in turn, is the smallest sized fraction. Mineralogically and physically, silt particles generally resemble sand particles, but since they are smaller and have a greater surface area per unit mass and are often coated with strongly adherent clay, they may exhibit, to a limited degree, some of the physicochemical attributes of clay.

The clay fraction, with particles ranging from 2 $\mu$m downwards, is the colloidal fraction. Clay particles are characteristically platelike or needlelike in shape and generally belong to a group of minerals called the *aluminosilicates*. These are *secondary minerals*, formed in the soil itself in the course of its evolution from the *primary minerals* contained in the original rock. In some cases, however, the clay fraction may include considerable concentrations of fine particles which do not belong to the aluminosilicate clay mineral category, e.g., iron oxide or calcium carbonate.

Because of its far greater surface area per unit mass and its resulting physicochemical activity, clay is the decisive fraction which has the most influence on soil behavior. Clay particles adsorb water and hydrate, thereby causing the soil to swell upon wetting and then shrink upon drying. Clay particles typically carry a negative charge and when hydrated form an *electrostatic double layer* with *exchangeable ions* in the surrounding solution. Another expression of surface activity is the heat which evolves when a dry clay is wetted, called the *heat of wetting*. A body of clay will typically exhibit plastic behavior and become sticky when moist and then cake up and crack to form cemented hard fragments when desiccated. (So important is the clay

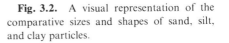

**Fig. 3.2.** A visual representation of the comparative sizes and shapes of sand, silt, and clay particles.

[1] Quartz particles have a hardness of 7 on the so-called *Mohs scale of hardness*, and will readily abrade steel (of hardness 5.5), as can be commonly observed with tillage implements.

fraction, in fact, that we have devoted our entire next chapter to it.) The relatively inert sand and silt fractions can be called the "soil skeleton," while the clay, by analogy, can be thought of as the "flesh" of the soil. Together, all three fractions of the solid phase, as they are combined in various configurations, constitute the *matrix* of the soil.

The expressions light soil and heavy soil are used in common parlance to describe the physical behavior of sandy versus clayey soils. Since a sandy soil tends to be loose, well drained, well aerated, and easy to cultivate, it is called light. A clayey soil, on the other hand, tends to absorb and retain much more water and to become plastic and sticky when wet, as well as tight and cohesive when dry, and is thus difficult to cultivate. It is therefore called heavy. These can be misleading expressions, however, since in actual fact it is the coarse-textured soils which are generally more dense (i.e., have a lower porosity) than the fine-textured soils and thus are heavier, rather than lighter, in weight (at least in the dry state).

## C. Nature of Clay

The fraction which influences the physical behavior of the soil most decisively is the colloidal clay, since it exhibits the greatest specific surface area and is therefore most active in physicochemical processes. Clay particles adsorb water and thus cause the soil to swell and shrink upon wetting and drying (Grim, 1958). Most of them are negatively charged and form an electrostatic double layer with exchangeable cations. Sand and silt have relatively small specific surface areas and consequently exhibit comparatively little physicochemical activity. These fractions can be termed the "skeleton," while the clay, by a similar analogy, can be thought of as the "flesh" of the soil. Together, they constitute the *solid matrix* of the soil.

The term clay designates not merely a range of particle sizes, but a large group of minerals, some of which are amorphous, but many of which occur in the form of highly structured microcrystals of colloidal size. The clay fraction thus differs mineralogically, as well as in particle sizes, from sand and silt, which are composed mainly of quartz and other primary mineral particles that have not been transformed chemically into secondary minerals as is the case with clay (Jenny, 1935; Jackson, *et al.,* 1948). The various clay minerals differ greatly in properties and prevalence. And while the ordinary measurement of soil texture does give an idea of the quantity of clay in the soil, it reveals very little of the specific character and activity of the clay.

The most prevalent clay minerals are the layered aluminosilicates. Their crystals are composed of two basic structural units (Grim, 1963; Marshall, 1964; Low, 1968; Hillel, 1980), namely: a tetrahedron of oxygen atoms surrounding a central cation, usually Si⁴⁺, and an octahedron of oxygen atoms or hydroxyl groups surrounding a larger cation usually Al³⁺ or Mg²⁺. The tetrahedra are joined at their basal corners and the octahedra are joined along their edges by means of shared oxygen atoms. Thus, tetrahedral and octahedral layers are formed (Fig. 3.3).

The layered aluminosilicate clay minerals are of two main types, depending upon the ratios of tetrahedral to octahedral layers, whether 1:1 or 2:1. In the 1:1 minerals like kaolinite, an octahedral layer is attached by the sharing of oxygens to a single tetrahedral layer. In the 2:1 minerals like montmorillonite, it is attached in the same way to two tetrahedral layers, one on each side. A clay particle is composed of multiply stacked composite layers (or unit cells) of this sort, called *lamellae*.

The structure described is an idealized one. Typically, some substitutions, or *isomorphous replacements*, of $Al^{3+}$ for $Si^{4+}$ occur in tetrahedral layers, and substitutions of $Mg^{2+}$ for $Al^{3+}$ occur in the octahedral layers. Hence, internally unbalanced negative charges occur at different sites in the lamellae. Another source of unbalanced charge on clay minerals is the incomplete charge neutralization of terminal atoms on lattice edges. These charges are balanced externally by exchangeable ions (mostly cations), which concentrate near the external surfaces of the particle and occasionally penetrate into interlamellar spaces. These cations are not an integral part of the lattice structure, and can be replaced, or exchanged, by other cations. The cation-exchange phenomenon is of great importance in soil physics as well as soil

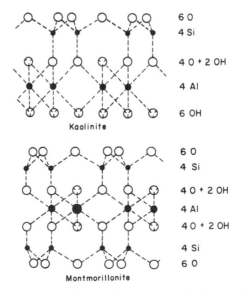

**Fig. 3.3.** Schematic representation of the structure of aluminosilicate minerals.

chemistry, since it affects the retention and release of nutrients and salts, and the flocculation–dispersion processes of soil colloids.

A hydrated clay particle forms a colloidal micelle, in which the excess negative charge of the particle is neutralized by a spatially separated swarm of cations. Together, the particle surface and the neutralizing cations form an *electrostatic double layer*. The cation swarm consists partly of a layer more or less fixed in the proximity of the particle surface (known as the Stern layer), and partly of a diffuse distribution extending some distance away from the particle surface. This distribution is illustrated schematically in Fig. 3.4. It results from an equilibrium between two opposing effects: the Coulomb (electrostatic) attraction of the clay particle, versus the Brownian (kinetic) motion of the liquid molecules, inducing outward diffusion of the cations toward the intermicellar solution. Just as cations are adsorbed positively toward the clay particles, so anions are repelled, or adsorbed negatively, and relegated from the micellar to the intermicellar solution (Kruyt, 1949).

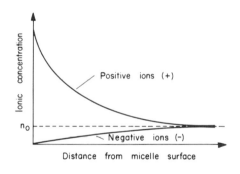

**Fig. 3.4.** Distribution of positive and negative ions in solution with distance from the surface of a clay micelle bearing net negative charge. Here, $n_0$ is the ionic concentration in the bulk solution outside the electrical double layer.

The quantity of cations adsorbed on soil-particle surfaces per unit mass of the soil under chemically neutral conditions is nearly constant and independent of the species of cation, and is generally known as the *cation exchange capacity*. Soils vary in cation exchange capacity from nil to perhaps 0.60 mEq per gm (Bear, 1955).

Clay minerals differ somewhat in *surface charge density* (i.e., the number of exchange sites per unit area of particle surface), and differ greatly in *specific surface area*. Hence, they differ also in their total *cation-exchange capacity*. Montmorillonite (also called smectite), with a specific surface area of nearly 800 m²/gm, has a cation-exchange capacity of about 0.95 mEq/gm, whereas kaolinite has an exchange capacity of only about 0.04—

0.09 mEq/gm. The greater specific surface area of montmorillonite is due to its lattice expansion and consequent exposure of internal (interlamellar) surfaces, which are not so exposed in the case of kaolinite. Other clay minerals (such as illite, micas, palygorskite, etc.) often exhibit properties intermediate between those of kaolinite and montmorillonite.

The attraction of a cation to a negatively charged clay micelle generally increases with increasing valency of the cation. Thus, monovalent cations are replaced more easily than divalent or trivalent cations. Highly hydrated cations, which tend to be farther from the surface, are also more easily replaced than less hydrated ones. The order of preference of cations in exchange reactions is generally as follows (Jenny, 1932, 1938):

$$Al^{3+} > Ca^{2+} > Mg^{2+} > K^+ > Na^+ > Li^+$$

When confined clays are allowed to sorb water, swelling pressures develop, which are related to the osmotic pressure difference between the double layer and the external solution (Aylmore and Quirk, 1959). Depending upon their state of hydration and the composition of their exchangeable cations, clay particles may either flocculate or disperse (Jenny and Reitemeier, 1935). Dispersion generally occurs with monovalent and highly hydrated cations (e.g., sodium). Conversely, flocculation occurs at high solute concentrations and/or in the presence of divalent and trivalent cations (e.g., $Ca^{2+}$, $Al^{3+}$) when the double layer is compressed so that its repulsive effect is lessened and any two micelles can approach each other more closely. Thus, the short-range attractive forces (known as the London–van der Waals forces) can come into play and join the individual micelles into *flocs*.

When a dispersed clay is dehydrated, it forms a dense and hard mass, or crust. On the other hand, when flocculated clay is dehydrated, it forms a crumbly and loose assemblage of small aggregates. Under rainfall action in the field, the dispersed clay will tend to become muddy, less pervious, and more highly erodible than flocculated clay. Thus, the desirable condition of a clayey soil is the flocculated one. Flocculation alone does not create an optimal structure, however, as will be explained in Chapter 4.

## D. Soil Classes

The overall textural designation of a soil, called the *textural class*, is conventionally determined on the basis of the mass ratios of the three fractions. Soils with different proportions of sand, silt, and clay are assigned to different classes, as shown in the triangular diagram of Fig. 3.5. To illustrate the use of the textural triangle, let us assume that a soil is composed of 50% sand, 20% silt, and 30% clay. Note that the lower left apex of the triangle represents 100% sand and the right side of the triangle represents 0% sand. Now find the point of 50% sand on the bottom edge of the triangle and follow the diagonally leftward line rising from that point and parallel to the zero line for sand. Next, identify the 20% line for silt, which is parallel to the zero line for silt, namely, the left edge of the triangle. Where the two lines intersect each other, as well as the 30% line for clay, is the point we are seeking. In this particular example, it happens to fall within the realm of "sandy clay loam."

Note that a class of soils called *loam* occupies a rather central location in the textural triangle. It refers to a soil which contains a "balanced" mixture of coarse and fine particles, so that its properties are intermediate among

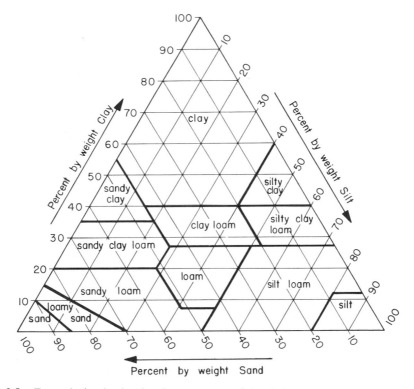

**Fig. 3.5.** Textural triangle, showing the percentages of clay (below 0.002 mm), silt (0.002–0.05 mm), and sand (0.05–2.0 mm) in the basic soil textural classes.

those of a sand, a silt, and a clay. As such, loam is often considered to be the optimal soil for plant growth and for agricultural production, as its capacity to retain water and nutrients is better than that of sand while its drainage, aeration, and tillage properties are more favorable than those of clay. This is an oversimplification, however, as under different environmental conditions and for different plant species a sand or a clay may be more suitable than a loam.

## E.  Particle Size Distribution

Any attempt to divide into distinct preconceived fractions what is usually in nature a continuous array of particle sizes is arbitrary to begin with, and the further classification of soils into discrete textural classes is doubly so. Although this approach is widely followed and evidently useful, it seems better to measure and display the complete distribution of particle sizes. This method of representing soil texture is used mostly by soil engineers.

Figure 3.6 presents typical *particle-size distribution curves*. The ordinate of the graph indicates the percentage of soil particles with diameters smaller than the diameter denoted in the abscissa, which is drawn on a logarithmic scale to encompass several orders of magnitude of particle diameters while allowing sufficient space for the representation of the fine particles. Note that this graph gives an integral, or cumulative, representation. In practice, the particle-size distribution curve is constructed by connecting a series of $n$ points, each expressing the cumulative fraction of particles finer than each of the $n$ diameters measured ($F_1, F_2, \ldots, F_i, \ldots, F_n$). Thus,

$$F_i = (M_s - \sum_1^i M_i)/M_s \qquad (3.1)$$

in which $M_s$ is the total mass of the soil sample analyzed and $\sum M_i$ is the cumulative mass of particles finer than the $i$th diameter measured.

The information obtainable from this representation of particle-size distribution includes the diameter of the largest grains in the assemblage, and the grading pattern, i.e., whether the soil is composed of distinct groups of particles each of uniform size or whether it consists of a more or less continuous array of particle sizes. Soils which contain a preponderance of particles of one or several distinct sizes, indicating a steplike distribution curve, are called *poorly graded*. Soils with a flattened and smooth distribution curve (without apparent discontinuities) are called *well graded*.

This aspect of the particle size distribution can be expressed in terms of the so-called *uniformity index* $I_u$, defined as the ratio of the diameter $d_{60}$ which includes 60% of the particles to the smaller diameter $d_{10}$ which includes 10% of the particles (as shown in Fig. 3.6). This index, also called the *uniformity coefficient*, is used mostly with coarse-grained soils. For a soil material consisting entirely of equal-sized particles, if such were to exist, $I_u$ would be unity. Some sand deposits may have uniformity indexes smaller than 10. Some well-graded soils, on the other hand, have $I_u$ values greater than 1000.

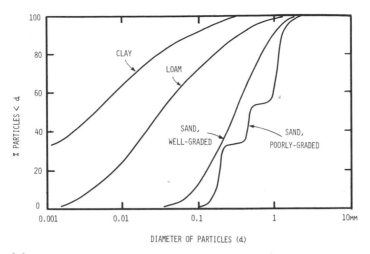

**Fig. 3.6.** Particle-size distribution curves for various types of soil material (schematic).

The particle size distribution curve can also be differentiated graphically to yield a frequency distribution curve for grain sizes, with a peak indicating the most prevalent grain size. Attempts have been made to correlate this index, as well as the harmonic mean diameter of grains, with various soil properties such as permeability.

## F. Mechanical Analysis

*Mechanical analysis* is the procedure for determining the particle-size distribution of a soil sample. The first step in this procedure is to disperse the soil sample in an aqueous suspension. The primary soil particles, often naturally aggregated, must be separated and made discrete by removal of cementing agents (such as organic matter, calcium carbonate, or iron oxides) and by deflocculating the clay. Removal of organic matter is usually achieved by oxidation with hydrogen peroxide, and calcium carbonate can be dissolved by addition of hydrochloric acid. Deflocculation is carried out by means of a chemical *dispersing agent* (e.g., sodium hexametaphosphate) and by mechanical agitation (shaking, stirring, or ultrasonic vibration). The function of the dispersing agent is to replace the cations adsorbed to the clay, particularly divalent or trivalent cations, with sodium, which has the effect of increasing the hydration of the clay micelles, thus causing them to repel each other rather than coalesce, as they do in the flocculated state. Failure to disperse the soil completely will result in flocs of clay or aggregates settling as if they were silt-sized or sand-sized primary particles, thus biasing the results of mechanical analysis to indicate an apparent content of clay lower than the real value.

Actual separation of particles into size groups can be carried out by passing the suspension through graded sieves, down to a particle diameter of approximately 0.05 mm. To separate and classify still finer particles, the method of sedimentation is usually used, based on measuring the relative settling velocity of particles of various sizes from the aqueous suspension.

According to *Stokes' law*, the terminal velocity of a spherical particle settling under the influence of gravity in a fluid of a given density and viscosity is proportional to the square of the particle's radius. We shall now proceed to derive this law, as it governs the method of sedimentation analysis.

A particle falling in a vacuum will encounter no resistance as it is accelerated by gravity and thus its velocity will increase as it falls. A particle falling in a fluid, on the other hand, will encounter a frictional resistance proportional to the product of its radius and velocity and to the viscosity of the fluid.

The resisting force due to friction $F_r$ was shown by Stokes in 1851 to be

$$F_r = 6\pi\eta r u \tag{3.2}$$

where $\eta$ is the viscosity of the fluid and $r$ and $u$ are the radius and velocity of the particle. Initially, as the particle begins its fall, its velocity increases. Eventually, a point is reached at which the increasing resistance force equals the constant downward force, and the particle then continues to fall without acceleration at a constant velocity known as the terminal velocity $u_t$.

The downward force due to gravity $F_g$ is

$$F_g = \tfrac{4}{3}\pi r^3 (\rho_s - \rho_f) g \tag{3.3}$$

where $\tfrac{4}{3}\pi r^3$ is the volume of the spherical particle, $\rho_s$ is its density, $\rho_f$ is the density of the fluid, and $g$ is the acceleration of gravity.

Setting the two forces equal, we obtain Stokes' law:

$$u_t = \frac{2}{9}\frac{r^2 g}{\eta}(\rho_s - \rho_f) = \frac{d^2 g}{18\eta}(\rho_s - \rho_f) \tag{3.4}$$

where $d$ is the diameter of the particle. Assuming that the terminal velocity is attained almost instantly, we can obtain the time $t$ needed for the particle to fall through a height $h$:

$$t = 18h\eta/d^2 g(\rho_s - \rho_f) \tag{3.5}$$

Rearranging and solving for the particle diameter gives

$$d = [18h\eta/tg(\rho_s - \rho_f)]^{1/2} \tag{3.6}$$

Since all terms on the right-hand side of the equation, except $t$, are constants, we can combine them all into a single constant $A$ and write

$$d = A/t^{1/2} \qquad \text{or} \qquad t = B/d^2 \tag{3.7}$$

where $B = A^2$.

One way of measuring particle-size distribution is to use a pipette to draw samples of known volume from a given depth in the suspension at regular times after sedimentation is begun. An alternative method is to use an *hydrometer* to measure the density of the suspension at a given depth as a function of time. With time this density decreases as the largest particles, and then progressively smaller ones, settle out of the region of the suspension being measured.

To visualize the process, let us focus upon some hypothetical plane at a depth $h_i$ below the surface of the suspension. Using Eq. (3.5) we can calculate the time $t_i$ necessary for all particles with a diameter equal to or greater than $d_i$ to fall below the given plane. A moment's contemplation will convince the reader that the concentration of all smaller particles (with a velocity smaller than $h_i/t_i$) will have remained constant in the plane under consideration, since the number of these particles reaching the plane from above must be equal to the number falling through it. Therefore, measuring the concentration of particles $d_i$ at this plane at time $t_i$ allows us to calculate the total mass of these particles originally present in the suspension from the known volume of water in the cylinder. With the standardized *Bouyoucos hydrometer*, for instance, a settling time of about 40 sec is needed at 20°C to measure the concentration of clay and silt (all the sand having settled through), and a time of about 8 hr is needed to measure the clay content alone. Correction factors must be applied whenever differences in temperature, or particle density, or initial suspension concentration, arise.

The use of Stokes' law for measurement of particle sizes is dependent upon certain simplifying assumptions which may not be in accord with reality. Among these are the following:

(1)   The particles are sufficiently large to be unaffected by the thermal (Brownian) motion of the fluid molecules;
(2)   the particles are rigid, spherical, and smooth;
(3)   all particles have the same density;
(4)   the suspension is sufficiently dilute that particles do not interfere with one another and each settles independently;
(5)   the flow of the fluid around the particles is laminar, i.e., no particle exceeds the critical velocity for the onset of turbulence.

In fact, we know that soil particles, while indeed rigid, are neither spherical nor smooth, and some may be platelike. Hence, the diameter calculated from the settlement velocity does not necessarily correspond to the actual dimensions of the particle. Rather, we should speak of an effective or equivalent settling diameter. The results of a mechanical analysis based on sieving may differ from those of a sedimentation analysis of the same particles. Moreover, soil particles are not all of the same density. Most silicates have values of 2.6–2.7 gm/cm³, whereas iron oxides and other heavy minerals may have density values of 5 gm/cm³ or even more. For all these reasons, the mechanical analysis of soils yields only approximate results. Its greatest shortcoming, however, is that it does not account for differences in *type* of clay, which can be of decisive importance in determining soil behavior.

## G. Specific Surface

The specific surface of a soil material can variously be defined as the total surface area of particles per unit mass ($a_m$), or per unit volume of particles ($a_v$), or per unit bulk volume of the soil as a whole ($a_b$):

$$a_m = A_s/M_s \tag{3.8}$$

$$a_v = A_s/V_s \tag{3.9}$$

$$a_b = A_s/V_t \tag{3.10}$$

where $A_s$ is the total surface area of a mass of particles $M_s$ having a volume $V_s$ and contained in a bulk volume $V_t$ of soil.

Specific surface is commonly expressed in terms of square meters per gram, or per cubic centimeter, of particles. It depends in the first place upon the size of the particles. It also depends upon their shape. Flattened or elongated particles obviously expose greater surface per volume or per mass than do equidimensional (e.g., cubical or spherical) particles. Since clay particles are generally platy, they contribute even more to the overall specific surface area of a soil than is indicated by their small size alone. In addition to their external surfaces, certain types of clay crystals exhibit internal surface areas, such as those which form when the open lattice of montmorillonite expands on imbibing water. Whereas the specific surface of sand is often less than 1 m²/gm, that of clay can be as high as several hundred square meters per gram. In fact, it is the clay fraction, its content and mineral composition, that largely determines the specific surface of a soil.

The specific surface of a soil material is a fundamental and intrinsic property which has been found to correlate with important phenomena such as cation exchange, retention and release of various chemicals (including nutrients and certain potential pollutants of the environment), swelling, retention of water, and such mechanical properties as plasticity, cohesion, and strength. Hence, it is a highly pertinent property to study, and its measurement can help provide a basis for evaluating and predicting soil behavior. It is probable that the measurement of soil specific surface, though not yet as common as the measurement of soil texture by the traditional methods, may eventually prove to be a more meaningful and pertinent index for characterizing a soil than are the percentages of sand, silt, and clay.

The usual procedure for determining surface area is to measure the amount of gas or liquid needed to form a *monomolecular layer* over the entire surface in a process of adsorption. The standard method is to use an inert gas such as nitrogen. Water vapor and organic liquids (e.g., glycerol and ethylene glycol) are also used.

The adsorption phenomenon was described by de Boer (1953). At low gas pressures, the amount of a gas adsorbed per unit area of adsorbing surface, $\sigma_a$, is related to the gas pressure $P$, the temperature $T$, and the heat of adsorption $Q_a$ by the equation

$$\sigma_a = k_i P \exp(Q_a/RT) \tag{3.11}$$

where $R$ is the gas constant and $k_i$ is also a constant. Thus, the amount of adsorption increases with pressure, but decreases with temperature.

The *equation of Langmuir* (1918) indicates the relation between the gas pressure $P$ and the volume of gas adsorbed per gram of adsorbent $v$ at constant temperature:

$$P/v = 1/k_2 v_m + P/v_m \tag{3.12}$$

where $v_m$ is the volume of adsorbed gas which forms a complete monomolecular layer over the adsorbent, and can be obtained by plotting $P/v$ versus $P$. The specific surface of the adsorbent can then be calculated by determining the number of molecules in $v_m$ and multiplying this by the cross-sectional area of these molecules. The Langmuir equation is based on the assumption that only one layer of molecules can be adsorbed, and that the heat of adsorption is uniform during the process.

Brunauer *et al.* (1938) derived what has come to be known as the *BET equation*, based on multilayer adsorption theory:

$$P/v(P_0 - P) = (1/v_m C) + (C - 1)P/v_m C P_0 \tag{3.13}$$

where $v$ is the volume of gas adsorbed at pressure $P$, $v_m$ is the volume of a single layer of adsorbed molecules over the entire surface of the adsorbent, $P_0$ is the gas pressure required for monolayer saturation at the temperature of the experiment, and $C$ is a constant for the particular gas, adsorbent, and temperature. The volume $v_m$ can be obtained from the BET theory by plotting $P/v(P_0 - P)$ versus $P/P_0$. The density of the adsorbed gas is usually assumed to be that of the liquefied or the solidified gas.

Polar adsorbents (such as water) may not obey the BET or Langmuir equations (which are similar at low pressures), since their molecules or ions may tend to cluster at charged sites rather than to spread out evenly over the adsorbent surface. The use of various adsorbents and techniques for the measurement of the specific surface area of soil materials was described by Mortland and Kemper (1965).

An estimation of the specific surface area can also be made by calculation based on the sizes and shapes of the particles. We shall proceed to give some examples of such calculations, for particles of definable geometry.

For a sphere of diameter $d$, the ratio of surface to volume is

$$a_v = \pi d^2/(\pi d^3/6) = 6/d \tag{3.14}$$

and the ratio of surface to mass is

$$a_m = 6/\rho_s d \tag{3.15}$$

Where the particles have a density $\rho_s$ of about 2.65 gm/cm$^3$, approximately,

$$a_m \approx 2.3/d \tag{3.16}$$

For a cube of edge $L$, the ratio of surface to volume is

$$a_v = 6L^2/L^3 = 6/L \tag{3.17}$$

and the ratio of surface to mass is, again,

$$a_m = 6/\rho_s L \tag{3.18}$$

Thus, the expressions for particles of nearly equal dimensions, such as most sand and silt grains, are similar, and knowledge of the particle size distribution can allow us to calculate the approximate specific surface by the summation equation:

$$a_m = (6/\rho_s)\sum c_i(d_i^2/d_i^3) = (6/\rho_s)\sum(c_i/d_i) \tag{3.19}$$

where $c_i$ is the mass fraction of particles of average diameter $d_i$.

Now let us consider a platy particle. For the sake of argument, we can assume that our plate is square shaped, with sides $L$ and thickness $l$. The surface-to-volume ratio is

$$a_v = (2L^2 + 4Ll)/L^2l \tag{3.20}$$

and the surface-to-mass ratio

$$a_m = 2(L + 2l)/\rho_s Ll \tag{3.21}$$

If the platelet is very thin, so that its thickness $l$ is negligible compared to principal dimension $L$, and if $\rho_s = 2.65$ gm/cm$^3$, then

$$a_m \approx 2/\rho_s l \approx 0.75/l \quad \text{cm}^2/\text{gm} \tag{3.22}$$

Thus, the specific surface area of a clay can be estimated if the thickness of its platelets is known. For example, the thickness of a platelet of fully dispersed montmorillonite is approximately 10 Å, or $10^{-7}$ cm. Therefore, $a_m = 0.75/10^{-7}$, or 750 m²/gm, which compares closely with the measured value of about 800 m²/gm. Often, however, montmorillonite particles are not in a state of ultimate dispersion, and their platelets are several unit-layers thick. The average platelet thickness for illite clay is about 50 Å, and for kaolinite clays it is a few hundred angstroms. The measured specific surface area for kaolinite is about 15 m²/gm, which is low among the clays. Although kaolinite seems to have a somewhat more active surface in that it is characterized by a higher density of electrostatic charges, by and large the activity of clay is more or less proportional to the specific surface.

## Sample Problems

**1.** Determine the textural class designations for soils with the following distributions of particle sizes:

| | <0.0002 | 0.0002–0.002 | 0.002–0.01 | 0.01–0.05 | 0.05–0.25 | 0.25–2.0 (mm) |
|---|---|---|---|---|---|---|
| (a) | 5% | 10% | 20% | 25% | 20% | 20% |
| (b) | 6% | 9% | 30% | 30% | 15% | 10% |
| (c) | 10% | 30% | 30% | 10% | 10% | 10% |
| (d) | 4% | 6% | 10% | 20% | 30% | 30% |

Using the USDA classification (Fig. 3.1) and the textural triangle (Fig. 3.3), we obtain the following textural classes:

(a) % sand = 40, % silt = 45, % clay = 15. Soil class: loam.
(b) % sand = 25, % silt = 60, % clay = 15. Soil class: silt-loam.
(c) % sand = 20, % silt = 30, % clay = 50. Soil class: clay.
(d) % sand = 60, % silt = 30, % clay = 10. Soil class: sandy loam.

**2.** Using Stokes' law, calculate the time needed for all sand particles (diameter $>50$ μm) to settle out of a depth of 20 cm in an aqueous suspension at 30°C. How long would it take for all silt particles to settle out? How long for "coarse" clay ($>1$ μm)?
We use Eq. (3.5):

$$t = 18 \, h\eta/d^2 g(\rho_s - \rho_f)$$

Substituting the appropriate values for depth $h$ (20 cm), viscosity $\eta$ (0.008 gm/cm sec, particle diameter $d$ (50 μm, 2 μm, and 1 μm for the lower limits of sand, silt, and coarse clay, respectively), gravitational acceleration $g$ (981 cm/sec²), average particle density $\rho_s$ (2.65 gm/cm³), and water density (1.0 gm/cm³), we can write:

(a) For all sand to settle out, leaving only silt and clay in suspension:

$$t = \frac{18 \times 20 \times (8 \times 10^{-3})}{(50 \times 10^{-4})^2 \times 981 \times (2.65 - 1.0)} \cong 71 \text{ sec.}$$

(b) For all silt to settle out, leaving only clay in suspension:

$$t = \frac{18 \times 20 \times (8 \times 10^{-3})}{(2 \times 10^{-4})^2 \times 981 \times 1.65} \cong 44500 \text{ sec} = 12.36 \text{ hr.}$$

(c) For all coarse clay to settle out, leaving only fine clay in suspension:

$$t = \frac{18 \times 20 \times (8 \times 10^{-3})}{(1 \times 10^{-4})^2 \times 981 \times 1.65} \cong 178000 \text{ sec} = 49.44 \text{ hr.}$$

**3.**   Calculate the approximate specific surface of a sand composed of the following array of particle sizes:

> Average diameter:      1 mm,     0.5 mm,     0.2 mm,     0.1 mm.
> Percentage by mass:   40%,       30%,         20%,         10%.

We use Eq. (3.19):

$$a_m = \frac{6}{2.65} \Sigma \frac{c_i}{d_i}$$
$$= 2.264 \left( \frac{0.4}{0.1} + \frac{0.3}{0.05} + \frac{0.2}{0.02} + \frac{0.1}{0.01} \right) = 67.92 \text{ cm}^2/\text{gm}$$

Note that the smallest-diameter fraction, constituting only a tenth of the mixture's mass, accounts for a third of its specific surface ($2.264 \times 0.1/0.01 = 22.64 \text{ cm}^2/\text{gm}$).

**4.**   Estimate the approximate specific surface ($m^2/\text{gm}$) of a soil composed of 10% coarse sand (average diameter 0.1 cm), 20% fine sand (average diameter 0.01 cm), 30% silt (average diameter 0.002 cm), 20% kaolinite clay (average platelet thickness 400 Å), 10% illite clay (average thickness 50 Å), and 10% montmorillonite (average thickness 10 Å).

For the sand and silt fractions, we use Eq. (3.19):

$$a_m = \frac{6}{2.65} \left( \frac{0.1}{0.1} + \frac{0.2}{0.01} + \frac{0.3}{0.002} \right) = 387 \text{ cm}^2/\text{gm} = 0.0387 \text{ m}^2/\text{gm.}$$

For the clay fraction, we use Eq. (3.22) in summation form to include the partial specific surface values for kaolinite, illite, and montmorillonite, respectively:

$$a_m = 0.2 \times 0.75/(400 \times 10^{-8}) + 0.1 \times 0.75/(50 \times 10^{-8})$$
$$+ 0.1 \times 0.75/(10 \times 10^{-8})$$
$$= 3.78 \text{ m}^2/\text{gm (kaol.)} + 15.09 \text{ m}^2/\text{gm (ill.)} + 75.45 \text{ m}^2/\text{gm (mont.)}$$
$$= 94.32 \text{ m}^2/\text{gm}$$

Total for the soil $= 0.0387 + 94.32 \cong 94.36 \text{ m}^2/\text{gm}$.

Note that the clay fraction, which constitutes 40% of the soil mass, accounts for 99.96% of the specific surface (i.e., 94.32/94.36). The montmorillonite constituent alone (10% of the mass) accounts for nearly 80% of the soil's specific surface.

# 4    *Soil Structure and Aggregation*

## A. Introduction

Bricks thrown haphazardly atop one another become an unsightly heap. The same bricks, only differently arranged and mutually bonded, can form a home or a factory. Similarly, a soil can be merely a loose and unstable assemblage of random particles, or it can consist of a distinctly structured pattern of interbonded particles associated into aggregates having regular sizes and shapes. Hence it is not enough to study the properties of individual soil particles. To understand how the soil behaves as a composite body, we must consider the manner in which the various particles are packed and held together in a continuous spatial network which is commonly called the *soil matrix*, or *fabric*.

The arrangement and organization of the particles in the soil is called *soil structure*. Since soil particles differ in shape, size, and orientation, and can be variously associated and interlinked, the mass of them can form complex and irregular configurations which are in general exceedingly difficult if not impossible to characterize in exact geometric terms. A further complication is the inherently unstable nature of soil structure and hence its inconstancy in time, as well as its nonuniformity in space. Soil structure is strongly affected by changes in climate, biological activity, and soil management practices, and it is vulnerable to destructive forces of a mechanical and physicochemical nature. For these various reasons, we have no truly objective or universally applicable method to measure soil structure per se, and the term soil structure therefore denotes a qualitative concept rather than a directly quantifiable property. The numerous methods which have been proposed for characterization of soil structure are in fact indirect methods which measure one soil attribute or another which are supposed to be dependent upon structure, rather than the structure itself, and many of these methods are at best specific to the purpose for which they were devised and at worst completely arbitrary.

The complexity and difficulty associated with soil structure notwithstanding, we can readily perceive and appreciate its critical importance, inasmuch as it determines the total porosity as well as the shapes of individual pores and their size distribution. Hence, soil structure affects the retention and transmission of fluids in the soil, including infiltration and aeration. Moreover, as soil structure influences the mechanical properties of the soil, it may also affect such disparate phenomena as germination, root growth, tillage, overland traffic, and erosion. Agriculturists are usually interested in having the soil, at least in its surface zone, in a loose and highly porous and permeable condition. Engineers, on the other hand, desire a dense and rigid structure so as to provide maximal stability and resistance to shear and minimal permeability. In either case, knowledge of soil structure relationships is essential for efficient management.

## B. Types of Soil Structure

In general, we can recognize three broad categories of soil structure—*single grained*, *massive*, and *aggregated*. When particles are entirely unattached to each other, the structure is completely loose, as it is in the case of coarse granular soils or unconsolidated deposits of desert dust. Such soils were labeled structureless in the older literature of soil physics, but, since even a loose arrangement is a structure of sorts, we prefer the term single grained structure. On the other hand, when the soil is tightly packed in large cohesive blocks, as is sometimes the case with dried clay, the structure can be called massive. Between these two extremes, we can recognize an intermediate condition in which the soil particles are associated in quasi-stable small clods known as *aggregates* or *peds*. This last type of structure, called aggregated, is generally the most desirable condition for plant growth, especially in the critical early stages of germination and seedling establishment. The formation and maintenance of stable aggregates is the essential feature of soil *tilth*, a qualitative term used by agronomists to describe that highly desirable, yet unfortunately elusive, physical condition in which the soil is an optimally loose, friable, and porous assemblage of aggregates permitting free movement of water and air, easy cultivation and planting, and unobstructed germination and root growth.

## C.  Structure of Granular Soils

The structure of most coarse-grained, or granular, soils is single grained, as there is little tendency for the grains to adhere to each other and to form aggregates. The actual arrangement and internal mode of packing of the grains depends upon the distribution of grain sizes and shapes, as well as upon the manner in which the material has been deposited or formed in place.

Hypothetically speaking, the two extreme cases of possible packing arrangements are, on the one hand, a system of uniform spherical grains in a state of open packing (and hence minimal density) and, on the other hand, a gradual distribution of grain sizes in which progressively smaller grains fill the voids between larger ones in an "ideal" succession which provides maximal density. A system of uniform grains is called *monodisperse*, and one which includes various grain sizes is *polydisperse* (Fig. 4.1).

Natural granular soils resemble collections of spheres in the sense that the particles are often rounded (though deposits of angular particles also occur). The actual porosities of natural sediments range within the theoretically derivable limits for ideal packings of monodisperse and polydisperse spheres; i.e., they lie between 25 and 50%, in general.

Loose deposits of granular material cannot be compacted very effectively by the application of static pressure (which, however, can be very effective in compacting unsaturated, clayey soils), unless, of course, the pressure

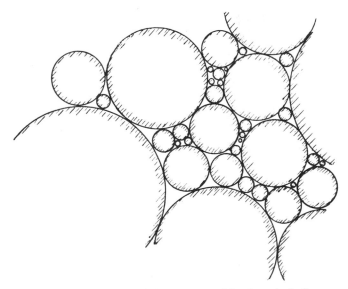

**Fig. 4.1.**  Packing of polydisperse particles (hypothetical).

exceeds the crushing strength of the individual grains. Short of that, however, static pressure is merely transmitted along the soil skeleton and borne by the high frictional resistance to the mutual sliding of the grains. Granular material can, however, be compacted by the application of vibratory action. The vibration pulsates the grains and allows smaller ones to enter between larger ones, thus increasing the packing density. The vibration of granular materials is effective in the dry and in the saturated states, whereas at unsaturated moisture contents surface tension forces of the water menisci wedged between the particles lend cohesiveness and hence greater rigidity to the matrix.

The structure of nonideal granular materials in which the grains are not spherical but oblong or angular is much more difficult to formulate theoretically, as the orientations of particles must be taken into account.

## D. Structure of Aggregated Soils

In soils with an appreciable content of clay, the primary particles tend, under favorable circumstances, to group themselves into structural units known as *secondary particles*, or *aggregates*. Such aggregates are not characterized by any universally fixed size, nor are they necessarily stable. The visible aggregates, which are generally of the order of several millimeters to several centimeters in diameter, are often called *peds*, or *macroaggregates*. These are usually assemblages of smaller groupings, or microaggregates, which themselves are associations of the ultimate structural units, being the flocs, clusters, or packets of clay particles. Bundles of the latter units attach themselves to, and sometimes engulf, the much larger primary particles of sand and silt. The internal organization of these various groupings can now be studied by means of electron microscopy, particularly with the use of newly developed scanning techniques.

A prerequisite for aggregation is that the clay be flocculated. However, flocculation is a necessary but not a sufficient condition for aggregation. As stated by Richard Bradfield of Cornell University some 40 years ago, "aggregation is flocculation—plus!" That "plus" is *cementation*.

The complex interrelationship of physical, biological, and chemical reactions involved in the formation and degradation of soil aggregates was reviewed by Harris *et al.* (1965). An important role is played by the extensive networks of roots which permeate the soil and tend to enmesh soil aggregates. Roots exert pressures which compress aggregates and separate between adjacent ones. Water uptake by roots causes differential dehydration, shrinkage, and the opening of numerous small cracks. Moreover, root exudations and the continual death of roots and particularly of root hairs promote microbial activity which results in the production of humic cements.

Since these binding substances are transitory, as they are susceptible to further microbial decomposition, organic matter must be replenished and supplied continually if aggregate stability is to be maintained in the long run.

Active humus is accumulated and soil aggregates are stabilized most effectively under perennial sod-forming herbage. Annual cropping systems, on the other hand, hasten the decomposition of humus and the destruction of soil aggregates. The foliage of close-growing vegetation, and its residues, also protect surface soil aggregates against slaking by water, particularly under raindrop impact, to which aggregates become especially vulnerable if exposed and desiccated in the absence of a protective cover.

The influence of a cropping system on soil aggregation is seen to be a function of root activity (density and depth of rooting and the rate of root proliferation), density and continuity of surface cover, and the mode and frequency of cultivation and traffic. Crops or cropping systems which develop sparse roots, provide little vegetative cover, and require intensive mechanical cultivation of the soil are the least likely to maintain optimal tilth. Moreover, soil wetness at the time of cultivation has a great bearing on whether aggregated structure is maintained or destroyed. Cultivation of excessively wet soil causes puddling (plastic remolding), whereas cultivation of excessively dry soil is likely to result in grinding or pulverizing the soil into dust. In between lies a rather narrow range of wetness considered optimal for cultivation, at which the soil will readily break up into clods of the desired size range with the least amount of effort. The precise value of "optimal wetness" must be determined for each soil specifically, and perhaps for each type of tillage implement as well.[1]

When we speak of *microbial activity* as affecting soil aggregation we are obliged to point out that this catch-all phrase in reality refers to the time-variable activity of numerous microorganisms, including thousands of species of bacteria, fungi, actinomycetes, etc. Especially important are rhizospheric bacteria, which flourish in direct association with roots of specific plants, as well as fungi, which often form extensive adhesive networks of fine filaments known as mycelia or hyphae. The composition of the soil microfauna and microflora depends on the thermal and moisture regimes, on soil pH and oxidation–reduction potential, the nutrient status of the soil substrate, and the type and quantity of organic matter present (Alexander, 1977).

Soil microorganisms bind aggregates by a complex of mechanisms, such as adsorption, physical entanglement and envelopment, and cementation by excreted mucilagenous products. Prominent among the many microbial products capable of binding soil aggregates are polysaccharides, hemicelluloses or uronides, levans, as well as numerous other natural polymers. Such materials are attached to clay surfaces by means of cation bridges, hydrogen bonding, van der Waals forces, and anion adsorption mechanisms.

Polysaccharides, in particular, consist of large, linear, and flexible molecules capable of forming multiple bonds with several particles at once. In some cases, organic polymers hardly penetrate between the individual clay particles but form a protective capsule around soil aggregates. In other cases, solutions of active organic agents penetrate into soil aggregates and then precipitate more or less irreversibly as insoluble (though still biologically decomposable) cements.

In addition to increasing the strength and stability of intra-aggregate bonding, organic products may further promote aggregate stability by reducing wettability and swelling. Some of the organic materials are inherently hydrophobic, or become so as they dehydrate, so that the organo–clay complex may have a reduced affinity for water.

Some inorganic materials can also serve as cementing agents. The importance of clay, and its state of flocculation, should be obvious by now. Cohesiveness between clay particles is, in fact, the ultimate internal binding force within microaggregates. Calcium carbonate, as well as iron and aluminum oxides, can impart considerable stability to otherwise weak soil aggregates. The latter, in particular, are responsible for the significant stability of aggregates in tropical soils which often contain little organic matter.

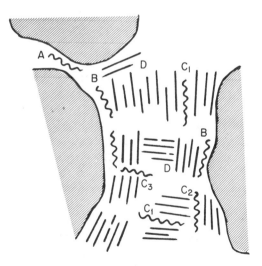

**Fig. 4.2.** Possible arrangements of quartz particles, clay domains, and organic matter in a soil aggregate. (A) Quartz–organic colloid–quartz, (B) quartz–organic colloid–clay domain, (C) clay domain–organic colloid–clay domain, (C$_1$) face–face, (C$_2$) edge–face, (C$_3$) edge–edge, (D) clay domain edge–clay domain face. (After Emerson, 1959.)

An interesting model of the internal bonding forms which can constitute a soil aggregate was provided by Emerson (1959) and is shown in Fig. 4.2. Emerson's model is composed of sand or silt-size quartz particles and of "domains" of oriented clay bonded by electrostatic forces. Stability of the aggregate is enhanced by linkage of organic polymers between the quartz particles and the faces or edges of clay crystals.

### E. Characterization of Soil Structure

Soil structure, or fabric, can be studied directly by microscopic observation of thin slices under polarized light. The structural associations of clay can be examined by means of electron microscopy, using either transmission or scanning techniques. The structure of single-grained soils, as well as of aggregated soils, can be considered quantitatively in terms of the total porosity and of the pore-size distribution. The specific structure of aggregated soils can, furthermore, be characterized qualitatively by specifying the typical shapes of aggregates found in various horizons within the soil profile, or quantitatively by measuring their sizes. Additional methods of characterizing soil structure are based on measuring mechanical properties and permeability to various fluids. None of these methods has been accepted universally. In each case, the choice of the method to be used depends upon the problem, the soil, the equipment available, and, not the least, upon the soil physicist.

*Total porosity* $f$ of a soil sample is usually computed on the basis of measured bulk density $\rho_b$, using the following equation (see Chapter 2):

$$f = 1 - \rho_b/\rho_s \tag{4.1}$$

where $\rho_s$ is the average particle density.

Bulk density is generally measured by means of a *core sampler* designed to extract "undisturbed" samples of known volume from various depths in the profile. An alternative is to measure the volumes and masses of individual clods (not including interclod cavities) by *immersion in mercury* or by coating with paraffin wax prior to *immersion in water*. Still other methods are the *sand-funnel* and *balloon techniques* used in engineering, and *gamma-ray attenuation densitometry* (Blake, 1965).

*Pore-size distribution* measurements can be made in coarse-grained soils by means of the *pressure-intrusion method* (Diamond, 1970), in which a nonwetting liquid, generally mercury, is forced into the pores of a predried sample. The pressure is applied incrementally, and the volume pentrated by the liquid is measured for each pressure step, equivalent (by the capillary theory) to a range of pore diameters. In the case of fine-grained soils, capillary condensation methods, or, more commonly, *desorption methods* are used (Vomocil, 1965). In the desorption method, a presaturated sample is subjected to a stepwise series of incremental suctions, and the capillary theory is used to obtain the equivalent pore-size distribution. Water is commonly used as the permeating liquid, though nonpolar liquids have also been tried in an attempt to assess, by the comparison, the possible effect of water saturation and desorption in modifying soil structure. Where the aggregates are fairly distinct, it is sometimes possible to divide pore-size distribution into two distinguishable ranges, namely *macropores* and *micropores*. The macropores are mostly the interaggregate cavities which serve as the principal avenues for the infiltration and drainage of water and for aeration. The micropores are the intraaggregate capillaries responsible for the retention of water and solutes. However, the demarcation is seldom truly distinct, and the separation between macropores and micropores is often arbitrary.

The shapes of aggregates observable in the field (illustrated in Fig. 4.3) can be classified as follows:

(1) *Platy:* Horizontally layered, thin and flat aggregates resembling wafers. Such structures occur, for example, in recently deposited clay soils.

(2) *Prismatic* or *columnar:* Vertically oriented pillars, often six sided, up to 15 cm in diameter. Such structures are common in the B horizon of clayey soils, particularly in semiarid regions. Where the tops are flat, these vertical aggregates are called prismatic, and where rounded—columnar.

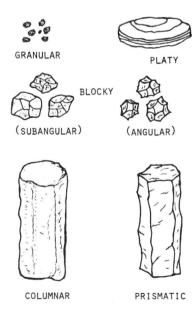

**Fig. 4.3.** Observable forms of soil aggregation.

(3) *Blocky:* Cubelike blocks of soil, up to 10 cm in size, sometimes angular with well-defined planar faces. These structures occur most commonly in the upper part of the B horizon.

(4) *Spherical:* Rounded aggregates, generally not much larger than 2 cm in diameter, often found in a loose condition in the A horizon. Such units are called granules, and where particularly porous, crumbs.

The shapes, sizes, and densities of aggregates generally vary within the profile. As the overburden pressure increases with depth, and inasmuch as the deeper layers do not experience such extreme fluctuations in moisture content (as does the alternately saturated and desiccated surface layer), the decrease of swelling and shrinkage activity causes the deeper aggregates to be larger. A typical structural profile in semiarid regions consists of a granulated A horizon underlain by a prismatic B horizon, whereas in humid temperate regions a granulated A horizon may occur with a platy or blocky B horizon. The number of variations found in nature is, however, legion.

Aggregate size distribution is a determinant of pore-size distribution and has a bearing on the erodibility of the surface, particularly by wind. In the field, adjacent particles often adhere to each other, though of course not as tenaciously as do the particles within each aggregate. Separating and classifying soil aggregates necessarily involves a disruption of the original, in situ structural arrangement. The application of too great a force may break up the aggregates themselves. Hence determination of aggregate size distribution depends on the mechanical means employed to separate the aggregates.

Aggregate screening methods were reviewed by Kemper and Chepil (1965). Screening through flat sieves is difficult to standardize and entails frequent clogging of the sieve openings. Chepil (1962) presented a detailed plan for a rotary sieve machine equipped with up to 14 concentrically nested sieves of various aperture sizes. The operation of this machine can be standardized, thus minimizing the arbitrary personal factor, and clogging is virtually eliminated. Samples for *dry sieving analysis* should be taken when the soil is reasonably dry, and care must be taken to avoid change of structure during handling The electrically rotated sieves are slanted downward so that the classified aggregates gradually tumble into separate bins for weighing.

Various indexes have been proposed for expressing the distribution of aggregate sizes. If a single characteristic parameter is desired, such as might allow correlation with various factors (e.g., erosion, infiltration, evaporation, or aeration), a method must be adopted for assigning an appropriate weighting factor to each size range of aggregates. A widely used index is the *mean weight diameter* (Van Bavel, 1949; Youker and McGuinness, 1956), based on weighting the masses of aggregates of the various size classes according to their respective sizes. The mean weight diameter $X$ is thus defined by the following equation:

$$X = \sum_{i=1}^{n} x_i w_i \qquad (4.2)$$

where $x_i$ is the mean diameter of any particular size range of aggregates separated by sieving, and $w_i$ is the weight of the aggregates in that size range as a fraction of the total dry weight of the sample analyzed. The summation accounts for all size ranges, including the group of aggregates smaller than the openings of the finest sieve.

An alternative index of aggregate size distribution is the *geometric mean diameter Y* calculated according to the equation (Mazurak, 1950):

$$Y = \exp\left[\left(\sum_{i=1}^{n} w_i \log x_i\right)\bigg/\left(\sum_{i=1}^{n} w_i\right)\right] \qquad (4.3)$$

wherein $w_i$ is the weight of aggregates in a size class of average diameter $x_i$, and $\sum_{i=1}^{n} w_i$ is the total weight of the sample.

Gardner (1956) reported that the aggregates of many soils exhibit a logarithmic–normal distribution which can be characterized in terms of two parameters, namely the *geometric mean diameter* and the *log standard deviation*. Other indexes proposed for characterizing aggregate size distribution are the so-called *coefficient of aggregation* (Retzer and Russell, 1941) and the *weighted mean diameter* and standard deviation (Puri and Puri, 1949). No universal prescription can be offered on which of these alternative indexes are to be preferred.

## F. Aggregate Stability

Determining the momentary state of aggregation of a soil at any particular time might not suffice to portray the soil's true structural characteristics as they may vary dynamically over a period of time. By any measure, the degree of aggregation is indeed a time-variable property, as aggregates form, disintegrate, and reform periodically. For instance, a newly cultivated field may for a time exhibit a nearly optimal array of aggregate sizes, with large interaggregate pores favoring high infiltration rates and unrestricted aeration. This blissful state often proves to be ephemeral, however, as many farmers have repeatedly discovered to their great chagrin. Soil structure may begin to deteriorate quite visibly and rapidly, as the soil is subject to destructive forces resulting from intermittent rainfall (causing slaking, swelling, shrinkage, and erosion) followed by dry spells (causing the soil to be vulnerable to deflation by wind). Repeated traffic, particularly by heavy machinery, furthermore tends to crush the aggregates remaining at the surface, and to compact the soil to some depth below the surface.

Soils vary, of course, in the degree to which they are vulnerable to externally imposed destructive forces. *Aggregate stability* is a measure of this vulnerability. More specifically, it expresses the resistance of aggregates to breakdown when subjected to potentially disruptive processes. Since the reaction of a soil to forces acting on it depends not only on the soil itself but also to a large degree upon the nature of the forces and the manner they are applied, aggregate stability is not measurable in absolute terms. Rather, it is a relative, and partly even subjective, concept. This, however, does not detract from its importance.

To test aggregate stability, soil physicists generally subject samples of aggregates to artificially induced forces designed to simulate phenomena which are likely to occur in the field. The nature of the forces applied during such testing depends on the investigator's perception of the natural phenomenon which he wishes to simulate, as well as on the equipment available and the mode of its employment. The degree of stability is then assessed by determining the fraction of the original sample's mass which has withstood destruction and retained its physical integrity, or, conversely, the fraction which appears, by some arbitrary but reproducible criterion, to have disintegrated.

If an indication of mechanical stability is sought, measurements can be made of the resistance of aggregates to prolonged dry sieving, or to crushing forces. The latter can be applied statically, as by compression between two plates; or dynamically, by dropping the aggregates from a given height onto a hard surface to observe the readiness with which they tend to shatter under repeated collisions. If the objective is to study the resistance to erosion by wind, the aggregates can be placed in a wind tunnel and tested for deflation under specified wind conditions (Chepil, 1958).

Most frequently, however, the concept of aggregate stability is applied in relation to the destructive action of water. Although mentioned before, it bears repeating that the very wetting of aggregates may cause their collapse, as the bonding substances dissolve or weaken and as the clay swells and possibly disperses. If wetting is nonuniform, one part of the aggregate will swell more than another, and the resulting stress, compounded during subsequent shrinkage, may fracture the aggregate. Aggregates are more vulnerable to sudden than to gradual wetting, owing to the air occlusion effect. Raindrops and flowing water provide the energy to detach particles and transport them away. Abrasion by particles carried as suspended matter in runoff water scours the surface and contributes to the overall breakdown of aggregated structure at the soil surface.

The classical and still most prevalent procedure for testing the water stability of soil aggregates is the *wet sieving method* (Tiulin, 1928; Yoder, 1936; De Boodt and De Leenheer, 1958). A representative sample of air-dry aggregates is placed on the uppermost of a set of graduated sieves and immersed in water to simulate flooding. The sieves are then oscillated vertically and rhythmically, so that water is made to flow up and down through the screens and the assemblage of aggregates. In this manner, the action of flowing water is simulated. At the end of a specified period of sieving (e.g., 20 min) the nest of sieves is removed from the water and the oven-dry weight of material left on each sieve is determined. As pointed out by Kemper (1965), the results should be corrected for the coarse primary particles retained on each sieve to avoid designating them falsely as aggregates. This is done by dispersing the material collected from each sieve, using a mechanical stirrer and a sodic dispersing agent, then washing the material back through the same sieve. The weight of sand retained after the second sieving is then subtracted from the total weight of undispersed material retained after the first sieving, and the percentage of stable aggregates %SA is given by

$$\%\text{SA} = 100 \times \frac{(\text{weight retained}) - (\text{weight of sand})}{(\text{total sample weight}) - (\text{weight of sand})} \qquad (4.4)$$

Obviously, the value of %SA depends upon the period of time the aggregates are shaken in water. Russell and Feng (1947) reported the following relationship:

$$\log(\%\text{SA}) = a - b \log t \qquad (4.5)$$

wherein $t$ is time, $a$ is the logarithm of the original weight of the sample, and $b$ is the slope of the $\log(\%\text{SA})$ versus $\log t$ curve. The difference in mean weight–diameter between dry-sieved and wet-sieved samples has also been used as an indicator of stability (De Leenheer and De Boodt, 1954).

An alternative approach is to subject soil aggregates to simulated rain. In the *drop method* (McCalla, 1944), individual aggregates are bombarded with drops of water in a standardized manner. The number of drops needed for total dissipation of the aggregate, or the fractional mass of the aggregate remaining after a given time, can be determined. A better way is to subject an aggregated soil surface in the field to periods of simulated rainstorms of controllable raindrop sizes and velocities (Morin *et al.*, 1967; Amerman *et al.*, 1970). The condition of the soil surface can then be compared to the initial condition, and the degree of aggregate stability thereby assessed.

## G. Soil Crusting

As we have already mentioned, it is the aggregates exposed at the soil surface which are most vulnerable to destructive forces. The surface aggregates which collapse and slake down during wetting may form a slick layer of dispersed mud, sometimes several centimeters thick, which clogs the surface macropores and thus tends to inhibit the infiltration of water into the soil and the exchange of gases between the soil and the atmosphere (aeration). Such a layer is often called a *surface seal*. Upon drying, this dispersed layer shrinks to become a dense, hard crust which impedes seedling emergence by its hardness and tears seedling roots as it cracks, forming a characteristic polygonal pattern.

The effect of soil crusting on seedlings depends on crust thickness and strength, as well as on the size of the seeds and vigor of the seedlings. Particularly sensitive are small seeded grasses and vegetables. In soils which exhibit a strong crusting tendency one can often observe that seedling emergence occurs only through the crust's cracks, while numerous unfortunate seedlings lie smothered under the hard crust fragments between cracks. Such an occurrence can doom an entire crop from the very outset.

Attempts have been made to characterize soil crusting, particularly with respect to its effect on seedling emergence, in terms of the resistance of the dry crust to the penetration of a probe (Parker and Taylor, 1965), as well as in terms of its strength as exhibited in the *modulus of rupture* test (Richards, 1953). These tests were designed to imitate the process by which a seedling forces its way upward by penetrating and rupturing the crust. Using the modulus of rupture technique, Richards (1953) reported that the emergence of bean seedlings in a fine sandy loam was reduced from 100 to 0% when crust strength increased from 108 to 273 mbar, whereas Allison (1956) found that the emergence of sweet corn was prevented only when crust strength exceeded 1200 mbar. However, the critical crust strength which prevents emergence obviously depends on crust thickness and soil wetness as well as on plant species and depth of seed placement (Hillel, 1972).

Crust strength increases as the rate of drying decreases and as the degree of colloidal dispersion increases. As water evaporates, the soil surface often becomes charged with a relatively high concentration of sodic salts and, consequently, with a high exchangeable sodium percentage. With subsequent infiltration of rain or irrigation water, the salts are leached but the exchangeable sodium percentage (ESP) remains high. The resulting combination of high ESP and low salt concentration induces colloidal dispersion, which contributes to the formation of a dense crust.

## Sample Problems

**1.** Calculate the mean weight diameters of the assemblages of aggregates given in the following tabulation. The percentages refer to the mass fractions of dry soil in each diameter range.

| Aggregate diameter range (mm) | Dry sieving | | Wet sieving | |
|---|---|---|---|---|
| | Virgin soil | Cultivated soil | Virgin soil | Cultivated soil |
| 0.0–0.5 | 10% | 25% | 30% | 50% |
| 0.5–1.0 | 10% | 25% | 15% | 25% |
| 1–2 | 15% | 15% | 15% | 15% |
| 2–5 | 15% | 15% | 15% | 5% |
| 5–10 | 20% | 10% | 15% | 4% |
| 10–20 | 20% | 7% | 5% | 1% |
| 20–50 | 10% | 3% | 5% | 0% |

First, we determine the mean diameters of the seven aggregate diameter ranges. These are

Range:     0–0.5,  0.5–1,  1–2,  2–5,  5–10,  10–20,  20–50 mm
Mean:      0.25,   0.75,   1.5,   3.5,   7.5,   15,    35 mm

Recall that the mean weight diameter $X$ is defined by Eq. (4.2):

$$X = \sum_{i=1}^{i=n} x_i w_i$$

Hence, for the dry-sieved virgin soil,

$$X = (0.25 \times 0.1) + (0.75 \times 0.1) + (1.5 \times 0.15) + (3.5 \times 0.15)$$
$$+ (7.5 \times 0.2) + (15 \times 0.2) + (35 \times 0.1) = 8.85 \text{ mm}$$

For the dry-sieved cultivated soil,

$$X = (0.25 \times 0.25) + (0.75 \times 0.25) + (1.5 \times 0.15) + (3.5 \times 0.15)$$
$$+ (7.5 \times 0.1) + (15 \times 0.07) + (35 \times 0.03) = 4.30 \text{ mm}$$

For the wet-sieved virgin soil,

$$X = (0.25 \times 0.3) + (0.75 \times 0.15) + (1.5 \times 0.15) + (3.5 \times 0.15)$$
$$+ (7.5 \times 0.15) + (15 \times 0.05) + (35 \times 0.05) = 4.56 \text{ mm}$$

For the wet-sieved cultivated soil,

$$X = (0.25 \times 0.5) + (0.75 \times 0.25) + (1.5 \times 0.15) + (3.5 \times 0.05)$$
$$+ (7.5 \times 0.04) + (15 \times 0.01) + (35 \times 0.0) = 1.16 \text{ mm}$$

*Note:* Wet sieving reduced the mean weight diameter from 8.85 to 4.56 mm in the virgin soil and from 4.30 to 1.16 mm in the cultivated soil. This indicates the degree of instability of the various aggregates under the slaking effect of immersion in water. The influence of cultivation is generally to reduce the water stability of soil aggregates and hence to render the soil more vulnerable to crusting and erosion processes.

# Part III     THE LIQUID PHASE

# 5  *Soil Water: Content and Potential*

## A. Introduction

The variable amount of water contained in a unit mass or volume of soil, and the energy state of water in the soil are important factors affecting the growth of plants. Numerous other soil properties depend very strongly upon water content. Included among these are mechanical properties such as consistency, plasticity, strength, compactibility, penetrability, stickiness, and trafficability. In clayey soils, swelling and shrinkage associated with addition or extraction of water change the overall specific volume (or bulk density) of the soil as well as its pore-size distribution. Soil water content also governs the air content and gas exchange of the soil, thus affecting the respiration of roots, the activity of microorganisms, and the chemical state of the soil (e.g., oxidation–reduction potential).

The per mass or per volume fraction of water in the soil can be characterized in terms of *soil wetness.* The physicochemical condition or state of soil water is characterized in terms of its free energy per unit mass, termed the *potential*. Of the various components of this potential, it is the *pressure* or *matric* potential which characterizes the tenacity with which soil water is held by the soil matrix. Wetness and matric potential are functionally related to each other, and the graphical representation of this relationship is termed the *soil-moisture characteristic* curve. The relationship is not unique, however, as it is affected by direction and rate of change of soil moisture and is sensitive to changes in soil volume and structure. Both wetness and matric potential vary widely in space and time as the soil is wetted by rain, drained by gravity, and dried by evaporation and root extraction.

The wettest possible condition of a soil is that of *saturation*, defined as the condition in which all the soil pores are filled with water. The saturation value is relatively easy to define in the case of nonswelling (e.g., sandy) soils. It can be difficult or even impossible to define in the case of swelling soils, as such soils may continue to imbibe water and swell even after all pores have been filled with water. The lowest wetness we are likely to encounter in nature is a variable state called *air dryness*, and in the laboratory it is an arbitrary state known as the *oven-dry* condition.

## B.  The Soil-Water Content (Wetness)[1]

The fractional content of water in the soil can be expressed in terms of either mass or volume ratios. As given in Chapter 2

$$w = M_w/M_s \tag{5.1}$$

$$\theta = V_w/V_t = V_w/(V_s + V_w + V_a) \tag{5.2}$$

where $w$, the mass wetness, is the dimensionless ratio of water mass $M_w$ to dry soil mass $M_s$, whereas $\theta$, the volume wetness, is the ratio of water volume $V_w$ to total (bulk) soil volume $V_t$. The latter is equal to the sum of the volumes of solids ($V_s$), water ($V_w$), and air ($V_a$). Both $\theta$ and $w$ are usually multiplied by 100 and reported as percentages by volume or mass.

The two expressions can be related to each other by means of the bulk density $\rho_b$ and the density of water:

$$\theta = w(\rho_b/\rho_w) = w\Gamma_b \tag{5.3}$$

where $\Gamma_b$ is the *bulk specific gravity* of the soil (a dimensionless ratio which usually lies in the range between 1.1 and 1.7). The conversion is relatively simple for nonswelling soils in which bulk density, and hence bulk specific gravity, are constant regardless of wetness, but it can be difficult in the case of swelling soils as the bulk density must be known as a function of mass wetness.

In many cases, it is useful to express the water content of a soil profile in terms of *depth*, i.e., as the volume of water contained in a specified total depth of soil $d_t$ per unit area of land. This indicates the equivalent depth

---

[1] The author prefers the term *wetness* to *soil water content* not only for reasons of verbal economy but also because wetness implies intensity whereas content implies extensity. One can speak of the water content when referring to the total amount of water in a bucket or in a defined volume of soil but the degree of wetness obviously pertains to the *relative* concentration (rather than the absolute amount) of water in a porous body, independent of the body's size.

$d_w$ soil water would have if it were extracted and then ponded over the soil surface. Thus

$$d_w = \theta d_t = w \Gamma_b d_t \tag{5.4}$$

Usually $d_w$ is given in millimeters, as are rainfall and evaporation.

Another expression of soil wetness is the *liquid ratio* $\vartheta$, defined as the volume of water per unit volume of the solid phase:

$$\vartheta = w \frac{\rho_s}{\rho_w} = \theta \frac{\rho_w}{\rho_b} \frac{\rho_s}{\rho_w} = \theta \frac{\rho_s}{\rho_b} \tag{5.5}$$

This expression is useful, along with the *void ratio* (defined in Chapter 2), in connection with soils having a nonrigid (swelling and shrinking) matrix.

## C. Measurement of Soil Wetness

The need to determine the amount of water contained in the soil arises frequently in many agronomic, ecological, and hydrological investigations aimed at understanding the soil's chemical, mechanical, hydrological, and biological relationships. There are direct and indirect methods to measure soil moisture (Gardner, 1965), and, as we have already pointed out, there are several alternative ways to express it quantitatively. As yet we have no universally recognized standard method of measurement and no uniform way to compute and present the results of soil moisture measurements. We shall proceed to describe, briefly, some of the most prevalent methods for this determination.

### 1. SAMPLING AND DRYING

The traditional (gravimetric) method of measuring mass wetness consists of removing a sample by augering into the soil and then determining its moist and dry weights. The *moist weight* is determined by weighing the sample as it is at the time of sampling, and the *dry weight* is obtained after drying the sample to a constant weight in an oven. The more or less standard method of drying is to place the sample in an oven at 105°C for 24 hr. An alternative method of drying, suitable for field use, is to impregnate the sample in a heat-resistant container with alcohol, which is then burned off, thus vaporizing the water (Bouyoucos, 1937). The mass wetness, also called *gravimetric wetness*, is the ratio of the weight loss in drying to the dry weight of the sample (mass and weight being proportional):

$$w = \frac{(\text{wet weight}) - (\text{dry weight})}{\text{dry weight}} = \frac{\text{weight loss in drying}}{\text{weight of dried sample}} \tag{5.6}$$

The gravimetric method, depending as it does on sampling, transporting, and repeated weighings, entails practically inevitable errors. It is also laborious and time consuming, since a period of at least 24 hr is usually considered necessary for complete oven drying. The standard method of oven drying is itself arbitrary. Some clays may still contain appreciable amounts of adsorbed water (Nutting, 1943) even at 105°C. On the other hand, some organic matter may oxidize and decompose at this temperature so that the weight loss may not be due entirely to the evaporation of water.

The errors of the gravimetric method can be reduced by increasing the sizes and number of samples. However, the sampling method is destructive and may disturb an observation or experimental plot sufficiently to distort the results. For these reasons, many workers prefer indirect methods, which permit making frequent or continuous measurements at the same points, and, once the equipment is installed and calibrated, with much less time, labor, and soil disturbance.

## 2. Electrical Resistance

The electrical resistance of a soil volume depends not only upon its water content, but also upon its composition, texture, and soluble-salt concentration. On the other hand, the electrical resistance of porous bodies placed in the soil and left to equilibrate with soil moisture can sometimes be calibrated against soil wetness. Such units (generally called *electrical resistance blocks*) generally contain a pair of electrodes embedded in gypsum (Bouyoucos and Mick, 1940), nylon, or fiberglass (Colman and Hendrix, 1949). (See Fig. 5.1.)

The electrical conductivity of moist porous blocks is due primarily to the permeating fluid rather than to the solid matrix. Thus it depends upon the electrolytic solutes present in the fluid as well as upon the volume content of the fluid. Blocks made of such inert materials as fiberglass, for instance, are highly sensitive to even small variations in salinity of the soil solution. On the other hand, blocks made of plaster of Paris (gypsum) maintain a nearly constant electrolyte concentration corresponding primarily to that

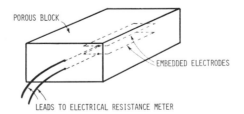

**Fig. 5.1.**  An electrical resistance block (schematic). The embedded electrodes may be plates, screens, or wires in a parallel or in a concentric arrangement.

of a saturated solution of calcium sulfate. This tends to mask, or buffer, the effect of small or even moderate variations in the soil solution (such as those due to fertilization or low levels of salinity). However, an undesirable consequence of the solubility of gypsum is that these blocks eventually deteriorate in the soil. Hence the relationship between electrical resistance and moisture suction varies not only from block to block but also for each block as a function of time, since the gradual dissolution of the gypsum changes the internal porosity and pore-size distribution of the blocks.

For these and other reasons (e.g., temperature sensitivity) the evaluation of soil wetness by means of electrical resistance blocks is likely to be of limited accuracy. On the other hand, an advantage of these blocks is that they can be connected to a recorder to obtain a continuous indication of soil moisture changes in situ.

## 3. NEUTRON SCATTERING

First developed in the 1950s, this method has gained widespread acceptance as an efficient and reliable technique for monitoring soil moisture in the field (Holmes, 1956; van Bavel, 1963). Its principal advantages over the gravimetric method are that it allows less laborious, more rapid, nondestructive, and periodically repeatable measurements, in the same locations and depths, of the volumetric wetness of a representative volume of soil. The method is practically independent of temperature and pressure. Its main disadvantages, however, are the high initial cost of the instrument, low degree of spatial resolution, difficulty of measuring moisture in the soil surface zone, and the health hazard associated with exposure to neutron and gamma radiation.

The instrument, known as a *neutron moisture meter* (Fig. 5.2) consists of two main components: (a) a *probe*, which is lowered into an access tube[2] inserted vertically into the soil, and which contains a *source of fast neutrons* and a *detector of slow neutrons*; (b) a *scaler* or ratemeter (usually battery powered and portable) to monitor the flux of slow neutrons scattered by the soil.

A source of fast neutrons is generally obtained by mixing a radioactive emitter of *alpha particles* (helium nuclei) with beryllium. Frequently used is a 2–5 millicurie pelletized mixture of radium and beryllium. An Ra–Be source emits about 16,000 neutrons per second per milligram (or millicurie) of radium. The energies of the neutrons emitted by this source vary from 1 to 15 MeV (million electron volts), with a preponderant energy range

---

[2] The purpose of the access tube is both to maintain the bore hole into which the probe is lowered and to standardize measuring conditions. Aluminum tubing is usually the preferred material for access tubes since it is nearly transparent to a neutron flux.

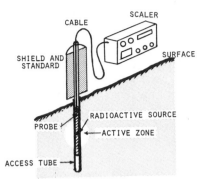

**Fig. 5.2.** Components of a portable neutron soil-moisture meter, including a probe (with a source of fast neutrons and a detector of slow neutrons) lowered from a shield containing hydrogenous material (e.g., paraffin, polyethylene, etc.) into the soil via an access tube. A scaler–ratemeter is shown alongside the probe. Recent models incorporate the scaler into the shield body and the integrated unit weights no more than 8 kgm.

of 2–4 MeV and an average speed of about 1600 km/sec. Hence, they are called *fast neutrons*. An alternative source of fast neutrons is a mixture of americium and beryllium. Both radium and americium incidentally also emit gamma radiation, but that of the americium is lower in energy and hence less hazardous than that of the radium. The source materials are chosen for their longevity (e.g., radium–beryllium has a half-life of about 1620 yr) so that they can be used for a number of years without an appreciable change in radiation flux.

The fast neutrons are emitted radially into the soil, where they encounter and collide elastically (as do billiard balls) with various atomic nuclei. Through repeated collisions, the neutrons are deflected and "scattered," and they gradually lose some of their kinetic energy. As the speed of the initially fast neutrons diminishes, it approaches a speed which is characteristic for particles at the ambient temperature. For neutrons this is about 2.7 km/sec, corresponding to an energy of about 0.03 eV. Neutrons that have been slowed to such a speed are said to be *thermalized* and are called *slow neutrons*. Such slow neutrons continue to interact with the soil and are eventually absorbed by the nuclei present.

The effectiveness of various nuclei present in the soil in moderating or thermalizing fast neutrons varies widely. The average loss of energy is maximal for collisions between particles of approximately the same mass. Of all nuclei encountered in the soil, the ones most nearly equal in mass to neutrons are the nuclei of hydrogen, which are therefore the most effective fast neutron moderators of all soil constituents. Thus, the average number of collisions required to slow a neutron from, say, 2 MeV to thermal energies is 18 for hydrogen, 114 for carbon, 150 for oxygen, and $9N + 6$ for nuclei of larger mass number $N$. If the soil contains an appreciable concentration of hydrogen, the emitted fast neutrons are thermalized before they get very far from the source, and the slow neutrons thus produced scatter randomly in the soil, quickly forming a swarm or cloud of constant density around the probe. The equilibrium density of the slow neutron cloud is determined by the rate of emission by the source and the rates of thermalization and absorption by the medium (i.e., soil) and is established within a small fraction of a second. Certain elements which might be present in the soil, incidentally, exhibit a high absorption capacity for slow neutrons (e.g., boron, cadmium, and chlorine), and their presence in nonnegligible concentrations might tend to decrease the density of slow neutrons. By and large, however, the density of slow neutrons formed around the probe is nearly proportional to the concentration of hydrogen in the soil, and therefore more or less proportional to the volume fraction of water present in the soil.

As the thermalized neutrons repeatedly collide and bounce about randomly, a number of them (proportional to the density of neutrons thus thermalized and scattered, and hence approximately linearly related to the concentration of soil moisture) return to the probe, where they are counted by the detector of slow neutrons. The detector cell is usually filled with $BF_3$ gas. When a thermalized neutron encounters a $^{10}B$ nucleus and is absorbed, an alpha particle is emitted, creating an electrical pulse on a charged wire. The number of pulses over a measured time interval is counted by a scaler, or indicated by a ratemeter.

The effective volume of soil in which the water content is measured depends upon the energy of emitted neutrons as well as upon the concentration of hydrogen nuclei; i.e., for a given source and soil, it depends by and large upon the volume concentration of soil moisture. If the soil is dry rather than wet, the cloud of slow neutrons surrounding the probe will be less dense and extend farther from the source, and vice versa for wet soil. With the commonly used radium–beryllium or americium–beryllium sources, the so-called *sphere of influence*, or effective volume of measurement, varies with a radius of less than 10 cm in a wet soil to 25 cm or more in a dry soil. This low and variable degree of spatial resolution makes the neutron moisture meter unsuitable for the detection of moisture profile discontinuities (e.g., wetting fronts or boundaries between distinct layers in the soil). Moreover, measurements close to the surface (say, within 20 cm of the surface, depending on soil wetness) are precluded[3] because of the escape of fast neutrons through the surface. However, the relatively large volume monitored can be an advantage in water balance studies, for instance, as such a volume is more truly representative of field soil than the very much smaller samples generally taken for the gravimetric measurement of soil moisture.

---

[3] A special surface probe is available commercially to allow measurement of the average moisture in the soil's top layer. It is used mostly in engineering practice in connection with soil compaction.

For the sake of safety, and also to provide a convenient means of making standard readings, the probe containing the fast neutron source is normally carried inside a protective *shield*, which is a cylindrical container filled with lead and some hydrogenous material such as paraffin or polyethylene designed to prevent the escape of fast neutrons. The shield should also protect users of the neutron soil moisture meter against emitted gamma radiation. Improper or excessive use of the equipment can be hazardous. The danger from exposure to radiation depends upon the strength of the source, the quality of the shield, the distance from source to operator, and the duration of contact. With strict observance of safety rules, the equipment can be used without undue risk. However, it is altogether too easy to become complacent and careless, since the radiation can be neither seen nor felt. A recent analysis of the radiation exposure hazard was given by Gee *et al.* (1976).

### 4.  OTHER METHODS

Additional approaches to the measurement of soil wetness include γ-ray absorption, the dependence of soil thermal properties upon water content, and the use of ultrasonic waves, radar waves, and dielectric properties. Some of these and other methods have been tried in connection with the remote sensing of land areas from aircraft or satellites. However, most of the methods currently under development are not yet practical for routine use in the field.

### D.  Energy State of Soil Water

Soil water, like other bodies in nature, can contain energy in different quantities and forms. Classical physics recognizes two principal forms of energy, *kinetic* and *potential*. Since the movement of water in the soil is quite slow, its kinetic energy, which is proportional to the velocity squared, is generally considered to be negligible. On the other hand, the potential energy, which is due to position or internal condition, is of primary importance in determining the state and movement of water in the soil.

The potential energy of soil water varies over a very wide range. Differences in potential energy of water between one point and another give rise to the tendency of water to flow within the soil. The spontaneous and universal tendency of all matter in nature is to move from where the potential energy is higher to where it is lower and for each parcel of matter to equilibrate with its surroundings. Soil water obeys the same universal pursuit of that elusive state known as *equilibrium*, definable as a condition of uniform potential energy throughout. In the soil, water moves constantly in the direction of decreasing potential energy. The rate of decrease of potential energy with distance is in fact the moving force causing flow. A knowledge of the relative potential energy state of soil water at each point within the soil can allow us to evaluate the forces acting on soil water in all directions, and to determine how far the water in a soil system is from equilibrium. This is analagous to the well-known fact that an object will tend to fall spontaneously from a higher to a lower elevation, but that lifting it requires work. Since potential energy is a measure of the amount of work a body can perform by virtue of the energy stored in it, knowing the potential energy state of water in the soil and in the plant growing in that soil can help us to estimate how much work the plant must expend to extract a unit amount of water.

Clearly, it is not the absolute amount of potential energy "contained" in the water which is important in itself, but rather the relative level of that energy in different regions within the soil. The concept of *soil-water potential*[4] is a criterion, or yardstick, for this energy. It expresses the specific potential energy of soil water relative to that of water in a standard reference state. The standard state generally used is that of a hypothetical reservoir of pure water, at atmospheric pressure, at the same temperature as that of soil water (or at any other specified temperature), and at a given and constant elevation. Since the elevation of this hypothetical reservoir can be set at will, it follows that the potential which is determined by comparison with this standard is not absolute, but by employing even so arbitrary a criterion we can determine the relative magnitude of the specific potential energy of water at different locations or times within the soil.

Just as an energy increment can be viewed as the product of a force and a distance increment, so the ratio of a potential energy increment to a distance increment can be viewed as constituting a force. Accordingly, the force acting on soil water, directed from a zone of higher to a zone of lower potential, is equal to the negative *potential gradient* ($-d\phi/dx$), which is the change of energy potential with distance $x$. The negative sign indicates that the force acts in the direction of decreasing potential.

The concept of soil-water potential is of great fundamental importance. This concept replaces the arbitrary categorizations which prevailed in the early stages of the development of soil physics and which purported to recognize and classify different forms of soil water: e.g., gravitational water, capillary water, hygroscopic water. The fact is that all of soil water, not merely a part of it, is affected by the earth's gravitational field, so that in

---

[4]The potential concept was first applied to soil water by Buckingham, in his classic and still-pertinent paper on the "capillary" potential (1907). Gardner (1920) showed how this potential is dependent upon the water content. Richards (1931) developed the tensiometer for measuring it in situ.

effect it is all gravitational. Furthermore, the laws of capillarity do not begin or cease at certain values of wetness or pore sizes.

In what way, then, does water differ from place to place and from time to time within the soil? Not in form, but in potential energy. The possible values of soil-water potential are continuous, and do not exhibit any abrupt discontinuities or changes from one condition to another (except in changes of phase). Rather than attempt to classify soil water, our task therefore is to obtain the measure of its potential energy state.

When the soil is saturated and its water is at a hydrostatic pressure greater than the atmospheric pressure (as, for instance, under a water table) the potential energy level of that water may be greater than that of the reference state reservoir described, and water will tend to move spontaneously from the soil into such a reservoir. If, on the other hand, the soil is moist but un-saturated, its water will no longer be free to flow out toward a reservoir at atmospheric pressure. On the contrary, the spontaneous tendency will be for the soil to draw water from such a reservoir if placed in contact with it, much as a blotter draws ink.

Under hydrostatic pressures greater than atmospheric, the potential of soil water (in the absence of osmotic effects) is greater than that of the reference state and therefore can be considered positive. In an unsaturated soil, the water is constrained by capillary and adsorptive forces, hence its energy potential is generally negative, since its equivalent hydrostatic pressure is less than that of the reference state. The potential energy of soil water is also reduced by the presence of solutes, i.e., by the osmotic effect.

Under normal conditions in the field, the soil is generally unsaturated and the soil-water potential is negative. Its magnitude at any point depends not only on hydrostatic pressure but also upon such additional physical factors as elevation (relative to that of the reference elevation), concentration of solutes, and temperature.

## E. Soil-Water Potential

We have already described the energy potential of soil water in a qualitative way. Thermodynamically, this energy potential can be regarded in terms of the difference in partial specific free energy between soil water and standard water. More explicitly, a soil physics terminology committee of the International Soil Science Society (Aslyng, 1963) defined the *total potential* of soil water as "the amount of work that must be done per unit quantity of pure water in order to transport reversibly and isothermally an infinitesimal quantity of water from a pool of pure water at a specified elevation at atmospheric pressure to the soil water (at the point under consideration)."

This is merely a formal definition, since in actual practice the potential is not measured by transporting water as per the definition, but by measuring some other property related to the potential in some known way (e.g., hydrostatic pressure, vapor pressure, elevation, etc.). The definition specifies transporting an infinitesimal quantity, in any case, to ensure that the determination procedure does not change either the reference state (i.e., the pool of pure, free water) or the soil-water potential being measured. It should be recognized that this definition provides a conceptual rather than an actual working tool.

Soil water is subject to a number of force fields, which cause its potential to differ from that of pure, free water. Such force fields result from the attraction of the solid matrix for water, as well as from the presence of solutes and the action of external gas pressure and gravitation. Accordingly, the *total potential* of soil water can be thought of as the sum of the separate contributions of these various factors, as follows:

$$\phi_t = \phi_g + \phi_p + \phi_o + \cdots \tag{5.7}$$

where $\phi_t$ is the total potential, $\phi_g$ the gravitational potential, $\phi_p$ the pressure (or matric) potential, $\phi_0$ the osmotic potential, and the ellipses signify that additional terms are theoretically possible.

Not all of the separate potentials given in Eq. (5.7) act in the same way, and their separate gradients may not always be equally effective in causing flow (for example, the osmotic potential gradient requires a semipermeable membrane to induce liquid flow). The main advantage of the total-potential concept is that it provides a unified measure by which the state of water can be evaluated at any time and everywhere within the soil–plant–atmosphere continuum.

## 1. GRAVITATIONAL POTENTIAL

Every body on the earth's surface is attracted toward the earth's center by a gravitational force equal to the weight of the body, that weight being the product of the mass of the body and the gravitational acceleration. To raise a body against this attraction, work must be expended, and this work is stored by the raised body in the form of *gravitational potential energy*. The amount of this energy depends on the body's position in the gravitational force field.

The gravitational potential of soil water at each point is determined by the elevation of the point relative to some arbitrary reference level. For the sake of convenience, it is customary to set the reference level at the elevation of a pertinent point within the soil, or below the soil profile being considered, so that the gravitational potential can always be taken as positive or zero.

On the other hand, if the soil surface is chosen as the reference level, as is often done, then the gravitational potential for all points below the surface is negative with respect to that reference level.

At a height $z$ above a reference, the gravitational potential energy $E_g$ of a mass $M$ of water occupying a volume $V$ is

$$E_g = Mgz = \rho_w Vgz \qquad (5.8)$$

where $\rho_w$ is the density of water and $g$ the acceleration of gravity. Accordingly, the gravitational potential in terms of the potential energy per unit mass is

$$\phi_g = gz \qquad (5.9)$$

and in terms of potential energy per unit volume is

$$\phi_{g,v} = \rho_w gz \qquad (5.10)$$

The gravitational potential is independent of the chemical and pressure conditions of soil water, and dependent only on relative elevation.

## 2. Pressure Potential

When soil water is at hydrostatic pressure greater than atmospheric, its pressure potential is considered positive. When it is at a pressure lower than atmospheric (a subpressure commonly known as *tension* or *suction*) the pressure potential is considered negative. Thus, water under a free-water surface is at positive pressure potential, while water at such a surface is at zero pressure potential, and water which has risen in a capillary tube above that surface is characterized by a negative pressure potential. This principle is illustrated in Fig. 5.3.

The positive pressure potential which occurs below the ground water level has been termed the *submergence potential*. The hydrostatic pressure $P$ of water with reference to the atmospheric pressure is

$$P = \rho g h \qquad (5.11)$$

where $h$ is the submergence depth below the free-water surface (called the *piezometric head.*

The potential energy    of this water is then

$$E = P \, dV \qquad (5.12)$$

and thus the submergence potential, taken as the potential energy per unit volume, is equal to the hydrostatic pressure, $P$:

$$\phi_{ps} = P \qquad (5.13)$$

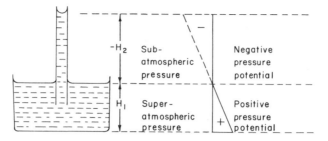

**Fig. 5.3.** Superatmospheric and subatmospheric pressures below and above a free-water surface.

A negative pressure potential has often been termed *capillary potential*, and more recently, *matric potential*.[11] This potential of soil water results from the capillary and adsorptive forces due to the soil matrix. These forces attract and bind water in the soil and lower its potential energy below that of bulk water. Capillarity results from the surface tension of water and its contact angle with the solid particles. In an unsaturated (three-phase) soil system, curved menisci form which obey the equation of capillarity equation of capillarity

$$P_0 - P_c = \Delta P = \gamma(1/R_1 + 1/R_2) \qquad (5.14)$$

where $P_0$ is the atmospheric pressure, conventionally taken as zero, $P_c$ the pressure of soil water, which can be smaller than atmospheric, $\Delta P$ is the pressure deficit, or subpressure, of soil water, $\gamma$ the surface tension of water, and $R_1$ and $R_2$ are the principal radii of curvature of a point on the meniscus.

If the soil were like a simple bundle of capillary tubes, the equations of capillarity might by themselves suffice to describe the relation of the negative pressure potential, or tension, to the radii of the soil pores in which the menisci are contained. However, in addition to the capillary phenomenon, the soil also exhibits adsorption, which forms hydration envelopes over the particle surfaces. These two mechanisms of soil–water interaction are illustrated in Fig. 5.4.

The presence of water in films as well as under concave menisci is most important in clayey soil and at high suctions, and it is influenced by the electric double layer and the exchangeable cations present. In sandy soils, adsorption is relatively unimportant and the capillary effect predominates. In general, however, the negative pressure potential results from the combined effect of the two mechanisms, which cannot easily be separated, since the capillary "wedges" are at a state of internal equilibrium with the adsorption "films," and the ones cannot be changed without affecting the others. Hence, the older term capillary potential is inadequate and the better term is *matric potential*, as it denotes the total effect resulting from the affinity of the water to the whole matrix of the soil, including its pores and particle surfaces together.

Some soil physicists prefer to separate the positive pressure potential from the matric potential, assuming the two to be mutually exclusive. Accordingly, soil water may exhibit either of the two potentials, but not both simultaneously. Unsaturated soil has no pressure potential, only a matric potential, expressible in negative pressure units. This, however, is really a matter of formality. There is an advantage in unifying the positive pressure potential and the matric potential (with the latter considered merely as a negative pressure potential) in that this unified concept allows one to consider the entire moisture profile in the field in terms of a single continuous potential extending from the saturated region into the unsaturated region, below and above the water table.

An additional factor which may affect the pressure of soil water is a possible change in the pressure of the ambient air. In general, this effect is negligible, as the atmospheric pressure remains nearly constant, small barometric pressure fluctuations notwithstanding. However, in the laboratory, the application of air pressure to change soil-water pressure or suction is a common practice. Hence, this effect has been recognized and termed the *pneumatic potential*. In an unsaturated soil $\phi_p$ can be taken as equal to the sum of the matric ($\phi_m$) and pneumatic ($\phi_a$) potentials.

In the absence of solutes, the liquid and vapor phases in an unsaturated porous medium are related at equilibrium by

$$\phi_m = RT \ln(p/p_0) \qquad (5.15)$$

where $p/p_0$ is the relative humidity, $R$ the gas constant for water vapor, and $T$ the absolute temperature. The term $p/p_0$ represents the ratio of the partial pressure of vapor in the air phase relative to the partial pressure at saturation.

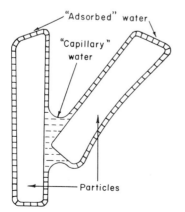

**Fig. 5.4.** Water in an unsaturated soil is subject to capillarity and adsorption, which combine to produce a matric suction.

## 3. OSMOTIC POTENTIAL

The presence of solutes in soil water affects its thermodynamic properties and lowers its potential energy. In particular, solutes lower the vapor pressure of soil water. While this phenomenon may not affect liquid flow in the soil significantly, it does come into play whenever a membrane or diffusion barrier is present which transmits water more readily than salts. The osmotic effect is important in the interaction between plant roots and soil, as well as in processes involving vapor diffusion.

Figure 5.5a is a schematic representation of a pure solvent separated from a solution by a semipermeable membrane. Solvent will pass through the membrane and enter the solution compartment, driving the solution level up the left-hand tube until the hydrostatic pressure of the column of dilute solution on the left is sufficient to counter the diffusion pressure of the solvent molecules drawn into the solution through the membrane. The hydrostatic pressure at equilibrium, when solvent molecules are crossing the membranes in both directions at equal rates, is the *osmotic pressure* of the solution. To measure the osmotic pressure of a solution, it is not necessary to wait until the flow stops and equilibrium is established. Theoretically, a counter pressure can be applied with a piston on the solution side, and when the counter pressure thus applied (Fig. 5.5b) is just sufficient to prevent the osmosis of water into the solution, it represents the osmotic pressure of the solution.[5]

The expression osmotic pressure of a solution can be misleading. What a solution exhibits relative to the pure solvent (say, water) is not an excess pressure but, on the contrary, a "suction" such that will draw water from a reservoir of pure water brought into contact with the solution across a semipermeable boundary. Hence we ought perhaps to speak of an osmotic suction as the characteristic property of a solution rather than of osmotic pressure, which, however, is the conventional term. The actual *process* of osmosis will obviously take place only in the presence of a semipermeable membrane separating a solution from its pure solvent, or from another solution of different concentration. In principle, however, the *property* of the solution which induces the process of osmosis exists whether or not a membrane happens to be present as it derives fundamentally from the decrease of potential energy of water in solution relative to that of pure water. The other manifestations of this property are a decrease in vapor pressure, a rise of the boiling point, and a depression of the freezing point (popularly known as the "antifreeze" effect).

---

[5] If a pressure greater than the osmotic pressure is applied on the solution side, some solvent will be forced out of the solution compartment through the membrane. This is the basis of the *reverse osmosis* method of water purification.

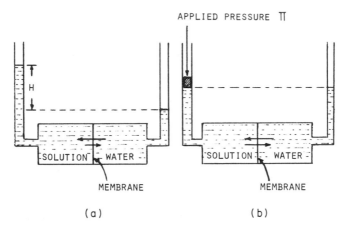

**Fig. 5.5.**  Osmosis and osmotic pressure. (a) Osmosis: the flow of water molecules through the membrane into the solution is at first greater than the reverse flow from the solution into the water compartment. The hydrostatic pressure due to the column of expanded solution increases the rate of water flow from the solution to the water compartment until at equilibrium, the opposite flows are equal. (b) The osmotic pressure of the solution is equal to the hydrostatic pressure ∏ which must be applied to the solution to equalize the rate of flow to and from the solution and produce a net flow of zero.

The term semipermeable membrane, first coined by van't Hoff in 1886 and since hallowed by tradition, is itself something of a misnomer. Membrane selectivity toward different species in a solution is not an absolute but a relative property. As such, it can be characterized by means of a parameter known as the selectivity or reflection coefficient, which varies from zero in the case of nonselectivity to unity in the (hypothetical) case of perfect selectivity. The nearest thing to a perfectly selective membrane is probably an air space separating two aqueous solutions of a completely nonvolatile solute. Such an air barrier allows free passage of solvent (water) molecules from liquid to vapor to liquid (from the more dilute to the more concentrated solution) while restricting the solute. However, most known porous membranes, including the various biological membranes, are "leaky," in the sense that they transmit molecules of the solute, as well as of the solvent, to a greater or lesser degree. The mechanism of selectivity is often related to pore size, as is evident from the often observed fact that solutes of small molecular weight tend to pass through membranes more readily than do solutes of large molecular weight (e.g., polymers). However, in many cases membranes are not merely molecular sieves. Often the molecules of solute are unable to penetrate the complex network of the membrane's pores owing to the preferential adsorption of solvent molecules on the membrane's inner surfaces, which makes the pores effectively smaller in diameter. In some cases the solute is subject to "negative adsorption" or repulsion by the membrane. In still other cases the semipermeable membrane may consist of a large number of fine capillaries not wetted by the liquid solution but through which molecules of the solvent can pass in the vapor phase. Osmosis can thus occur by distillation through the membrane from the region of higher to lower vapor pressure (i.e., from the more dilute to the more concentrated solution).

In dilute solutions, the osmotic pressure is proportional to the concentration of the solution and to its temperature according to the following equation:

$$\Pi = MRT \tag{5.16}$$

where $\Pi$ is the osmotic pressure in atmospheres, $M$ the total molar concentration of solute particles (whether molecules or dissociated ions), $T$ the temperature in degrees Kelvin, and $R$ the gas constant (0.08205 liter atm/deg mole).

## F. Quantitative Expression of Soil-Water Potential

The soil-water potential (Table 5.1) is expressible physically in at least three ways.

1. *Energy per unit mass:* This is often taken to be the fundamental expression of potential, using units of ergs per gram or joules per kilogram. The dimensions of energy per unit mass are $L^2T^{-2}$.

2. *Energy per unit volume:* Since water is a practically incompressible liquid, its density is almost independent of potential. Hence, there is a direct proportion between the expression of the potential as energy per unit mass and its expression as energy per unit volume. The latter expression yields the dimensions of pressure (for, just as energy can be expressed as the product of pressure and volume, so the ratio of energy to volume gives a pressure). This equivalent pressure can be measured in terms of dynes per square centimeter, Newtons per square meter, bars, or atmospheres. The basic dimensions are those of force per unit area: $ML^{-1}T^{-2}$. This method of expression is convenient for the osmotic and pressure potentials, but is seldom used for the gravitational potential.

3. *Energy per unit weight* (*hydraulic head*): Whatever can be expressed in units of hydrostatic pressure can also be expressed in terms of an equivalent head of water, which is the height of a liquid column corresponding to the given pressure. For example, a pressure of 1 atm is equivalent to a vertical water column, or hydraulic head, of 1033 cm, and to a mercury head of 76 cm. This method of expression is certainly simpler, and often

**Table 5.1**

ENERGY LEVELS OF SOIL WATER

| Soil-water potential | | | | Soil-water suction[a] | | Vapor pressure (torr) 20 C (%) | Relative humidity[b] at 20 C (%) |
|---|---|---|---|---|---|---|---|
| Per unit mass | | Per unit volume | | Pressure (bar) | Head (cm H$_2$O) | | |
| erg/gm | joule/kg | bar | cm H$_2$O | | | | |
| 0 | 0 | 0 | 0 | 0 | 0 | 17.5350 | 100.00 |
| $-1 \times 10^4$ | $-1$ | $-0.01$ | $-10.2$ | 0.01 | 10.2 | 17.5349 | 100.00 |
| $-5 \times 10^4$ | $-5$ | $-0.05$ | $-51.0$ | 0.05 | 51.0 | 17.5344 | 99.997 |
| $-1 \times 10^5$ | $-10$ | $-0.1$ | $-102.0$ | 0.1 | 102.0 | 17.5337 | 99.993 |
| $-2 \times 10^5$ | $-20$ | $-0.02$ | $-204.0$ | 0.2 | 204.0 | 17.5324 | 99.985 |
| $-3 \times 10^5$ | $-30$ | $-0.3$ | $-306.0$ | 0.3 | 306.0 | 17.5312 | 99.978 |
| $-4 \times 10^5$ | $-40$ | $-0.4$ | $-408.0$ | 0.4 | 408.0 | 17.5299 | 99.971 |
| $-5 \times 10^5$ | $-50$ | $-0.5$ | $-510.0$ | 0.5 | 510.0 | 17.5286 | 99.964 |
| $-6 \times 10^5$ | $-60$ | $-0.6$ | $-612.0$ | 0.6 | 612.0 | 17.5273 | 99.965 |
| $-7 \times 10^5$ | $-70$ | $-0.7$ | $-714.0$ | 0.7 | 714.0 | 17.5260 | 99.949 |
| $-8 \times 10^5$ | $-80$ | $-0.8$ | $-816.0$ | 0.8 | 816.0 | 17.5247 | 99.941 |
| $-9 \times 10^5$ | $-90$ | $-0.9$ | $-918.0$ | 0.9 | 918.0 | 17.5234 | 99.934 |
| $-1 \times 10^6$ | $-100$ | $-1.0$ | $-2040$ | 1.0 | 2040 | 17.5222 | 99.927 |
| $-2 \times 10^6$ | $-200$ | $-2$ | $-1040$ | 2 | 1040 | 17.5089 | 99.851 |
| $-3 \times 10^6$ | $-300$ | $-3$ | $-3060$ | 3 | 3060 | 17.4961 | 99.778 |
| $-4 \times 10^6$ | $-400$ | $-4$ | $-4080$ | 4 | 4080 | 17.4833 | 99.705 |
| $-5 \times 10^6$ | $-500$ | $-5$ | $-5100$ | 5 | 5100 | 17.4704 | 99.637 |
| $-6 \times 10^6$ | $-600$ | $-6$ | $-6120$ | 6 | 6120 | 17.4572 | 99.556 |

[a] In the absence of osmotic effects (soluble salts), soil-water suction equals matric suction; otherwise, it is the sum of matric and osmotic suctions.

[b] Relative humidity of air in equilibrium with the soil at different suction values.

more convenient, than the previous methods. Hence, it is common to characterize the state of soil water in terms of the total potential head, the gravitational potential head, and the pressure potential head, which are usually expressible in centimeters of water. Accordingly, instead of

$$\phi = \phi_g + \phi_p \tag{5.17}$$

one could write

$$H = H_g + H_p$$

which reads: The total potential head of soil water ($H$) is the sum of the gravitational ($H_g$) and pressure ($H_p$) potential heads. $H$ is commonly called, simply, the *hydraulic head*.

In attempting to express the negative pressure potential of soil water (relative to atmospheric pressure) in terms of an equivalent hydraulic head, we must contend with the fact that this head may be of the order of $-10,000$ or even $-100,000$ cm of water. To avoid the use of such cumbersomely large numbers, Schofield (1935) suggested the use of pF (by analogy with the pH acidity scale) which he defined as the logarithm of the negative pressure (tension, or suction) head in centimeters of water. A pF of 1 is, thus, a tension head of 10 cm $H_2O$, a pF of 3 is a tension head of 1000 cm $H_2O$, and so forth.

The use of various alternative methods for expressing the soil-water potential can be perplexing to the uninitiated. It should be understood that these alternative expressions are in fact equivalent, and each method of expression can be translated directly into any of the other methods. If we use $\phi$ to designate the potential in terms of energy per mass, $P$ for the potential in terms of pressure, and $H$ for the potential head, then

$$\phi = P/\rho_w \qquad (5.18)$$

$$H = P/\rho_w g = \phi/g \qquad (5.19)$$

where $\rho_w$ is the density of liquid water and $g$ the acceleration of gravity.

A remark is in order concerning the use of the synonymous terms tension and suction in lieu of negative or subatmospheric pressure. Tension and suction are merely semantic devices to avoid the use of the unesthetic negative sign which generally characterizes the pressure of soil water, and to allow us to speak of the osmotic and matric potentials in positive terms.

## G. Soil-Moisture Characteristic Curve

In a saturated soil at equilibrium with free water at the same elevation, the actual pressure is atmospheric, and hence the hydrostatic pressure and the suction (or tension) are zero.

If a slight suction, i.e., a water pressure slightly subatmospheric, is applied to water in a saturated soil, no outflow may occur until, as suction is increased, a certain critical value is exceeded at which the largest pore of entry begins to empty. This critical suction is called the *air-entry suction*. Its value is generally small in coarse-textured and in well-aggregated soils. However, since in coarse-textured soils the pores are often more nearly uniform in size, these soils may exhibit critical air-entry phenomena more distinctly and sharply than do fine-textured soils.

As suction is further increased, more water is drawn out of the soil and more of the relatively large pores, which cannot retain water against the suction applied, will empty out. Recalling the capillary equation[6] ($-P = 2\gamma/r$), we can readily predict that a gradual increase in suction will result in the emptying of progressively smaller pores, until, at high suction values, only the very narrow pores retain water. Similarly, an increase in soil-water

---

[6]The use of this equation assumes that the contact angle is zero and that soil pores are approximately cylindrical.

suction is associated with a decreasing thickness of the hydration envelopes covering the soil-particle surfaces. Increasing suction is thus associated with decreasing soil wetness. The amount of water remaining in the soil at equilibrium is a function of the sizes and volumes of the water-filled pores and hence it is a function of the matric suction. This function is usually measured experimentally, and it is represented graphically by a curve known as the soil-moisture retention curve, or the soil-moisture characteristic (Childs, 1940).

The amount of water retained at relatively low values of matric suction (say, between 0 and 1 bar of suction) depends primarily upon the capillary effect and the pore-size distribution, and hence is strongly affected by the structure of the soil. On the other hand, water retention in the higher suction range is due increasingly to adsorption and is thus influenced less by the structure and more by the texture and specific surface of the soil material. According to Gardner (1968), the water content at a suction of 15 bar (often taken to be the lower limit of soil moisture availability to plants) is fairly well correlated with the surface area of a soil and would represent, roughly, about 10 molecular layers of water if it were distributed uniformly over the particle surfaces.

It should be obvious from the foregoing that the soil-moisture characteristic curve is strongly affected by soil texture. The greater the clay content, in general, the greater the water retention at any particular suction, and the more gradual the slope of the curve. In a sandy soil, most of the pores are relatively large, and once these large pores are emptied at a given suction, only a small amount of water remains. In a clayey soil, the pore-size distribution is more uniform, and more of the water is adsorbed, so that increasing the matric suction causes a more gradual decrease in water content (Fig. 5.6).

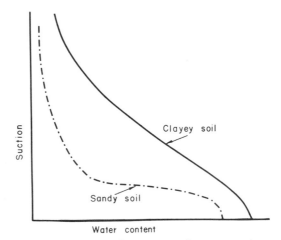

**Fig. 5.6.**   The effect of texture on soil-water retention.

Soil structure also affects the shape of the soil-moisture characteristic curve, particularly in the low-suction range. The effect of compaction upon a soil is to decrease the total porosity, and, especially, to decrease the volume of the large interaggregate pores. This means that the saturation water content and the initial decrease of water content with the application of low suction are reduced. On the other hand, the volume of intermediate-size pores is likely to be somewhat greater in a compact soil (as some of the originally large pores have been squeezed into intermediate size by compaction), while the intraaggregate micropores remain unaffected and thus the curves for the compacted and uncompacted soil may be nearly identical in the high suction range (Fig. 5.7). In the very high suction range, the predominant mechanism of water retention is adsorptive rather than capillary, and hence the retention capacity becomes more of a textural than a soil-structural attribute.

If two soil bodies differing in texture or structure are brought into direct physical contact, they will tend toward a state of potential-energy equilibrium in which the water potential would become equal throughout, but each of the two bodies will retain an amount of water determined by its own soil-moisture characteristic. Two soil bodies or layers can thus attain equilibrium and yet exhibit a marked nonuniformity, or discontinuity, in wetness. We can easily envision a situation in which a drier soil layer will contribute water to a wetter one simply because the water potential of the former is higher than that of the latter, owing to textural, structural, elevational, osmotic, or thermal differences.

In a nonshrinking soil, the soil-moisture characteristic curve, once obtained, allows calculation of the effective pore-size distribution (i.e., the

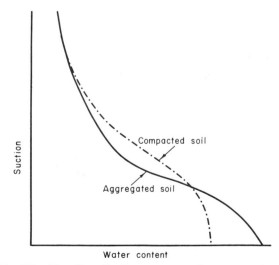

**Fig. 5.7.** The effect of soil structure on soil-water retention.

volumes of different classes of pore sizes). If an increase in matric suction from $\psi_1$ to $\psi_2$ results in the release of a certain volume of water, then that volume is evidently equal to the volume of pores having a range of effective radii between $r_1$ and $r_2$, where $\psi_1$ and $r_1$, and $\psi_2$ and $r_2$, are related by the equation of capillarity, namely $\psi = 2\gamma/r$.

An important and as yet incompletely understood phenomenon is the possible change in soil-moisture characteristics caused by the swelling and shrinkage of clay, which in turn is affected by the composition and concentration of the soil solution (Russell, 1941). Swelling is generally suppressed when the soil solution is fairly concentrated with electrolytes, particularly in the presence of a preponderance of divalent cations such as calcium. On the other hand, swelling—and hence water retention at any suction value— can be much more pronounced when the soil solution is dilute and with a preponderance of monovalent cations such as sodium (Dane and Klute, 1977). Other factors that affect the soil-moisture characteristic are the entrapment and persistence of air bubbles (Peck, 1969) and the change in soil structure resulting from sudden wetting, as well as from prolonged saturation (Hillel and Mottes, 1966).

## H. Hysteresis

The relation between matric potential and soil wetness can be obtained in two ways: (1) in *desorption*, by taking an initially saturated sample and applying increasing suction to gradually dry the soil while taking successive measurements of wetness versus suction; and (2) in *sorption*, by gradually

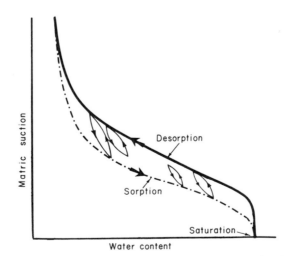

**Fig. 5.8.** The suction–water content curves in sorption and desorption. The intermediate loops are *scanning curves*, indicating transitions between the main branches.

wetting up an initially dry soil sample while reducing the suction. Each of these two methods yields a continuous curve, but the two curves will in general not be identical. The equilibrium soil wetness at a given suction is greater in desorption (drying) than in sorption (wetting). This dependence of the equilibrium content and state of soil water upon the direction of the process leading up to it is called *hysteresis*[16] (Haines, 1930; Miller and Miller, 1955a,b, 1956; Philip, 1964; Topp and Miller, 1966; Bomba, 1968; Topp, 1969).

Figure 5.8 shows a typical soil-moisture characteristic curve and illustrates the hysteresis effect in the soil–water equilibrium relationship.

The hysteresis effect may be attributed to several causes:

(1)  the geometric nonuniformity of the individual pores (which are generally irregularly shaped voids interconnected by smaller passages), resulting in the "inkbottle" effect, illustrated in Fig. 5.9:

(2)  the contact-angle effect, resulting from the fact that the contact angle of water on the solid walls of pores tends to be greater and hence the radius of curvature is greater, in an advancing meniscus than in the case of a receding one.[7] A given water content will tend therefore to exhibit greater suction in desorption than in sorption.

(3)  entrapped air, which further decreases the water content of newly wetted soil. Failure to attain true equilibrium can accentuate the hysteresis effect.

(4)  swelling, shrinking, or aging phenomena, which result in differential changes of soil structure, depending on the wetting and drying history of the sample (Hillel and Mottes, 1966). The gradual solution of air, or the release of dissolved air from soil water, can also have a differential effect upon the suction–wetness relationship in wetting and drying systems.

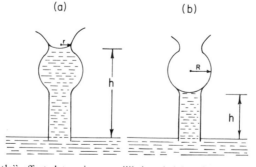

(a)                    (b)

**Fig. 5.9.** "Ink-bottle" effect determines equilibrium height of water in a variable-width pore: (a) in capillary drainage (desorption) and (b) in capillary rise (sorption).

[7] Contact-angle hysteresis can arise because of surface roughness, the presence and distribution of adsorbed impurities on the solid surface, and the mechanism by which liquid molecules adsorb or desorb when the interface is displaced.

Of particular interest is the ink bottle effect. Consider the hypothetical pore shown in Fig. 5.9. This pore consists of a relatively wide void of radius $R$, bounded by narrow channels of radius $r$. If initially saturated, this pore will drain abruptly the moment the suction exceeds $\psi_r$, where $\psi_r = 2\gamma/r$. For this pore to rewet, however, the suction must decrease to below $\psi_R$, where $\psi_R = 2\gamma/R$, whereupon the pore abruptly fills. Since $R > r$, it follows that $\psi_r > \psi_R$. Desorption depends on the narrow radii of the connecting channels, whereas sorption depends on the maximum diameter of the large pores. These discontinuous spurts of water, called *Haines jumps*, can be observed readily in coarse sands. The hysteresis effect is in general more pronounced in coarse-textured soils in the low-suction range, where pores may empty at an appreciably larger suction than that at which they fill.

In the past, hysteresis was generally disregarded in the theory and practice of soil physics. This may be justifiable in the treatment of processes entailing monotonic wetting (e.g., infiltration) or drying (e.g., evaporation). But the hysteresis effect may be important in cases of composite processes in which wetting and drying occur simultaneously or sequentially in various parts of the soil profile (e.g., redistribution). It is possible to have two soil layers of identical texture and structure at equilibrium with each other (i.e., at identical energy states) and yet they may differ in wetness or water content if their wetting histories have been different. Furthermore, hysteresis can affect the dynamic, as well as the static, properties of the soil (i.e., hydraulic conductivity and flow phenomena).

The two complete characteristic curves, from saturation to dryness and vice versa, are called the *main branches* of the hysteretic soil moisture characteristic. When a partially wetted soil commences to drain, or when a partially desorbed soil is rewetted, the relation of suction to moisture content follows some intermediate curve as it moves from one main branch to the other. Such intermediate spurs are called *scanning curves*. Cyclic changes often entail wetting and drying scanning curves, which may form loops between the main branches (Fig. 5.8). The $\psi - \theta$ relationship can thus become very complicated. Because of its complexity, the hysteresis phenomenon is too often ignored, and the soil moisture characteristic which is generally reported is the *desorption curve*, also known as the soil-moisture release curve. The *sorption curve*, which is equally important but more difficult to determine, is seldom even attempted.

## I. Measurement of Soil-Moisture Potential

The measurement of soil wetness, described earlier in this chapter, though essential in many soil physical and engineering investigations, is obviously not sufficient to provide a description of the state of soil water. To obtain such a description, evaluation of the energy status of soil water (soil-moisture potential, or suction) is necessary. In general, the twin variables, wetness and potential, should each be measured directly, as the translation of one to the other on the basis of calibration curves of soil samples is too often unreliable.

Total soil-moisture potential is often thought of as the sum of matric and osmotic (solute) potentials and is a useful index for characterizing the energy status of soil water with respect to plant water uptake. The sum of the matric and gravitational (elevation) heads is generally called the hydraulic head (or hydraulic potential) and is useful in evaluating the directions and magnitudes of the water-moving forces throughout the soil profile. Methods are available for measuring matric potential as well as total soil moisture potential, separately or together (Black, 1965). To measure matric potential in the field, an instrument known as the tensiometer is used, whereas in the laboratory use is often made of tension plates and of air-pressure extraction cells. Total soil moisture potential can be obtained by measuring the equilibrium vapor pressure of soil water by means of thermocouple psychrometers.

We shall now describe the tensiometer, which has won widespread acceptance as a practical device for the in situ measurement of matric suction, hydraulic head, and hydraulic gradients.

## 1. THE TENSIOMETER

The essential parts of a tensiometer are shown in Fig. 5.10. The tensiometer consists of a porous cup, generally of ceramic material, connected through a tube to a manometer, with all parts filled with water. When the cup is placed in the soil where the suction measurement is to be made, the bulk water inside the cup comes into hydraulic contact and tends to equilibrate with soil water through the pores in the ceramic walls. When initially placed in the soil, the water contained in the tensiometer is generally at atmospheric pressure. Soil water, being generally at subatmospheric pressure, exercises a suction which draws out a certain amount of water from the rigid and airtight tensiometer, thus causing a drop in its hydrostatic pressure. This pressure is indicated by a manometer, which may be a simple water- or mercury-filled U tube, a vacuum gauge, or an electrical transducer.

A tensiometer left in the soil for a long period of time tends to follow the changes in the matric suction of soil water. As soil moisture is depleted by drainage or plant uptake, or as it is replenished by rainfall or irrigation, corresponding readings on the tensiometer gauge occur. Owing to the hydraulic resistance of the cup and the surrounding soil, or of the contact zone between the cup and the soil, the tensiometer response may lag behind suction changes

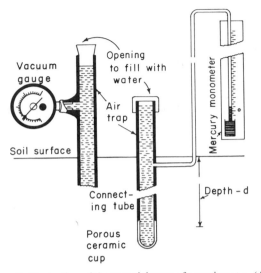

**Fig. 5.10.** Schematic illustration of the essential parts of a tensiometer. (After S. J. Richards, 1965.)

in the soil. This lag time can be minimized by the use of a null-type device or of a transducer-type manometer with rigid tubing, so that practically no flow of water need take place as the tensiometer adjusts to changes in the soil matric suction.

Since the porous cup walls of the tensiometer are permeable to both water and solutes, the water inside the tensiometer tends to assume the same solute composition and concentration as soil water, and the instrument does not indicate the osmotic suction of soil water (unless equipped with some type of an auxiliary salt sensor).

Suction measurements by tensiometry are generally limited to matric suction values of below 1 atm. This is due to the fact that the vacuum gauge or manometer measures a partial vacuum relative to the external atmospheric pressure, as well as to the general failure of water columns in macroscopic systems to withstand tensions exceeding 1 bar.[8] Furthermore, as the ceramic material is generally made of the most permeable and porous material possible, too high a suction may cause air entry into the cup, which would equalize the internal pressure to the atmospheric. Under such conditions, soil suction will continue to increase even though the tensiometer fails to show it.

---

[8] It is interesting that water in capillary systems can maintain continuity at tensions of many bars. Witness, for example, the continuity of liquid water in the xylem vessels of tall trees. The ultimate tensile strength of water is apparently equivalent to several hundred bars.

In practice, the useful limit of most tensiometers is at about 0.8 bar of maximal suction. To measure higher suctions, the use of an osmometer with a semipermeable membrane at the wall has been proposed, but this instrument is still in the experimental stage. However, the limited range of suction measurable by the tensiometer is not as serious a problem as it may seem at first sight. Though the suction range of 0–0.8 bar is but a small part of the total range of suction variation encountered in the field, it generally encompasses the greater part of the soil wetness range. In many agricultural soils the tensiometer range accounts for more than 50% (and in coarse-textured soils 75% or more) of the amount of soil water taken up by plants. Thus, where soil management (particularly in irrigation) is aimed at maintaining low-suction conditions which are most favorable for plant growth, tensiometers are definitely useful.

Despite their many shortcomings, tensiometers are practical instruments, available commercially, and, when operated and maintained by a skilled worker, are capable of providing reliable data on the in situ state of soil-moisture profiles and their changes with time.

Tensiometers have long been used in guiding the timing of irrigation of field and orchard crops, as well as of potted plants (Richards and Weaver, 1944). A general practice is to place tensiometers at one or more soil depths representing the root zone, and to irrigate when the tensiometers indicate that the matric suction has reached some prescribed value. The use of several tensiometers at different depths can indicate the amount of water needed in irrigation, and can also allow calculation of the hydraulic gradients in the soil profile. If $\psi_1, \psi_2, \psi_3, \ldots, \psi_n$ are the matric suction values in centimeters of water head ($\cong$ millibars) at depths $d_1, d_2, d_3, \ldots, d_n$ measured in centimeters below the surface, the average hydraulic gradient $i$ between depths $d_n$ and $d_{n+1}$ is

$$i = [(\psi_{n+1} + d_{n+1}) - (\psi_n + d_n)]/(d_{n+1} - d_n) \qquad (5.20)$$

Measurement of the hydraulic gradient is particularly important in the region below the root zone, where the direction and magnitude of water movement cannot easily be ascertained otherwise.

The still considerable cost of tensiometers may limit the number of instruments used below the number needed for characterizing the often highly variable distribution of moisture and hence the pattern of suction in heterogeneous soils. Air diffusion through the porous cup into the vacuum gauging system requires frequent purging with deaired water. Tensiometers are also sensitive to temperature gradients between their various parts. Hence the above-ground parts should prefereably be shielded from direct exposure to the sun. When installing a tensiometer, it is important that good contact be made between the cup and the soil so that equilibration is not hindered by contact-zone impedance to flow.

## 2. The Thermocouple Psychrometer

At equilibrium, the potential of soil moisture is equal to the potential of the water vapor in the ambient air. If thermal equilibrium is assured and the gravitational effect is neglected, the vapor potential can be taken to be equal to the sum of the matric and osmotic potentials, since air acts as an ideal semipermeable membrane in allowing only water molecules to pass (provided that the solutes are nonvolatile).

Fortunately, recent years have witnessed the development of highly precise, miniaturized *thermocouple psychrometers* which indeed make possible the in situ measurement of soil moisture potential (Dalton and Rawlins, 1968; Brown, 1970). A thermocouple is a double junction of two dissimilar metals. If the two junctions are subjected to different temperatures, they will generate a voltage difference. If, on the other hand, an electromotive force (emf) is applied between the junctions, a difference in temperature will result; depending on which way a direct current is applied, one junction can be heated while the other is cooled, and vice versa. The *soil psychrometer* (Fig. 5.11) consists of a fine wire thermocouple, one junction of which is equilibrated with the soil atmosphere by placing it inside a hollow porous cup embedded in the soil, while the other junction is kept in an insulated medium to provide a temperature lag. During operation, an emf is applied so that the junction exposed to the soil atmosphere is cooled to a temperature below the dew point of that atmosphere, at which point a droplet of water condenses on the junction, allowing it to become, in effect, a wet bulb thermometer. This is a consequence of the so-called *Peltier effect* (see, for example, Yavorsky and Detlaf, 1972). The cooling is then stopped, and as

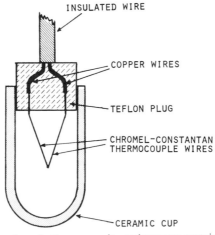

**Fig. 5.11.**   Cross section of a thermocouple psychrometer contained in an air-filled ceramic cup.

the water from the droplet reevaporates the junction attains a wet bulb temperature which remains nearly constant until the junction dries out, after which it returns to the ambient soil temperature. While evaporation takes place, the difference in temperature between the wet bulb and the insulated junction serving as dry bulb generates an emf which is indicative of the soil moisture potential. The relative humidity (i.e., the vapor pressure depression relative to that of pure, free water) is related to the soil-water potential according to

$$\phi = \bar{R}T \ln(p/p_0) \qquad (5.21)$$

where $p$ is the vapor pressure of soil water, $p_0$ the vapor pressure of pure, free water at the same temperature and air pressure, and $\bar{R}$ is the specific gas constant for water vapor.

### 3. MEASUREMENT OF THE SOIL-MOISTURE CHARACTERISTIC CURVES

The fundamental relation between soil wetness and matric suction is often determined by means of a tension plate assembly (Fig. 5.12) in the low suction (<1 bar) range, and by means of a pressure plate or pressure membrane apparatus (Fig. 5.13) in the higher suction range. These instruments allow the application of successive suction values and the repeated measurement of the equilibrium soil wetness at each suction.

The maximum suction value obtainable by porous-plate devices is limited to 1 bar if the soil air is kept at atmospheric pressure and the pressure difference across the plate is controlled either by vacuum or by a hanging water column. Matric suction values considerably greater than 1 bar (say, 20 bar or even more) can be obtained by increasing the pressure of the air phase. This requires placing the porous-plate assembly inside a pressure chamber, as shown in Fig. 5.13. The limit of matric suction obtainable with such a

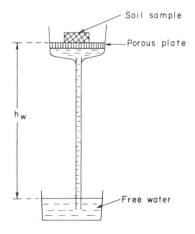

**Fig. 5.12.** Tension plate assembly for equilibrating a soil sample with a known matric suction value. This assembly is applicable in the range of 0–1 bar only.

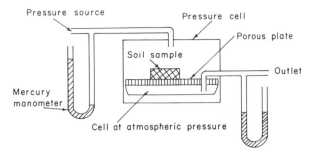

**Fig. 5.13.**  Pressure plate apparatus for moisture characteristic measurements in the high-suction range. The lower side of the porous plate is in contact with water at atmospheric pressure. Air pressure is used to extract water from initially saturated soil samples.

device is determined by the design of the chamber (i.e., its safe working pressure) and by the maximal air-pressure difference the saturated porous plate can bear without allowing air to bubble through its pores. Ceramic plates generally do not hold pressures greater than about 20 bar, but cellulose acetate membranes can be used with pressures exceeding 100 bar.

Soil moisture retention in the low-suction range (0–1 bar) is strongly influenced by soil structure and pore-size distribution. Hence, measurements made with disturbed samples (e.g., dried, screened, and artificially packed samples) cannot be expected to represent field conditions. The use of undisturbed soil cores is therefore preferable. Even better, in principle, is the in situ determination of the soil moisture characteristic by making simultaneous measurements of wetness (e.g., with the neutron moisture meter) and suction (using tensiometers) in the field. Unfortunately, this approach has often been frustrated by soil heterogeneity and by uncertainties over hysteretic phenomena as they occur in the field.

We have already mentioned that the soil-moisture characteristic is hysteretic. Ordinarily, the desorption curve is measured by gradually and monotonically extracting water from initially saturated samples. The resulting curve, often called the soil-moisture release curve, is applicable to processes involving drainage, evaporation, or plant extraction of soil moisture. On the other hand, the sorption curve is needed whenever infiltration or wetting processes are studied. Modified apparatus is required for the measurement of wetness versus suction during sorption (Tanner and Elrick, 1958). Both primary curves and knowledge of scanning patterns in transition from wetting to drying soil (and vice versa) are needed for a complete description.

**Sample Problems**

**1.**  The accompanying data were obtained by gravimetric sampling just prior to and two days following an irrigation.

| Sampling time | Sample number | Depth (cm) | Bulk density (gm/cm$^3$) | Gross weight of sample plus container | | Weight container (gm) |
|---|---|---|---|---|---|---|
| | | | | Wet sample | Dried sample | |
| Before | 1 | 0–40 | 1.2 | 160 | 150 | 50 |
| irrigation | 2 | 40–100 | 1.5 | 146 | 130 | 50 |
| After | 3 | 0–40 | 1.2 | 230 | 200 | 50 |
| irrigation | 4 | 40–100 | 1.5 | 206 | 170 | 50 |

From these data, calculate the mass and volume wetness values of each layer before and after the irrigation, and determine the amount of water (in millimeters) added to each layer and to the profile as a whole.

Using Eq. (5.6), we obtained the following mass wetness values:

$$w_1 = \frac{160 - 150}{150 - 50} = \frac{10 \text{ gm}}{100 \text{ gm}} = 0.1, \qquad w_2 = \frac{146 - 130}{130 - 50} = \frac{16 \text{ gm}}{80 \text{ gm}} = 0.2$$

$$w_3 = \frac{230 - 200}{200 - 50} = \frac{30 \text{ gm}}{150 \text{ gm}} = 0.2, \qquad w_4 = \frac{206 - 170}{170 - 50} = \frac{36 \text{ gm}}{120 \text{ gm}} = 0.3$$

Using Eq. (5.3), we obtain the following volume wetness values:

$$\theta_1 = 1.2 \times 0.1 = 0.12, \qquad \theta_2 = 1.5 \times 0.2 = 0.30$$

$$\theta_3 = 1.2 \times 0.2 = 0.24, \qquad \theta_4 = 1.5 \times 0.3 = 0.45$$

Using Eq. (5.4), we obtain the following water depths per layer:

$$d_{w_1} = 0.12 \times 400 \text{ mm} = 48 \text{ mm}, \qquad d_{w_2} = 0.30 \times 600 \text{ mm} = 180 \text{ mm}$$

$$d_{w_3} = 0.24 \times 400 \text{ mm} = 96 \text{ mm}, \qquad d_{w_4} = 0.45 \times 600 \text{ mm} = 270 \text{ mm}$$

Depth of water in profile before irrigation = 48 + 180 = 228 mm.
Depth of water in profile after irrigation = 96 + 270 = 366 mm.
Depth of water added to top layer = 96 − 48 = 48 mm.
Depth of water added to bottom layer = 270 − 180 = 90 mm.
Depth of water added to entire profile = 48 + 90 = 138 mm.

**2.** From calibration of a neutron probe we know that when a soil's volumetric wetness is 15% we get a reading of 24,000 cpm (counts per minute), and at a wetness of 40% we get 44,000 cpm. Find the equation of the straight line defining the calibration curve (in the form of $Y = mX + b$, where $Y$ is counts per minute, $X$ is volumetric wetness, $m$ is the slope of the line, and $b$ is the intercept on the $Y$ axis). Using the equation derived, find the wetness value corresponding to a count rate of 30,000 cpm.

We first obtain the slope $m$:

$$m = (Y_2 - Y_1)/(X_2 - X_1) = (44{,}000 - 24{,}000)/(40 - 15)$$
$$= 800 \text{ cpm per } 1\% \text{ wetness}$$

We next obtain the $Y$ intercept $b$:

$$Y = 800X + b, \qquad b = Y - 800X = 44{,}000 - 800 \times 40 = 12{,}000 \text{ cpm}$$

(or $\quad 24{,}000 - 800 \times 15 = 12{,}000 \text{ cpm})$

The complete equation is therefore

$$Y = 800X + 12{,}000$$

Now, to find the wetness value corresponding to 30,000 cpm, we set

$$30{,}000 = 800X + 12{,}000, \qquad X = (30{,}000 - 12{,}000)/800 = 22.5\%$$

**3.**   The accompanying data were obtained with tension plate and pressure plate extraction devices from two soils of unknown texture.

| Suction head | | Volumetric wetness (%) | |
|---|---|---|---|
| (bar) | (cm) | Soil A | Soil B |
| 0 | 0 | 44 | 52 |
| 0.01 | 10 | 44 | 52 |
| 0.02 | 20 | 43.9 | 52 |
| 0.05 | 50 | 38 | 51 |
| 0.1 | 100 | 22.5 | 48 |
| 0.3 | 300 | 12.5 | 32 |
| 1 | 1000 | 7 | 20 |
| 10 | 10,000 | 5.2 | 13.5 |
| 20 | 20,000 | 5.1 | 13 |
| 100 | 100,000 | 4.9 | 12.8 |

Plot the two soil-moisture characteristic curves on a semilog scale (logarithm of matric suction versus wetness). Estimate the bulk density, assuming that the soils do not swell or shrink. Estimate the volume and mass wetness values at suctions of $\frac{1}{3}$ bar and at 15 bar. How much water in depth units (mm) can each soil release per one meter depth of profile in transition from $\frac{1}{3}$ bar to 15 bar of suction?

The curves shown in Fig. 5.14 are obtained. *Bulk density* can be estimated from the volumetric wetness at saturation (zero suction) assuming it to be equal to the porosity $f$ (i.e., no occluded air):

$$f = 1 - \rho_b/\rho_s, \qquad \rho_b = \rho_s(1 - f)$$

For Soil A: $\rho_b$ = 2.65(1 − 0.44) = 1.48 gm/cm³.
For Soil B: $\rho_b$ = 2.65(1 − 0.52) = 1.27 gm/cm³.

*Volume wetness* values for different suctions can be read directly from the soil moisture characteristic curves:

For Soil A: 12% at ⅓ bar; 5% at 15 bar.
For Soil B: 31% at ⅓ bar; 13% at 15 bar.

**Fig. 5.14.**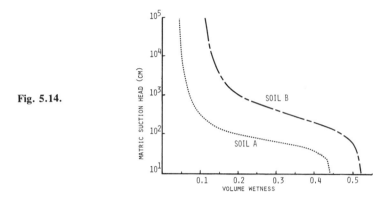

*Mass wetness w* values can be calculated from the volumetric wetness $\theta$ values and the bulk density $\rho_b$ values by the equation

$$\theta = w(\rho_b/\rho_w) \quad \text{or} \quad w = \theta(\rho_w/\rho_b) \quad \text{where } \rho_w \text{ is the density of water)}$$

For Soil A at ⅓ bar: $w$ = 12% × (1/1.48) = 8.1%; at 15 bar: $w$ = 5% × (1/1.48) = 3.4%.

For Soil B at ⅓ bar: $w$ = 31% × (1/1.27) = 24.4%; at 15 bar: $w$ = 13% × (1/1.27) = 10.2%.

*Water released* per 1 meter depth in transition from ⅓ to 15 bar:

For Soil A: (12 − 5)% × 1000 mm = 70 mm.
For Soil B: (31 − 13)% × 1000 mm = 180 mm.

# 6    *Flow of Water in Saturated Soil*

## A. Introduction

Before we enter into a discussion of flow in so complex a medium as soil, it might be helpful to consider some basic physical phenomena associated with fluid flow in narrow tubes.

Early theories of fluid dynamics were based on the hypothetical concepts of a "perfect" fluid, i.e., one that is both frictionless and incompressible. In the flow of a perfect fluid, contacting layers can exhibit no tangential forces (shearing stresses), only normal forces (pressures). Such fluids do not in fact exist. In the flow of real fluids, adjacent layers do transmit tangential stresses (drag), and the existence of intermolecular attractions causes the fluid molecules in contact with a solid wall to adhere to it rather than slip over it. The flow of a real fluid is associated with the property of viscosity, which characterizes the fluid's resistance to flow.

We can visualize the nature of viscosity by considering the motion of a fluid between two parallel plates, one at rest, the other moving at a constant velocity (Fig. 6.1). Experience shows that the fluid adheres to both walls, so that its velocity at the lower plate is zero, and that at the upper plate is equal to the velocity of the plate. Furthermore, the velocity distribution in the fluid between the plates is linear, so that the fluid velocity is proportional to the distance $y$ from the lower plate.

To maintain the relative motion of the plates at a constant velocity, it is necessary to apply a tangential force, that force having to overcome the frictional resistance in the fluid. This resistance, per unit area of the plate, is proportional to the velocity of the upper plate $U$ and inversely proportional to the distance $h$. The shearing stress at any point, is proportional to the

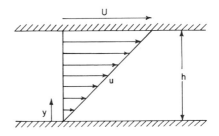

**Fig. 6.1.**   Velocity distribution in a viscous
fluid between two parallel flat plates, with the
upper plate moving at a velocity $U$ relative to
the lower plate.

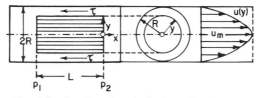

**Fig. 6.2.**   Laminar flow through a cylindrical tube.

velocity gradient $du/dy$. The viscosity $\eta$ is the proportionality factor between
$\tau_s$ and $du/dy$[1]:

$$\tau_s = \eta \, du/dy \qquad (6.1)$$

We can now apply these relationships to describe flow through a straight,
cylindrical tube with a constant diameter $D = 2R$ (Fig. 6.2). The velocity is
zero at the wall (because of adhesion), maximal on the axis, and constant on
cylindrical surfaces which are concentric about the axis. Adjacent cylindrical *laminae*, moving at different velocities, slide over each other. A parallel
motion of this kind is called *laminar*. Fluid movement in a horizontal tube is
generally caused by a pressure gradient acting in the axial direction. A fluid
"particle," therefore, is accelerated by the pressure gradient and retarded by
the frictional resistance.

Now let us consider a coaxial fluid cylinder of length $L$ and radius $y$. For
flow velocity to be constant, the pressure force acting on the face of the
cylinder $\Delta p \pi y^2$, where $\Delta p = p_1 - p_2$, must be equal to the frictional resistance due to the shear force $2\pi y L \tau_s$ acting on the circumferential area. Thus,

$$\tau_s = (\Delta p/L)(y/2)$$

Recalling Eq. (6.1)

$$\tau_s = -\eta \, du/dy$$

---

[1]Equation (6.1) bears an analogy to Hooke's law of elasticity. In an elastic solid, the shearing stress is proportional to the strain, whereas in a viscous fluid, the shearing stress is proportional to the time rate of the strain.

(the negative sign arises because in this case $u$ decreases with increasing $y$), we obtain

$$du/dy = -(\Delta p/\eta L)(y/2)$$

which, upon integration, gives

$$u(y) = (\Delta p/\eta L)(c - y^2/4)$$

The constant of integration $c$ is evaluated from the boundary condition of no slip at the wall; that is, $u = 0$ at $y = R$, so that $c = R^2/4$.
    Therefore,

$$u(y) = (\Delta p/4\eta L)(R^2 - y^2) \tag{6.2}$$

    Equation (6.2) indicates that the velocity is distributed parabolically over the radius, with the maximum velocity $u_{max}$ being on the axis ($y = 0$):

$$u_{max} = \Delta p\ R^2/4\eta L$$

The discharge $Q$, being the volume flowing through a section of length $L$ per unit time, can now be evaluated. The volume of a paraboloid of revolution is $\frac{1}{2}$(base $\times$ height), hence

$$Q = \tfrac{1}{2}\pi R^2 u_{max} = \pi R^4\ \Delta p/8\eta L \tag{6.3}$$

    This equation, known as *Poiseuille's law*, indicates that the volume flow rate is proportional to the pressure drop per unit distance ($\Delta p/L$) and the fourth power of the radius of the tube.
    The mean velocity over the cross section is

$$\bar{u} = \Delta p\ R^2/8\eta L = (R^2/a\eta)\ \nabla p \tag{6.4}$$

where $\nabla p$ is the pressure gradient. Parameter $a$, equal to 8 in a circular tube, varies with the shape of the conducting passage.
    Laminar flow prevails only at relatively low flow velocities and in narrow tubes. As the radius of the tube and the flow velocity are increased, the point is reached at which the mean flow velocity is no longer proportional to the pressure drop, and the parallel *laminar flow* changes into a *turbulent flow* with fluctuating eddies. Conveniently, however, laminar flow is the rule rather than the exception in most water flow processes taking place in soils, because of the narrowness of soil pores.

## B.  Darcy's Law

    Were the soil merely a bundle of straight and smooth tubes, each uniform in radius, we could assume the overall flow rate to be equal to the sum of the separate flow rates through the individual tubes. Knowledge of the size distribution of the tube radii could then enable us to calculate the total flow through a bundle caused by a known pressure difference, using Poiseuille's equation.

Unfortunately from the standpoint of physical simplicity, however, soil pores do not resemble uniform, smooth tubes, but are highly irregular, tortuous, and intricate. Flow through soil pores is limited by numerous constrictions, or "necks," and occasional "dead end" spaces. Hence, the actual geometry and flow pattern of a typical soil specimen is too complicated to be described in microscopic detail, as the fluid velocity varies drastically from point to point, even along the same passage. For this reason, flow through complex porous media is generally described in terms of *a macroscopic flow velocity vector*, which is the overall average of the microscopic velocities over the total volume of the soil. The detailed flow pattern is thus ignored, and the conducting body is treated as though it were a uniform medium, with the flow spread out over the entire cross section, solid and pore space alike.[2]

Let us now examine the flow of water in a macroscopically uniform, saturated soil body, and attempt to describe the quantitative relations connecting the rate of flow, the dimensions of the body, and the hydraulic conditions at the inflow and outflow boundaries.

Figure 6.3 shows a horizontal column of soil, through which a steady flow of water is occurring from left to right, from an upper reservoir to a lower one, in each of which the water level is maintained constant.

Experience shows that the discharge rate $Q$, being the volume $V$ flowing through the column per unit time, is directly proportional to the cross-sectional area and to the hydraulic head drop $\Delta H$, and inversely proportional to the length of the column $L$:

$$Q = V/t \propto A \, \Delta H/L$$

The usual way to determine the hydraulic head drop across the system is to measure the head at the inflow boundary $H_i$ and at the outflow boundary $H_o$, relative to some reference level. $\Delta H$ is the difference between these two heads:

$$\Delta H = H_i - H_o$$

Obviously, no flow occurs in the absence of a hydraulic head difference, i.e., when $\Delta H = 0$.

---

[2] An implicit assumption here is that the soil volume taken is sufficiently large relative to the pore sizes and microscopic heterogeneities to permit the averaging of velocity and potential over the cross section.

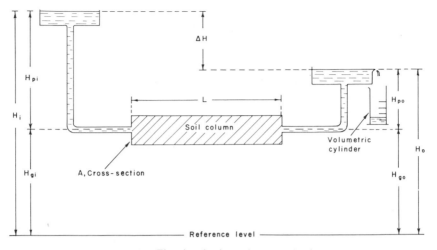

**Fig. 6.3.** Flow in a horizontal saturated column.

The head drop per unit distance in the direction of flow ($\Delta H/L$) is the *hydraulic gradient*, which is, in fact, the driving force. The specific discharge rate $Q/A$ (i.e., the volume of water flowing through a unit cross-sectional area per unit time $t$) is called the *flux density* (or simply the *flux*) and is indicated by $q$. Thus, the flux is proportional to the hydraulic gradient:

$$q = Q/A = V/At \propto \Delta H/L$$

The proportionality factor $K$ is generally designated as the *hydraulic conductivity*:

$$q = K\,\Delta H/L \tag{6.5}$$

This equation is known as *Darcy's law*, after Henri Darcy, the French engineer who discovered it over a century ago in the course of his classic investigation of seepage rates through sand filters in the city of Dijon (Darcy, 1856; Hubbert, 1956).

Where flow is unsteady (i.e., the flux changing with time) or the soil nonuniform, the hydraulic head may not decrease linearly along the direction of flow. Where the hydraulic head gradient or the conductivity is variable, we must consider the localized gradient, flux, and conductivity values rather than overall values for the soil system as a whole. A more exact and generalized expression of the Darcy law is, therefore, in differential form. Slichter (1899) generalized Darcy's law for saturated porous media into a three-dimensional macroscopic differential equation of the form[3]

[3] $K\nabla H$ is the product of a scalar $K$ and a vector $\nabla H$, and hence the flux $\mathbf{q}$ is a vector, the direction of which is determined by $\nabla H$. This direction in an isotropic medium is orthogonal to surfaces of equal hydraulic potential $H$.

$$\mathbf{q} = -K \nabla H \tag{6.6}$$

where $\nabla H$ is the gradient of the hydraulic head in three-dimensional space.

Stated verbally, this law indicates that the flow of a liquid through a porous medium is in the direction of, and at a rate proportional to, the *driving force* acting on the liquid (i.e., the *hydraulic gradient*) and also proportional to the property of the conducting medium to transmit the liquid (namely, the *conductivity*).

In a one-dimensional system, Eq. (6.6) takes the form

$$q = -K \, dH/dx \tag{6.7}$$

Mathematically, Darcy's law is similar to the linear transport equations of classical physics, including *Ohm's law* (which states that the current, or flow rate of electricity, is proportional to the electrical potential gradient), *Fourier's law* (the rate of heat conduction is proportional to the temperature gradient), and *Fick's law* (the rate of diffusion is proportional to the concentration gradient).

## C. Gravitational, Pressure, and Total Hydraulic Heads

The water entering the column of Fig. 6.3 is under a pressure $P_i$, which is the sum of the hydrostatic pressure $P_s$ and the atmospheric pressure $P_a$ acting on the surface of the water in the reservoir. Since the atmospheric pressure is the same at both ends of the system, we can disregard it and consider only the hydrostatic pressure. Accordingly, the water pressure at the inflow boundary is $\rho_w g H_{pi}$. Since $\rho_w$ and $g$ are both nearly constant, we can express this pressure in terms of the pressure head $H_{pi}$.

Water flow in a horizontal column occurs in response to a pressure head gradient. Flow in a vertical column can be caused by gravitation as well as pressure.[4] The *gravitational head* $H_g$ at any point is determined by the height of the point relative to some reference plane, while the *pressure head* is determined by the height of the water column resting on that point.

---

[4]In classical hydraulics, the fluid potential $\Phi$ (the mechanical energy per unit mass) is generally stated in terms of the *Bernoulli equation*

$$\Phi = \int_{P_0}^{P} \frac{dP}{\rho} + gz + \frac{v^2}{2}$$

wherein $P$ is pressure ($P_0$ being the pressure at the standard state), $\rho$ is density of the fluid, $g$ is gravitational acceleration, $z$ is elevation above a reference level, and $v$ is velocity. The three terms thus represent the pressure, gravity, and velocity potentials, respectively. Since flow is a porous medium is generally extremely slow, the third term can almost always be neglected (Freeze and Cherry, 1979). For an incompressible liquid (with $\rho$ independent of $P$), moreover, the first term can be written as $(P - P_0)/\rho$. Finally, if $P_0$ is assumed equal to zero, we get

$$\Phi = P/\rho + gz$$

The total hydraulic head $H$ is composed of the sum of these two heads:

$$H = H_p + H_g \tag{6.8}$$

To apply Darcy's law to vertical flow, we must consider the total hydraulic head at the inflow and at the outflow boundaries ($H_i$ and $H_o$, respectively):

$$H_i = H_{pi} + H_{gi}, \qquad H_o = H_{po} + H_{go}$$

Darcy's law thus becomes

$$q = K[(H_{pi} + H_{gi}) - (H_{po} + H_{go})]/L$$

The gravitational head is often designated as $z$, which is the vertical distance in the rectangular coordinate system $x, y, z$. It is convenient to set the reference level $z = 0$ at the bottom of a vertical column, or at the center of a horizontal column. However, the exact elevation of this hypothetical level is unimportant, since the absolute values of the hydraulic heads determined in reference to it are immaterial and only their differences from one point in the soil to another affect flow.

The pressure and gravity heads can be represented graphically in a simple way. To illustrate this, we shall immerse and equilibrate a vertical soil column in a water reservoir, so that the upper surface of the column will be level with the water surface, as shown in Fig. 6.4.

The coordinates of Fig. 6.4 are arranged so that the height above the bottom of the column is indicated by the vertical axis $z$; and the pressure, gravity, and hydraulic heads are indicated on the horizontal axis. The gravity head is determined with respect to the reference level $z = 0$, and increases with height at the ratio of $1:1$. The pressure head is determined with reference to the free-water surface, at which the hydrostatic pressure is zero. Accordingly, the hydrostatic pressure head at the top of the column

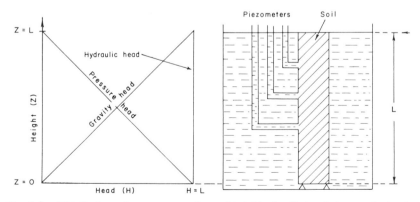

**Fig. 6.4.** Distribution of pressure, gravity, and total hydraulic heads in a vertical column immersed in water, at equilibrium.

is zero, and at the bottom of the column it is equal to $L$, the column length. Just as the gravity head decreases from top to bottom, so the pressure head increases; thus, their sum, which is the hydraulic head, remains constant all along the column. This is a state of equilibrium in which no flow occurs.

This statement should be further elaborated. The water pressure is not equal along the column, being greater at the bottom than at the top of the column. Why, then, will the water not flow from a zone of higher to one of lower pressure? If the pressure gradient were the only force causing flow (as it is, in fact, in a horizontal column), the water would tend to flow upward. However, opposing the pressure gradient is a gravitational gradient of equal magnitude, resulting from the fact that the water at the top is at a higher gravitational potential than at the bottom. Since these two opposing gradients in effect cancel each other, the total hydraulic head is constant, as indicated by the standpipes (piezometer) connected to the column at the left.

As we have already point out, the reference level is generally set at the bottom of the column, so that the gravitational potential can always be positive. On the other hand, the pressure head of water is positive under a free-water surface (i.e., a water table) and negative above it. A "negative" hydraulic head signifies a pressure smaller than atmospheric, and it occurs whenever the soil becomes unsaturated. Flow under these conditions will be dealt with in the next chapter.

## D. Flow in a Vertical Column

Figure 6.5 shows a uniform, saturated vertical column, the upper surface of which is ponded under a constant head of water $H_1$, and the bottom surface of which is set in a lower, constant-level reservoir. Flow is thus taking place from the higher to the lower reservoir through a column of length $L$.[5]

In order to calculate the flux according to Darcy's law, we must know the hydraulic head gradient, which is the ratio of the hydraulic head drop (between the inflow and outflow boundaries) to the column length as shown in the accompanying tabulation.

| | | Pressure head | | Gravity head |
|---|---|---|---|---|
| Hydraulic head at inflow boundary | $H_i =$ | $H_1$ | $+$ | $L$ |
| Hydraulic head at outflow boundary | $H_o =$ | $0$ | $+$ | $0$ |
| Hydraulic head difference $\Delta H = H_i - H_o =$ | | $H_1$ | $+$ | $L$ |

[5]This is the same system that Darcy considered in his classic filter bed analysis.

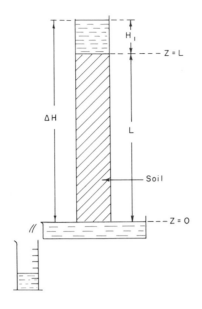

**Fig. 6.5.** Downward flow of water in a vertical saturated column.

The Darcy equation for this case is

$$q = K\,\Delta H/L = K(H_1 + L)/L, \qquad q = K\,H_1/L + K \qquad (6.9)$$

Comparison of this case with the horizontal one shows that the rate of downward flow of water in a vertical column is greater than in a horizontal column by the magnitude of the hydraulic conductivity. It is also apparent that, if the ponding depth $H_1$ is negligible, the flux is equal to the hydraulic conductivity. This is due to the fact that, in the absence of a pressure gradient, the only driving force is the gravitational head gradient, which, in a vertical column, has the value of unity (since this head varies with height at the ratio of $1:1$).

We shall now examine the case of upward flow in a vertical column, as shown in Fig. 6.6.

In this case, the direction of flow is opposite to the direction of the gravitational gradient, and the hydraulic gradient becomes

|  |  | Pressure head | Gravity head |
|---|---|---|---|
| Hydraulic head at inflow boundary | $H_i =$ | $H_1$ + | 0 |
| Hydraulic head at outflow boundary | $H_o =$ | 0 + | $L$ |
| Hydraulic head difference $\Delta H = H_i - H_o =$ | | $H_1$ − | $L$ |

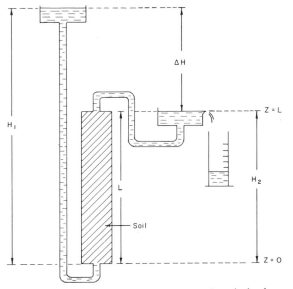

**Fig. 6.6.** Steady upward flow in a saturated vertical column.

The Darcy equation is thus

$$q = K(H_1 - L)/L = KH_1/L - K, \qquad q = K\,\Delta H/L$$

### E. Flux, Flow Velocity, and Tortuosity

As stated above, the *flux density* (hereafter, simply flux) is the volume of water passing through a unit cross-sectional area (perpendicular to the flow direction) per unit time. The dimensions of the flux are:

$$q = V/At = L^3/L^2T = LT^{-1}$$

i.e., length per time (in cgs units, centimeters per second). These are the dimensions of velocity, yet we prefer the term flux to flow velocity, the latter being an ambiguous term. Since soil pores vary in shape, width, and direction, the actual flow velocity in the soil is highly variable (e.g., wider pores conduct water more rapidly, and the liquid in the center of each pore moves faster than the liquid in close proximity to the particles). Strictly speaking, therefore, one cannot refer to a single velocity of liquid flow, but at best to an average velocity.

Yet, even the average velocity of the flowing liquid differs from the flux, as we have defined it. Flow does not in fact take place through the entire cross-sectional area $A$, since part of this area is plugged by particles and only the porosity fraction is open to flow. Since the real area through which flow takes place is smaller than $A$, the actual average velocity of the liquid must be greater than the flux $q$. Furthermore, the actual length of the path traversed by an average parcel of liquid is greater than the soil column length $L$, owing to the labyri⸱⸱hine, or tortuous, nature of the pore passages, as shown in Fig. 6.7.

*Tortuosity* can be defined as the average ratio of the actual roundabout path to the apparent, or straight, flow path; i.e., it is the ratio of the average length of the pore passages (as if they were stretched out in the manner one can stretch out a coiled or tangled telephone wire) to the length of the soil specimen. Tortuosity is thus a dimensionless geometric parameter of porous media which, though difficult to measure precisely, is always greater than 1 and may exceed 2. The *tortuosity factor* is sometimes defined as the inverse of what we defined as the tortuosity.

## F. Hydraulic Conductivity, Permeability, and Fluidity

The hydraulic conductivity, again, is the ratio of the flux to the hydraulic gradient, or the slope of the flux versus gradient curve (Fig. 8.8).

With the dimensions of flux being $LT^{-1}$, those of hydraulic conductivity depend on the dimensions assigned to the driving force (the potential gradient). In the last chapter, we showed that the simplest way to express the potential gradient is by use of length, or head, units. The hydraulic head gradient $H/L$, being the ratio of a length to a length, is dimensionless.[6] Accordingly, the dimensions of hydraulic conductivity are the same as the dimensions of flux, namely $LT^{-1}$. If, on the other hand, the hydraulic gradient is expressed in terms of the variation of pressure with length, then the hydraulic conductivity assumes the dimensions of $M^{-1}L^3T$. Since the latter is cumbersome, the use of head units is generally preferred.

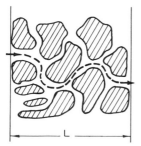

**Fig. 6.7.**  Flow path tortuosity in the soil.

[6]Though, strictly speaking, $H$ is not a true length, but a pressure equivalent in terms of a water column height; $H = P/\rho g$, and its gradient should be assigned the units of $cm_{H_2O}/cm$.

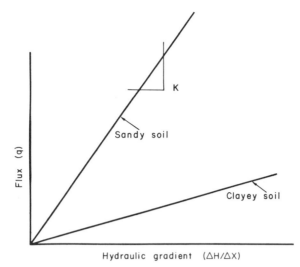

**Fig. 6.8.** The linear dependence of flux upon hydraulic gradient, the hydraulic conductivity being the slope (i.e., the flux per unit gradient).

In a saturated soil of stable structure, as well as in a rigid porous medium such as sandstone, for instance, the hydraulic conductivity is characteristically constant. Its order of magnitude is about $10^{-2}$–$10^{-3}$ cm/sec in a sandy soil and $10^{-4}$—$10^{-7}$ cm/sec in a clayey soil.

To appreciate the practical significance of these values in more familiar terms, consider the hypothetical case of an unlined (earth-bottom) reservoir or pond in which one wishes to retain water against losses caused by downward seepage. If the seepage into and through the underlying soil is by gravity alone (i.e., no pressure or suction gradients in the soil), we can assume it will take place at a rate approximately equal to the hydraulic conductivity. A coarse sandy soil might have a $K$ value of, say, $10^{-2}$ cm/sec and would therefore lose water at the enormous rate of nearly 10 m/day. A fine loam soil with a $K$ value of $10^{-4}$ cm/sec would lose "only" about 10 cm/day. Finally, and in contrast, a bed of clay with a conductivity of $10^{-6}$ cm/sec would allow the seepage of no more than 1 mm/day, much less than the expectable rate of evaporation. So, the retention of water in earthen dams and reservoirs and the prevention of seepage from unlined canals can be greatly aided by a bed of clay, particularly if the clay is dispersed to further reduce its hydraulic conductivity.

In many soils, the hydraulic conductivity does not in fact remain constant. Because of various chemical, physical, and biological processes, the hydraulic conductivity may change as water permeates and flows in a soil. Changes occurring in the composition of the exchangeable-ion complex, as when the water entering the soil has a different composition or concentration of solutes than the original soil solution, can greatly change the hydraulic conductivity. In general, the conductivity decreases with decreasing concentration of electrolytic solutes, due to swelling and dispersion phenomena, which are also affected by the species of cations present. Detachment and migration of clay particles during prolonged flow may result in the clogging of pores. The interactions of solutes with the soil matrix, and their effect on hydraulic conductivity are particularly important in saline and sodic soils.

In practice, it is extremely difficult to saturate a soil with water without trapping some air. Entrapped air bubbles may block pore passages, as shown in Fig. 6.9. Temperature changes may cause the flowing water to dissolve or to release gas, and will also cause a change in the volume of the gas phase, thus affecting conductivity.

The hydraulic conductivity $K$ is not an exclusive property of the soil alone, since it depends upon the attributes of the soil and of the fluid together. The soil characteristics which affect $K$ are the total porosity, the distribution of pore sizes, and tortuosity—in short, the pore geometry of the soil. The fluid attributes which affect conductivity are fluid density and viscosity.

It is possible in theory, and sometimes in practice, to separate $K$ into two factors: *intrinsic permeability* of the soil $k$ and *fluidity* of the liquid or gas $f$:

$$K = kf \qquad (6.10)$$

When $K$ is expressed in terms of cm/sec $(LT^{-1})$, $k$ is expressed in cm$^2$ $(L^2)$ and $f$ in $1/(cm\ sec)(L^{-1}T^{-1})$.

Fluidity is inversely proportional to viscosity:

$$f = \rho g/\eta \qquad (6.11)$$

hence,

$$k = K\eta/\rho g \qquad (6.12)$$

where $\eta$ is the viscosity in poise units (dyn sec/cm$^2$), $\rho$ is the fluid density (gm/cm$^3$), and $g$ is the gravitational acceleration (cm/sec$^2$).

In an ordinary liquid, the density is nearly constant (though it varies somewhat with temperature and solute concentration), and changes in fluidity are likely to result primarily from changes in viscosity. In compressible fluids such as gases, on the other hand, changes in density due to pressure and temperature variation can also be considerable.

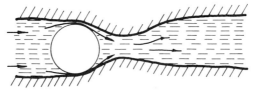

**Fig. 6.9.** An entrapped air bubble plugging flow.

The use of the term permeability has in the past been a source of some confusion, as it has often been applied synonymously with hydraulic conductivity. Permeability has also been used in a loosely qualitative sense to describe the readiness with which a porous medium transmits water or various other fluids. For this reason, the use of *permeability* in a strict, quantitative sense with the dimensions of length squared as previously defined in Eq. (6.12) may require the use of some such qualifying adjective as "intrinsic" permeability (Richards, 1954) or "specific" permeability (Scheidegger, 1957). For convenience, however, we shall henceforth refer to $k$ simply as permeability.

It should be clear from the foregoing that, while fluidity varies with composition of the fluid and with temperature, the permeability is ideally an exclusive property of the porous medium and its pore geometry alone — provided the fluid and the solid matrix do not interact in such a way as to change the properties of either. In a completely stable porous body, the same permeability will be obtained with different fluids, e.g., with water, air, or oil.[8] In many soils, however, matrix–water interactions are such that conductivity cannot be resolved into separate and exclusive properties of water and of soil, and Eq. (6.10) is impractical to apply.

## G. Measurement of Hydraulic Conductivity of Saturated Soils

Methods for measuring hydraulic conductivity in the laboratory were reviewed by Klute (1965a), and for measurement in the field by Talsma (1960) and by Boersma (1965a,b). The use of permeameters for laboratory determinations is illustrated in Figs. 6.10 and 6.11. Such determinations can

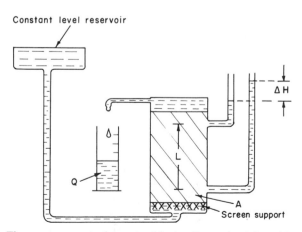

**Fig. 6.10.** The measurement of saturated hydraulic conductivity with a constant head permeameter; $K = VL/At\Delta H$.

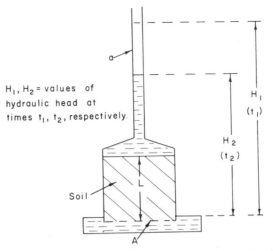

$H_1, H_2$ = values of hydraulic head at times $t_1$, $t_2$, respectively.

**Fig. 6.11.**   The measurement of saturated hydraulic conductivity with a falling head perme-ameter; $K = [2.3aL/A(t_2 - t_1)](\log H_1 - \log H_2)$. $H_1$ and $H_2$ are the values of hydraulic head at times $t_1$ and $t_2$, respectively.

be made with dried and fragmented specimens, which then must be packed into the flow cells in a standard manner, or, preferably, with undisturbed core samples taken directly from the field. In either case, provision must be made to avoid boundary flow along the walls of the container. Field measure-ments can be made most conveniently below the water table, as by the *augerhole method* (Luthin, 1957) or by the *piezometer method* (Johnson *et al.*, 1952). Techniques have also been proposed for measurements above the water table, as by the double-tube method (Bouwer, 1961, 1962), the shallow-well pump-in method, and the field-permeameter method (Winger, 1960). A more recent review of methods for measuring the hydraulic con-ductivity of saturated soils was given by van Schilfgaarde (1974).

### Sample Problems

**1.**   Water in an irrigation hose is kept at an hydrostatic pressure of 1 bar. Five drip-irrigation emitters are inserted into the wall of the hose. Calculate the drip rate (liter/hr) from the emitters if each contained a coiled capillary tube 1 m long and the capillary diameters are 0.2, 0.4, 0.6, 0.8, and 1.0 mm. (Assume laminar flow.) What percentage of the total discharge is due to the single largest emitter?

Use Poiseuille's law to calculate the discharge $Q$:

$$Q = \pi R^4 \, \Delta P / 8\eta L$$

Substituting the values for $\pi$ (3.14), the pressure differential $\Delta P$ ($10^6$ dyn/cm$^2$), the viscosity $\eta$ [$10^{-2}$ gm/cm sec at 20°C], the capillary tube length (100 cm), and the appropriate tube radii (0.01, 0.02, 0.03, 0.04, and 0.05 cm), we obtain

Emitter #1: $Q_1 = \dfrac{3.14 \times (10^{-2})^4 \times 10^6}{8 \times 10^{-2} \times 10^2} = 3.91 \times 10^{-3}$ cm$^3$/sec

$= 0.014$ liter/hr

Emitter #2: $Q_2 = \dfrac{3.14 \times (2 - 10^{-2})^4 \times 10^6}{8 \times 10^{-2} \times 10^2} = 6.28 \times 10^{-2}$ cm$^3$/sec

$= 0.226$ liter/hr

Emitter #3: $Q_3 = \dfrac{3.14 \times (3 \times 10^{-2})^4 \times 10^6}{8 \times 10^{-2} \times 10^2} = 3.18 \times 10^{-1}$ cm$^3$/sec

$= 1.14$ liter/hr

Emitter #4: $Q_4 = \dfrac{3.14 \times (4 \times 10^{-2})^4 \times 10^6}{8 \times 10^{-2} \times 10^2} = 1$ cm$^3$/sec

$= 3.6$ liter/hr

Emitter #5: $Q_5 = \dfrac{3.14 \times (5 \times 10^{-2})^4 \times 10^6}{8 \times 10^{-2} \times 10^2} = 2.45$ cm$^3$/sec

$= 8.83$ liter/hr

Total discharge from all five emitters:

$$Q_{tot} = 0.014 + 0.226 + 1.14 + 3.6 + 8.83 = 13.81 \text{ liter/hr}$$

Percentage contribution of the largest emitter:

$$\frac{Q_5}{Q_{tot}} \times 100 = \frac{8.83}{13.81} \times 100 = 63.9\%$$

The single largest emitter thus accounts for nearly two-thirds of the total discharge, while the smallest emitter accounts for only 0.1% (though its diameter is only $\frac{1}{5}$ that of the largest emitter).

*Note:* Modern drip-irrigation emitters generally depend on partly turbulent (rather than completely laminar) flow, to reduce sensitivity to pressure fluctuations and vulnerability to clogging by particles.

   **2.** Over a century ago, the population of the French mustard making town of Dijon was, perhaps, 10,000. Since they drank mostly wine, their daily water requirements were, say, no more than 20 liter per person. Then the denizens of the town began to notice that their water supply had somehow become polluted. Since they were unable at the time to find any bona fide soil physicists, they invited an engineer named Darcy to design a filtration system. He must have looked for a textbook or a handbook on the topic but found none, so he had to experiment from scratch. Darcy ended up promulgating a new law and achieving immortal, if posthumous, fame. Now supposing we were given the same task today (with the benefits of hindsight) and we knew that a column thickness of 30 cm was needed for adequate filtration and that the hydraulic conductivity of the available sand was $2 \times 10^{-3}$ cm/sec. Could we calculate the area of filter bed needed under an hydrostatic pressure (ponding) head of 0.7 m? Consider the flow to be vertically downward to a fixed drainage surface. We begin by calculating the discharge $Q$ needed:

$$Q = \frac{10^4 \text{ persons} \times 20 \text{ liter/person day} \times 10^3 \text{ cm}^3/\text{liter}}{8.64 \times 10^4 \text{ sec/day}}$$
$$= 2.31 \times 10^3 \text{ cm}^3/\text{sec}$$

We recall Darcy's law:

$$Q = AK \, \Delta H/L$$

Hence, the area $A$ needed is

$$A = QL/(K \, \Delta H)$$

The hydraulic head drop $\Delta H$ equals the sum of the pressure head and gravitational head drops. Hence

$$\Delta H = 70 + 30 = 100 \text{ cm}$$

We now substitute the appropriate values for $L$ (30 cm), $\Delta H$ (100 cm), and $K$ ($2 \times 10^{-3}$ cm/sec), to obtain

$$A = \frac{2.31 \times 10^3 \text{ cm}^3/\text{sec} \times 30 \text{ cm}}{2 \times 10^{-3} \text{ cm/sec} \times 10^2 \text{ cm}} = 3.5 \times 10^5 \text{ cm}^2 = 35 \text{ m}^2$$

*Note:* Since populations and per capita water use tend to increase whereas filter beds tend to clog, it might behoove use to apply a factor of safety to our calculations and increase the filtration capacity severalfold (particularly to accommodate peak demand periods). Per capita water use in the U.S., incidentally, ranges from 100 to 400 liter/day.

# 7     *Flow of Water in Unsaturated Soil*

## A. Introduction

Most of the processes involving soil–water interactions in the field, and particularly the flow of water in the rooting zone of most crop plants, occur while the soil is in an unsaturated condition. Unsaturated flow processes are in general complicated and difficult to describe quantitatively, since they often entail changes in the state and content of soil water during flow. Such changes involve complex relations among the variable soil wetness, suction, and conductivity, whose interrelations may be further complicated by hysteresis. The formulation and solution of unsaturated flow problems very often require the use of indirect methods of analysis, based on approximations or numerical techniques. For this reason the development of rigorous theoretical and experimental methods for treating these problems was rather late in coming. In recent decades, however, unsaturated flow has become one of the most important and active topics of research in soil physics, and this research has resulted in significant theoretical and practical advances.

## B. Comparison of Flow in Unsaturated versus Saturated Soil

In the previous chapter, we stated that soil-water flow is caused by a driving force resulting from a potential gradient, that flow takes place in the direction of decreasing potential, and that the rate of flow (flux) is proportional to the potential gradient and is affected by the geometric properties of the pore channels through which flow takes place. These principles apply in unsaturated, as well as saturated, soils.

The moving force in a saturated soil is the gradient of a positive pressure potential.[1] On the other hand, water in an unsaturated soil is subject to a subatmospheric pressure, or suction, which is equivalent to a negative pressure potential. The gradient of this potential likewise constitutes a moving force. Matric suction is due, as we have pointed out, to the physical affinity of water to the soil-particle surfaces and capillary pores. Water tends to be drawn from a zone where the hydration envelopes surrounding the particles are thicker to where they are thinner, and from a zone where the capillary menisci are less curved to where they are more highly curved.[2] In other words, water flows spontaneously from where matric suction is lower to where it is higher. When suction is uniform all along a horizontal column of soil the column is at equilibrium and there is no moving force. Not so when a suction *gradient* exists. In that case, water will flow in the pores which are water filled at the existing suction and will creep along the hydration films over the particle surfaces, in a tendency to equilibrate the potential. Vapor transfer is an additional mechanism of water movement in unsaturated soils. In the absence of temperature gradients, it is likely to be much slower than liquid flow as long as the soil is fairly moist. In the surface zone, however, where the soil becomes desiccated and strong temperature gradients occur, vapor transfer can become the dominant mechanism of water movement.

The moving force is greatest at the *wetting front* zone, where water invades and advances into an originally dry soil (see Fig. 7.3). In this zone, the suction gradient can amount to many bars per centimeter of soil. Such a gradient constitutes a moving force thousands of times greater than the gravitational force. As we shall see later on, such strong forces are sometimes required for water movement to take place in the face of the extremely low hydraulic conductivity which a relatively dry soil often exhibits.

Perhaps the most important difference between unsaturated and saturated flow is in the hydraulic conductivity. When the soil is saturated, all of the pores are water filled and conducting, so that continuity and hence con-

---

[1] We shall disregard, for the moment, the gravitational force, which is completely unaffected by the saturation or unsaturation of the soil.

[2] The question of how water-to-air interfaces behave in a conducting porous medium that is unsaturated is imperfectly understood. It is generally assumed, at least implicitly, that these interfaces, or menisci, are anchored rigidly to the solid matrix so that, as far as the flowing water is concerned, air-filled pores are like solid particles. The presence of organic surfactants which adsorb to these surfaces is considered to increase their rigidity or viscosity. Even if the air–water interfaces are not entirely stationary, however, the drag, or momentum transfer, between flowing water and air appears to be very small.

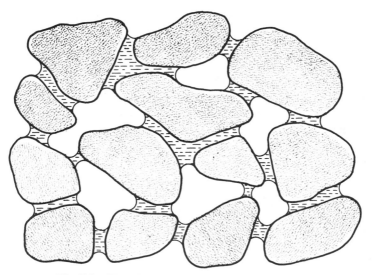

**Fig. 7.1.** Water in an unsaturated coarse-textured soil.

ductivity are maximal. When the soil desaturates, some of the pores become air filled and the conductive portion of the soil's cross-sectional area decreases correspondingly. Furthermore, as suction develops, the first pores to empty are the largest ones, which are the most conductive,[3] thus leaving water to flow only in the smaller pores. The empty pores must be circumvented, so that, with desaturation, tortuosity increases. In coarse-textured soils, water sometimes remains almost entirely in capillary wedges at the contact points of the particles, thus forming separate and discontinuous pockets of water (see Fig. 7.1). In aggregated soils, too, the large interaggregate spaces which confer high conductivity at saturation become (when emptied) barriers to liquid flow from one aggregate to its neighbors.

For these reasons, the transition from saturation to unsaturation generally entails a steep drop in hydraulic conductivity, which may decrease by several orders of magnitude (sometimes down to 1/100,000 of its value at saturation) as suction increases from 0 to 1 bar. At still higher suctions, or lower wetness values, the conductivity may be so low that very steep suction gradients, or very long times, are required for any appreciable flow to occur.

At saturation, the most conductive soils are those in which large and continuous pores constitute most of the overall pore volume, while the least conductive are the soils in which the pore volume consists of numerous

---

[3] By Poiseuille's law, the total flow rate of water through a capillary tube is proportional to the fourth power of the radius, while the flow rate per unit cross-sectional area of the tube is proportional to the square of the radius. A 1-mm-radius pore will thus conduct as 10,000 pores of radius 0.1 mm.

micropores. Thus, as is well known, a saturated sandy soil conducts water more rapidly than a clayey soil. However, the very opposite may be true when the soils are unsaturated. In a soil with large pores, these pores quickly empty and become nonconductive as suction develops, thus steeply decreasing the initially high conductivity. In a soil with small pores, on the other hand, many of the pores retain and conduct water even at appreciable suction, so that the hydraulic conductivity does not decrease as steeply and may actually be greater than that of a soil with large pores subjected to the same suction.

Since in the field the soil is unsaturated much, and perhaps most, of the time, it often happens that flow is more appreciable and persists longer in clayey than in sandy soils. For this reason, the occurrence of a layer of sand in a fine-textured profile, far from enhancing flow, may actually impede unsaturated water movement until water accumulates above the sand and suction decreases sufficiently for water to enter the large pores of the sand. This simple principle is all too often misunderstood.

## C. Relation of Conductivity to Suction and Wetness

Let us consider an unsaturated soil in which water is flowing under suction. Such flow is illustrated schematically in the model of Fig. 7.2. In this model, the potential difference between the inflow and outflow ends is maintained not by different heads of positive hydrostatic pressure, but by different imposed suctions. In general, as the suction varies along the sample, so will the wetness and conductivity. If the suction head at each end of the sample is maintained constant, the flow process will be steady but the suction gradient will vary along the sample's axis. Since the product of gradient and conductivity is constant for steady flow, the gradient will increase as the conductivity decreases with the increase in suction along the length of the sample. This phenomenon is illustrated in Fig. 7.3.

Since the gradient along the column is not constant, as it is in uniform saturated systems, it is not possible, strictly speaking, to divide the flux by the overall ratio of the head drop to the distance $(\Delta H / \Delta x)$ to obtain the conductivity. Rather, it is necessary to divide the flux by the exact gradient at each point to evaluate the exact conductivity and its variation with suction. In the following treatment, however, we shall assume that the column of Fig. 7.2 is sufficiently short to allow us to evaluate at least an average conductivity for the sample as a whole (i.e., $K = q \, \Delta x / \Delta H$).

The average negative head, or suction, acting in the column is

$$-\bar{H} = \bar{\psi} = -\tfrac{1}{2}(H_1 + H_2)$$

We shall further assume that the suction everywhere exceeds the air-entry value so that the soil is unsaturated throughout.

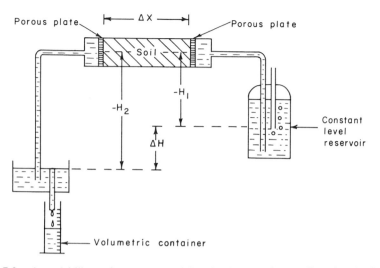

**Fig. 7.2.** A model illustrating unsaturated flow (under a suction gradient) in a horizontal column.

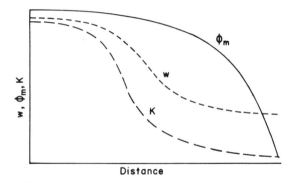

**Fig. 7.3.** The variation of wetness $w$, matric potential $\phi_m$, and conductivity $K$ along a hypothetical column of unsaturated soil conducting a steady flow of water.

Let us now make successive and systematic measurements of flux versus suction gradient for different values of average suction. The results of such a series of measurements are shown schematically in Fig. 7.4. As in the case of saturated flow, we find that the flux is proportional to the gradient. However, the slope of the flux versus gradient line, being the hydraulic conductivity, varies with the average suction. In a saturated soil, by way of contrast, the hydraulic conductivity is generally independent of the magnitude of the water potential, or pressure.

Figure 7.5 shows the general trend of the dependence of conductivity on suction[4] in soils of different texture. It is seen that, although the saturated

[4] $K$ versus suction curves are usually drawn on a log–log scale, as both $K$ and $\psi$ vary over several orders of magnitude within the suction range of general interest (say, 0– 10,000 cm of suction head).

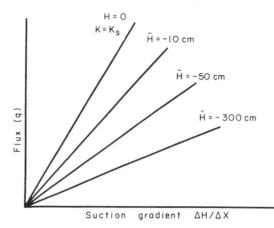

**Fig. 7.4.**  Hydraulic conductivity, being the slope of the flux versus gradient relation, depends upon the average suction in an unsaturated soil.

conductivity of the sandy soil $K_{s1}$ is typically greater than that of the clayey soil $K_{s2}$, the unsaturated conductivity of the former decreases more steeply with increasing suction and eventually becomes lower.

Although attempts have been made to develop theoretically based equations for the relation of conductivity to suction or to wetness, existing knowledge still does not allow reliable a priori prediction of the unsaturated conductivity function from basic soil properties such as texture. Various empirical equations have been proposed, however, including (Gardner, 1960)

$$K(\psi) = a / \psi^m \tag{7.1a}$$

$$K(\psi) = a / (b + \psi^m) \tag{7.1b}$$

$$K(\psi) = K_s / [1 + (\psi / \psi_c)^m] \tag{7.1c}$$

$$K(\theta) = a \theta^m \tag{7.1d}$$

$$K(\theta) = K_s s^m = K_s (\theta / f)^m \tag{7.1e}$$

where $K$ is the hydraulic conductivity at any degree of saturation (or unsaturation): $K_s$ is the saturated conductivity of the same soil; $a$, $b$, and $m$ are empirical constants (different in each equation); $\psi$ is matric suction head; $\theta$ is volumetric wetness; $s$ is the degree of saturation; and $\psi_c$ is the suction head at which $K = \frac{1}{2} K_s$. Note that $s = \theta / f$, where $f$ is porosity.

Of these various equations, the most commonly employed are the first two (of which the first is the simplest, but cannot be used in the suction range approaching zero). In all of the equations, the most important parameter is the exponential constant, since it controls the steepness with which conductivity decreases with increasing suction or with decreasing water content. The $m$ value of the first two equations is about 2 or less for clayey soils and

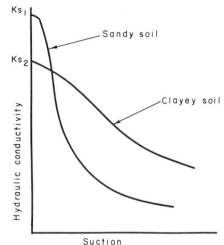

**Fig. 7.5.** Dependence of conductivity on suction in soils of different texture (log–log scale).

may be 4 or more for sandy soils. For each soil, the equation of best fit, and the values of the parameters, must be determined experimentally.

The relation of conductivity to suction depends upon hysteresis, and is thus different in a wetting than in a drying soil. At a given suction, a drying soil contains more water than a wetting one. The relation of conductivity to water content, however, appears to be affected by hysteresis to a much lesser degree.

## D. General Equation of Unsaturated Flow

Darcy's law, though originally conceived for saturated flow only, was extended by Richards (1931) to unsaturated flow, with the provision that the conductivity is now a function of the matric suction head (i.e., $K = K(\psi)$]:

$$\mathbf{q} = -K(\psi)\,\nabla H \tag{7.2}$$

where $\nabla H$ is the hydraulic head gradient, which may include both suction and gravitational components.

As pointed out by Miller and Miller (1956), this formulation fails to take into account the hysteresis of soil-water characteristics. In practice, the hysteresis problem can sometimes be evaded by limiting the use of Eq. (7.1) to cases in which the suction (or wetness) change is monotonic—that is, either increasing or decreasing continuously. In processes involving both wetting and drying phases, Eq. (7.1) is difficult to apply, as the $K(\psi)$ function may be highly hysteretic. As mentioned in the previous section, however, the relation of conductivity to volumetric wetness $K(\theta)$ or to degree of saturation $K(s)$ is affected by hysteresis to a much lesser degree than is the $K(\psi)$

function. Thus, Darcy's law for unsaturated soil can also be written

$$\mathbf{q} = -K(\theta)\,\mathbf{\nabla}H \tag{7.3}$$

which, however, still leaves us with the problem of dealing with the hysteresis between $\psi$ and $\theta$.

To obtain the general flow equation and account for transient, as well as steady, flow processes, we must introduce the continuity equation, which embodies the conservation of mass law in mathematical form:

$$\partial\theta/\partial t = -\mathbf{\nabla}\cdot\mathbf{q}$$

Thus,

$$\partial\theta/\partial t = \mathbf{\nabla}\cdot[K(\psi)\,\mathbf{\nabla}H] \tag{7.4}$$

Remembering that the hydraulic head is, in general, the sum of the pressure head (or its negative, the suction head $\psi$) and the gravitational head (or elevation) $z$, we can write

$$\partial\theta/\partial t = -\mathbf{\nabla}\cdot[K(\psi)\,\mathbf{\nabla}(\psi - z)] \tag{7.5}$$

Since $\mathbf{\nabla}z$ is zero for horizontal flow and unity for vertical flow, we can rewrite (9.6) as follows:

$$\frac{\partial\theta}{\partial t} = -\mathbf{\nabla}\cdot(K(\psi)\,\mathbf{\nabla}\psi) + \frac{\partial K}{\partial z} \tag{7.5a}$$

or

$$\frac{\partial\theta}{\partial t} = -\frac{\partial}{\partial x}\left(K\frac{\partial\psi}{\partial x}\right) - \frac{\partial}{\partial y}\left(K\frac{\partial\psi}{\partial y}\right) - \frac{\partial}{\partial z}\left(K\frac{\partial\psi}{\partial z}\right) + \frac{\partial K}{\partial z} \tag{7.6}$$

Processes may also occur in which $\mathbf{\nabla}z$ (the gravity gradient) is negligible compared to the strong matric suction gradient $\mathbf{\nabla}\psi$. In such cases,

$$\partial\theta/\partial t = \mathbf{\nabla}\cdot[K(\psi)\,\mathbf{\nabla}\psi] \tag{7.7}$$

or, in a one-dimensional horizontal system,

$$\frac{\partial\theta}{\partial t} = \frac{\partial}{\partial x}\left[K(\psi)\frac{\partial\psi}{\partial x}\right] \tag{7.8}$$

### E.  Hydraulic Diffusivity

To simplify the mathematical and experimental treatment of unsaturated flow processes, it is often advantageous to change the flow equations into a form analogous to the equations of diffusion and heat conduction, for which ready solutions are available (e.g., Carslaw and Jaeger, 1959; Crank, 1956) in some cases involving boundary conditions applicable to soil-water flow processes. To transform the flow equation, it is sometimes possible to relate the flux to the water content (wetness) gradient rather than to the suction gradient.

The matric suction gradient $\partial\psi/\partial x$ can be expanded by the chain rule as follows

$$\frac{\partial\psi}{\partial x} = \frac{d\psi}{d\theta}\frac{\partial\theta}{\partial x} \tag{7.9}$$

where $\partial\theta/\partial x$ is the wetness gradient and $d\psi/d\theta$ is the reciprocal of the specific water capacity $c(\theta)$:

$$c(\theta) = d\theta/d\psi \tag{7.10}$$

which is the slope of the soil-moisture characteristic curve at any particular value of wetness $\theta$.

We can now rewrite the Darcy equation as follows:

$$q = K(\theta)\frac{\partial\psi}{\partial x} = -\frac{K(\theta)}{c(\theta)}\frac{\partial\theta}{\partial x} \tag{7.11}$$

To cast this equation into a form analogous to Fick's law of diffusion, a function was introduced (Childs and Collis-George, 1950), originally called the *diffusivity D*, where

$$D(\theta) = K(\theta)/c(\theta) = K(\theta)\,d\psi/d\theta \tag{7.12}$$

$D$ is thus defined as the ratio of the hydraulic conductivity $K$ to the specific water capacity $c$, and since both of these are functions of soil wetness, $D$ must also be so. To avoid any possibility of confusion between the classical concept of diffusivity pertaining to the *diffusive transfer* of components in the gaseous and liquid phases (see, for example, our chapters on solute movement and on gas exchange in the soil) and this borrowed application of the same term to describe *convective flow*, we propose to qualify it with the adjective hydraulic. In this book, therefore, we shall henceforth employ the term *hydraulic diffusivity* when referring to $D$ of Eq. (7.12). Now we can rewrite Eq. (7.12)

$$\mathbf{q} = -D(\theta)\,\nabla\theta \tag{7.13}$$

or, in one dimension,

$$q = -D(\theta)\,\partial\theta/\partial x \tag{7.14}$$

which is mathematically identical with Fick's first equation of diffusion. Hydraulic diffusivity can thus be viewed as the ratio of the flux to the soil-water content (wetness) gradient. As such, $D$ has dimensions of length squared per unit time ($L^2T^{-1}$), since $K$ has the dimensions of volume per unit area per time ($LT^{-1}$) and the specific water capacity $c$ has dimensions of volume of water per unit volume of soil per unit change in matric suction head ($L^{-1}$). In the use of Eq. (7.14), the gradient of wetness is taken to represent, implicitly, a gradient of matric potential, which is the true driving force.

Introducing the hydraulic diffusivity into Eq. (7.8), for one-dimensional flow in the absence of gravity, we obtain[5]

$$\frac{\partial \theta}{\partial t} = \frac{\partial}{\partial x}\left[ D(\theta) \frac{\partial \theta}{\partial x} \right] \tag{7.15}$$

which has only one dependent variable ($\theta$) rather than the two ($\theta$ and $\psi$) of Eq. (7.8). This is analogous to Fick's second diffusion equation.

A word of caution is now in order. In employing the diffusivity concept, and all relationships derived from it, we must remember that the process of liquid water movement in the soil is not one of diffusion but of mass flow, or convection. As we have already suggested, the borrowed term diffusivity, if taken literally, can be misleading. Furthermore, the diffusivity equations become awkward whenever the hysteresis effect is appreciable or where the soil is layered, or in the presence of thermal gradients, since under such conditions flow bears no simple or consistent relation to the decreasing water-content gradient and may actually be in the opposite direction to it. On the other hand, an advantage in using the hydraulic diffusivity is in the fact that its range of variation is smaller than that of hydraulic conductivity.[5]

The relation of hydraulic diffusivity to wetness is shown in Fig. 7.6. This relation is sometimes expressed in the empirical equation (Gardner and Mayhugh, 1958)

$$D(\theta) = ae^{b\theta} \tag{7.16}$$

This equation applies only to the right-hand section of the curve showing a rise in diffusivity with wetness. In the very dry range, the diffusivity often indicates an opposite trend—namely, a rise with decreasing soil wetness. This is apparently due to the contribution of vapor movement (Philip, 1955). In the very wet range, as the soil approaches complete saturation, the diffusivity becomes indeterminate as it tends to infinity [since $c(\theta)$ tends to zero].

---

[5]The maximum value of $D$ found in practice is of the order of $10^4$ cm²/day. $D$ generally decreases to about $1-10$ cm²/day at the lower limit of wetness normally encountered in the root zone. It thus varies about a thousandfold rather than about a millionfold, as does the hydraulic conductivity in the same wetness range.

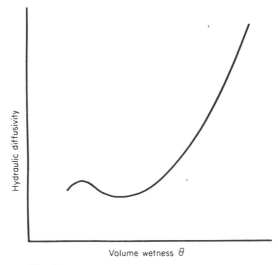

**Fig. 7.6.** Relation of diffusivity to soil wetness.

## F. Measurement of Unsaturated Hydraulic Conductivity and Diffusivity in the Laboratory

Knowledge of the unsaturated hydraulic conductivity and diffusivity at different suction and wetness values is generally required before any of the mathematical theories of water flow can be applied in practice. Since there is as yet no universally proven way to predict these values from more basic or more easily obtainable soil properties, $K$ and $D$ must be measured experimentally. In principle, $K$ and $D$ can be obtained from either *steady-state* or *transient-state* flow systems. In *steady flow systems*, flux, gradient, and water content are constant in time, while in *transient flow systems*, they vary. In general, therefore, measurements based on steady flow are more convenient to carry out and often more accurate. The difficulty, however, lies in setting up the flow system, which may take a very long time to stabilize.

Techniques for measurement of conductivity and diffusivity of soil samples or models in the laboratory were described by Klute (1965b). The conductivity is usually measured by applying a constant hydraulic head difference across the sample and measuring the resulting steady flux of water. Soil samples can be desaturated either by tension-plate devices or in a pressure chamber. Measurements are made at successive levels of suction and wetness, so as to obtain the functions $K(\psi)$, $K(\theta)$, and $D(\theta)$. The $K(\psi)$ relationship is hysteretic, and therefore, to completely describe it, measurements should be made both in desorption and in sorption, as well, perhaps, as in intermediate scanning. This is difficult, however, and requires specialized apparatus (Tanner and Elrick, 1958), so that all too often only the desorption curve is measured (starting at saturation and proceeding to increase the suction in increments).

Such laboratory techniques can also be applied to the measurement of undisturbed soil cores taken from the field. This is certainly preferable to measurements taken on fragmented and artificially packed samples, though it should be understood that no field sampling technique yet available provides truly undisturbed samples. Moreover, any attempt to represent a field soil by means of extracted samples incurs the problem of field soil heterogeneity as well as the associated problem of determining the appropriate scale (i.e., the representative volume) for realistic measurement of parameters.

A widely used transient flow method for measurement of conductivity and diffusivity in the laboratory is the *outflow method* (Gardner, 1956). It is based on measuring the falling rate of outflow from a sample in a pressure cell when the pressure is increased by a certain increment. One problem encountered in the application of this method is that of the hydraulic resistance (also called impedance) of the porous plate or membrane upon which the sample is placed, and of the soil-to-plate contact zone. Techniques to account for this resistance were proposed by Miller and Elrick (1958), Rijtema (1959), and Kunze and Kirkham (1962).

Laboratory measurements of conductivity and diffusivity can also be made on long columns of soil, not only on small samples contained in cells. In such columns, steady-state flow can be induced by evaporation (e.g., Moore, 1939) or by infiltration (Youngs, 1964). If the column is long enough to allow the measurement of suction gradients (e.g., by a series of tensiometers) and of wetness gradients (by sectioning, or, preferably, by some nondestructive technique such as gamma-ray scanning), the $K(\theta)$ and $K(\psi)$ relationship can be obtained for a range of $\theta$ with a single column or with a series of similarly packed columns.

Measurements in columns under transient flow conditions have also been made (e.g., the horizontal infiltration technique of Bruce and Klute, 1956). If periodic suction and wetness profiles are measured, the flux values at different time and space intervals can be evaluated by graphic integration between successive moisture profiles. This procedure has been called the *instantaneous profile* technique (Watson, 1966) and it can be applied in the field as well (Rose *et al.*, 1965).

## G. Measurement of Unsaturated Hydraulic Conductivity of Soil Profiles *in Situ*

In recent years research in soil physics has resulted in the development of mathematical theories and models describing the state and movement of water in both saturated and unsaturated soil bodies. Moreover, experimental work has resulted in the development of more precise and reliable techniques for the measurement of flow phenomena and of pertinent soil parameters. All too often, however, our theories and models have been validated, if at all, only in highly artificial sets of laboratory produced conditions. Yet too rare are the instances in which theories and models have been applied and found to be valid and useful under realistic conditions, in actual soil management practice. In fact, there is still a great dichotomy in this area between fundamental knowledge and practical application.

Application of the theories of soil physics to the description or prediction of actual processes in the field (e.g., processes involved in irrigation, drainage, water conservation, groundwater recharge and pollution, as well as infiltration and runoff control) depends upon knowledge of the pertinent hydraulic characteristics of the soil, including the functional relation of hydraulic conductivity and of matric suction to soil wetness as well as the spatial and temporal variation of these in the often heterogeneous field situation.

The problem of field soil heterogeneity relates to a fundamental theoretical question which is too often ignored, namely the characteristic scale of the system. Obviously such soil properties as conductivity, porosity, and pore-size distribution are scale dependent and their magnitudes should be considered in relation to some specified or implied size of sample. All soils are inherently inhomogeneous in that their primary and secondary particles and pore spaces differ from point to point and their geometry is too complicated to characterize in microscopic detail. For this reason, the soil is generally characterized in macroscopic terms based on the gross averaging of microscopic and sometimes of mesoscopic heterogeneities. An implicit assumption is that the physical properties are measured on a volume of soil sufficiently large relative to the microscopic heterogeneities to permit such an averaging. Yet how large must the measured volume be? This is generally left unspecified. For instance, in the case of a uniform sand soil, a measurement made on a cubic decimeter may be sufficient to characterize the entire soil. On the other hand, in some cases the soil may be layered, aggregated, and fissured, with relatively large cracks present, so that the hydraulic properties of the medium as a whole can be represented only by a volume as great perhaps as several cubic meters.

From the considerations given, it seems basically unrealistic to try to measure the unsaturated hydraulic conductivity of field soil by making laboratory determinations on discrete and small samples removed from their natural continuum, particularly when such samples are fragmented or otherwise disturbed. Hence it is necessary to devise and test practical methods for measuring soil hydraulic conductivity on a macroscale in situ.

We shall now proceed to give a brief description of several of these methods.

## 1. Method I: Sprinkling Infiltration

This method has been described in principle by Youngs (1964), who, however, tested it only under uniform laboratory conditions. The principle of the method is that the continued supply of water to the soil, as under sprinkling, at a constant rate lower than the effective hydraulic conductivity of the soil, eventually results in the establishment of a steady moisture distribution in the conducting profile. Once steady-state conditions are established, a constant flux exists. In a uniform soil the suction gradients will tend to zero, and with only a unit gravitational gradient in effect the hydraulic conductivity becomes essentially equal to the flux. If this test is carried out on an initially dry soil with a series of successively increasing application rates (sprinkling intensities), it becomes possible to obtain different values of hydraulic conductivity corresponding to different values of soil wetness. The theoretical relationship between rain intensity and the soil moisture profile during infiltration was described by Rubin (1966).

The difficulty of the steady sprinkling infiltration test in the field is that it requires rather elaborate equipment which must be maintained in continuous operation for considerable periods of time. The requirement of maintaining continuous operation becomes increasingly important, and difficult, as one attempts to extend the test toward the greater suction range by reducing the application rate below 1 mm/hr. Another difficulty is to avoid the raindrop impact effect which can cause the exposed surface soil to disperse and seal, thus reducing infiltrability. Variable intensity sprinkling infiltrometers adaptable to field use have been described by Steinhardt and Hillel (1966), Morin *et al.* (1967), and more recently by Amerman *et al.* (1970) and Rawitz *et al.* (1972). The field plot should normally be at least 1 m² in size surrounded by a buffer area under the same sprinkling so as to minimize the lateral flow component in the test plot. The assumption of one-dimensional (vertical) flow becomes questionable and perhaps invalid in the case of a soil profile with distinct impeding layers, over which temporary perched water table conditions can develop, leading to subsurface lateral flows beyond the boundaries of the test plot. A comparison between the steady sprinkling method and the instantaneous-profile unsteady drainage method for determination of $K(\theta)$ was presented by Hillel and Benyamini (1974).

## 2. Method II: Infiltration through an Impeding Layer

This method was suggested by Hillel and Gardner (1970), who showed that an impeding layer at the surface of the soil can be used to achieve the desired boundary conditions to allow measurements of the unsaturated hydraulic conductivity and diffusivity as a function of soil wetness. The effect of an impeding layer present over the top of the profile during infiltration is to decrease the hydraulic potential in the profile under the impeding layer. Thus, the soil wetness, and correspondingly the conductivity and diffusivity values of the infiltrating profile, are reduced. These authors held that their proposed test can be applied during the transient stage of infiltrations, but it becomes most reliable if the test is continued long enough to attain steady state. Qualitatively, when the surface is covered by an impeding layer (crust) with the saturated conductivity smaller than that of the soil, the steady flow conditions which develop are such that the head gradient through the impeding layer is necessarily greater than unity. If the ponding depth is negligible, such an impeding layer thus induces the development of suction in the subsoil, the magnitude of which will increase with increasing hydraulic resistance of the crust. When steady infiltration is achieved, the flux and the conductivity of the subcrust soil become equal.

The advantage of this procedure over those using constant application rates is the simplicity of the experimental system. It may require a relatively long time at high values of soil moisture suction. However, there is no a priori theoretical limit to the suction or conductivity range measurable, which depends only on the range of soil moisture suction one can find in the field at the outset of the test. In principle, one could use a nearly impervious plate and continue the experiment almost indefinitely, provided the flux can be measured accurately while evaporation is prevented. The test procedure obviously involves a series of infiltration trials through capping plates (or crusts) of different hydraulic resistance values. The surface crust can be made artificially by dispersing the surface of the soil itself or by applying a puddled slurry of dispersed soil or clay of different hydraulic resistance. The use of a series of crusts of progressively lower hydraulic resistance can give progressively high $K$ values corresponding to higher water content up to saturation. Such a series of tests can be carried out if the soil is initially fairly dry, either successively in the same location or concurrently on adjacent locations.

This method was applied to field use by Bouma *et al.* (1971). They used ring infiltrometers and puddled the surface soil to obtain the necessary boundary condition.[6] They also made tensiometric measurements in the infiltrating profile to monitor the vertical and horizontal gradients. Each infiltration run yielded a point on the $K$ versus suction or $K$ versus $\theta$ curves. A serious problem which has come to light recently and which may invalidate this test in some cases is related to the so-called unstable flow phenomenon. It has been observed that in transition from a fine-textured zone to a coarse-textured zone during infiltration, the advance of the water may not be even, but that sudden breakthrough flows may occur in specific locations where fingerlike intrusions take place. Despite several theoretical studies (e.g.,

[6]Attempts have also been made to form a crust by using gypsum (calcium sulfate). However, the addition of electrolytes into the soil solution may affect and modify soil properties and therefore seems undesirable.

Raats, 1974; Philip, 1974; Parlange, 1975) this phenomenon is still insufficiently understood.

### 3. METHOD III: INTERNAL DRAINAGE

This method is based upon monitoring the transient state internal drainage of a profile (Fig. 7.7) as suggested originally by Richards and Weeks (1953) and Ogata and Richards (1957), and later by Rose et al. (1965) and Watson (1966). This method, when carried out in the field, can help to eliminate the possible alteration of soil hydraulics due to disturbance of structure as well as the doubtful procedure of applying steady-state methods to transient state processes. The use of strain gauge pressure transducers in tensiometry make possible the rapid and automatic acquisition of soil moisture suction data while the soil wetness data required for the method are obtainable with a neutron moisture gauge.

Work along these lines has been carried out by Rose and Stern (1967), Van Bavel et al. (1968a,b), Davidson et al. (1969), Gardner (1970), and Giesel et al. (1970). These workers have proven the feasibility of determining the unsaturated hydraulic conductivity function of soils in the field. More recently, Hillel et al. (1972) gave a detailed description of a simplified procedure for determining the intrinsic hydraulic properties of a layered soil profile in situ.

The method requires frequent and simultaneous measurements of the soil wetness and matric suction profiles under conditions of drainage alone (evapotranspiration prevented). From these measurements it is possible to obtain instantaneous values of the potential gradients and fluxes operating within the profile and hence also of hydraulic conductivity values. Once the hydraulic conductivity at each elevation within the profile is known in relation to wetness, the data can be applied to the analysis of drainage and evapotranspiration in a vegetated field. To apply this method in the field, one must choose a characteristic fallow plot that is large enough (say, 10 × 10 m, or at least 5 × 5 m) so that processes at its center are unaffected by boundaries. Within this plot at least one neutron access tube is installed as deeply as possible through and below the root zone. The desirable depth will sometimes exceed 2 m. A series of tensiometers is installed at various depths near the access tube so as to represent profile horizons. The depth intervals between succeeding tensiometers should not exceed 30 cm. Water is then ponded on the surface and the plot is irrigated long enough so that the entire profile becomes as wet as it can be. The soil is then covered by a sheet of plastic so as to prevent any water flux across the surface. As internal drainage proceeds, periodic measurements are made of water content distribution and tension throughout the profile. The handling of the data and a sample calculation of the hydraulic conductivity values of a soil were described by Hillel et al. (1972).

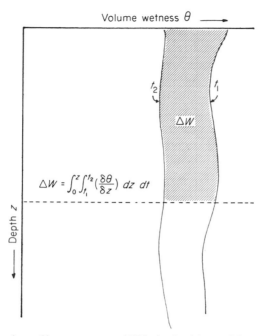

**Fig. 7.7.** Change in profile water content $\Delta W$ in the depth interval from $Z = 0$ (soil surface) to a given depth $Z_1$, in the time interval from $t_1$ to $t_2$, during internal drainage in the absence of either evaporation or lateral flow (schematic).

This method appears to be the most practical one at the present time for obtaining the unsaturated hydraulic conductivity function of soils in situ. The only instruments required are a neutron moisture meter (or some other nondestructive method for repeated determination of the soil moisture profile) and a set of tensiometers. As such it is easier to carry out than alternative methods based on controlled infiltration. Measurements can be made over several succeeding irrigation-drainage cycles in the same location. Moreover, the method does not assume previous knowledge of the soil moisture characteristic (matric suction versus water content) and in fact can yield information of this function in situ for the range of change occurring during the process of soil moisture redistribution (internal drainage) in the field. The method as described is not applicable where lateral movement of soil moisture is appreciable. This movement is not normally significant when the soil profile is unsaturated, but lateral movement can become significant wherever an impeding layer occurs upon which saturated conditions might prevail for some time. In practice, the moisture range for which conductivity can be measured by the internal drainage method is generally limited to suctions not exceeding about 0.5 bar, as the drainage process often slows down within a few days or weeks to become practically imperceptible.

## H. Vapor Movement

We have already stated that liquid water moves in the soil by *mass flow*, a process by which the entire body of a fluid flows in response to differences in hydraulic potential. In certain special circumstances, water vapor movement can also occur as mass flow; for instance, when wind gusts induce bulk movement of air and vapor mixing in the surface zone of the soil. In general, however, vapor movement through most of the soil profile occurs by *diffusion*, a process in which different components of a mixed fluid move independently, and at times in opposite directions, in response to differences in concentration (or partial pressure) from one location to another. Water vapor is always present in the gaseous phase of an unsaturated soil, and vapor diffusion occurs whenever differences in vapor pressure develop within the soil.

The diffusion equation for water vapor is

$$q_v = -D_v \, \partial \rho_v / \partial x \qquad (7.17)$$

wherein $\rho_v$ is the vapor density (or concentration) in the gaseous phase and $D_v$ the diffusion coefficient for water vapor. $D_v$ in the soil is lower than in open air because of the restricted volume and the tortuosity of air-filled pores (Currie, 1961).

By considering that the liquid water serves as a source and sink for water vapor and assuming that changes in liquid water content with time are much greater than vapor density changes with time, Jackson (1964) derived the equation

$$\frac{\partial \theta}{\partial t} = \frac{\partial}{\partial x}\left[ D_v \frac{\partial \rho_v}{\partial \theta} \frac{\partial \theta}{\partial x} \right] \qquad (7.18)$$

which describes nonsteady vapor transfer in terms of the liquid water content $\theta$ (the soil's volumetric wetness). For the simultaneous transfer of both liquid and vapor, the following equation applies:

$$\frac{\partial \theta}{\partial t} = \frac{\partial}{\partial x}\left[ \left( D_v \frac{\partial \rho_v}{\partial \theta} + D_\theta \right) \frac{\partial \theta}{\partial x} \right] \qquad (7.19)$$

in which $D_\theta$ is the hydraulic diffusivity for liquid water described in Section E.

The foregoing equations consider water vapor diffusion as an isothermal process, assuming that both viscous flow in the liquid phase and diffusion of vapor are impelled by the force fields of soil particle surface and capillarity. No explicit account has been taken of osmotic or solute effects on vapor pressure, though they can obviously be significant. More importantly, the foregoing discussion disregards the simultaneous and interactive transport of both water and heat in nonisothermal situations, which will be described in a subsequent chapter.

At constant temperature, the vapor-pressure differences which may develop in a nonsaline soil are likely to be very small. For example, a change in matric suction between 0 and 100 bar is accompanied by a vapor pressure change of only 17.54–16.34 torr,[7] a difference of only 1.6 mbar. For this reason, it is generally assumed that under normal field conditions soil air is nearly vapor saturated at almost all times. Vapor-pressure gradients can be caused by differences in the concentration of dissolved salts, but this effect is probably appreciable only in soils which contain zones of high salt concentration.

When temperature differences occur, however, they might cause considerable differences in vapor pressure. For example, a change in water temperature from 19 to 20°C results in an increase in vapor pressure of 1.1 torr. In other words, a change in temperature of 1°C has nearly the same effect upon vapor pressure as a change in suction of 100 bar! In the range of temperatures prevailing in the field, the variation of saturated vapor pressure (that is, the vapor pressure in equilibrium with pure, free water) is as follows:

| Temperature (°C) | 0 | 20 | 30 | 40 |
|---|---|---|---|---|
| Vapor pressure (torr) | 4.58 | 17.5 | 38.0 | 55.8 |

Vapor movement tends to take place from warm to cold parts of the soil. Since during the daytime the soil surface is warmer, and during the night colder, than the deeper layers, vapor movement tends to be downward during the day and upward during the night. Temperature gradients can also induce liquid flow.

Since liquid movement includes the solutes, while vapor flow does not, there have been attempts to separate the two mechanisms by monitoring salt movement in the soil (Gurr *et al.*, 1952; Deryaguin and Melnikova, 1958). It has been observed that the rate of vapor movement often exceeds the rate which could be predicted on the basis of diffusion alone (Cary and Taylor, 1962). It appears to be impossible to separate absolutely the liquid from the vapor movement, as overall flow can consist of a complex sequential process of evaporation, short-range liquid flow, reevaporation, etc. (Philip and de Vries, 1957). The two phases apparently move simultaneously and interdependently as a consequence of the suction and vapor-pressure gradients in the soil. It is commonly assumed, however, that liquid flow is the dominant mode in moist, nearly isothermal soils (Miller and Klute, 1967) and hence that the contribution of vapor diffusion to overall water movement is negligible in the main part of the root zone, where diurnal temperature fluctuations are slight.

[7] Torr is a unit of pressure equal to $1.316 \times 10^{-3}$ atm (i.e., 1 mm of mercury under standard conditions).

## Sample Problems

**1.** The hydraulic conductivity versus matric suction functions of two hypothetical soils, a sandy and a clayey soil, conform to the empirically based equation

$$K = a/[b + (\psi - \psi_a)^n] \qquad \text{for} \quad \psi \geq \psi_a$$

where $K$ is hydraulic conductivity (cm/sec); $\psi$ is suction head (cm $H_2O$) and $\psi_a$ is air-entry suction; and $a$, $b$, $n$ are constants, with $a/b$ representing the saturated soil's hydraulic conductivity $K_s$. The exponential parameter $n$ characterizes the steepness with which $K$ decreases with increasing $\psi$. Assume that in the sandy soil $K_s = 10^{-3}$ cm/sec, $a = 1$, $b = 10^3$, $\psi_a = 10$ cm, and $n = 3$; whereas in the clayey soil $K_s = 2 \times 10^{-5}$, $a = 0.2$, $b = 10^4$, $\psi_a = 20$, and $n = 2$.

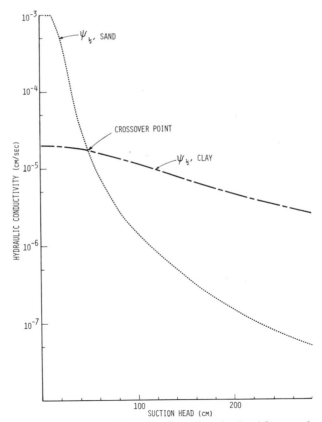

**Fig. 7.8.**   Hydraulic conductivities as functions of suction head for a sand and a clay.

Plot the $K$ versus $\psi$ curves. Note that the curves cross, and the relative conductivities are reversed: the sandy soil with the higher $K_s$ exhibits a steeper decrease of $K$ and falls below the clayey soil beyond a certain suction value $\psi_c$. Calculate the values of $\psi$ (designated $\psi_{1/2}$) at which each $K$ equals $\frac{1}{2}K_s$, and estimate the common value of $\psi_c$, at which the two curves intersect.

*For the sandy soil:*

$$K = [10^3 + (\psi - 10)^3]^{-1}$$

The suction $\psi_{1/2}$ at which $K = 0.5\ K_s = a/2b$ can be obtained by substituting $\tilde{\psi}$ for $(\psi - \psi_a)$ and setting $\tilde{\psi}^3 = 10^3$ (thus doubling the denominator). Therefore,

$$\tilde{\psi} = 10 \qquad \text{and} \qquad \psi_{1/2} = \tilde{\psi} + \psi_a = 10 + 10 = 20 \text{ cm}$$

*For the clayey soil:*

$$K = 0.2/[10^4 + (\psi - 20)^2]$$

Following the previous procedure, and setting $\tilde{\psi}^2 = 10^4$, we obtain

$$\tilde{\psi} = 10^2 \qquad \text{and} \qquad \psi_{1/2} = \tilde{\psi} + \psi_a = 100 + 20 = 120 \text{ cm}$$

"*Crossover suction value* ($\psi_c$): We can attempt to obtain this value algebraically (by setting the two expressions for $K$ equal to each other and solving for the common $\psi$ value) or graphically (Fig. 7.8) by reading the $\psi$ value at which the two $K(\psi)$ curves intersect. The latter procedure is easier in this case, and it shows $\psi_c$ to be about 48 cm.

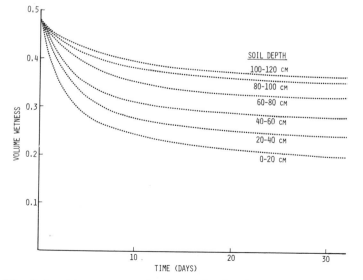

**Fig. 7.9.** Volumetric wetness as a function of time for different depth layers in a draining profile.

*Note:* The soils depicted are completely hypothetical and are not to be taken as typical of real sandy and clayey soils.

**2.** Using the "instantaneous profile" internal drainage method (Section G.3 of this chapter) compute the hydraulic conductivity $K$ as a function of volume wetness $\theta$ for the set of data shown in the following two graphs: (1) Figure 7.9 shows a plot of volumetric wetness variation with time for each of five depth layers in a 1-m deep profile (neutron meter data); (2) Fig. 7.10 shows a plot of matric suction variation with time for each depth (tensiometric data).

From the first graph, we calculate the soil moisture flux through the bottom of each depth increment (or layer) by integrating moisture–time curves with respect to depth. First, the slopes of the wetness curves ($-\partial\theta/\partial t$) are measured at days 1, 2, 4, 8, 16, and 32. Second, these slopes are multiplied by their respective depth increments to obtain the per layer rate of water content change $dz(\partial\theta/\partial t)$. Then the flux $q$ through the bottom of each depth is obtained by accumulating the water content increments of all layers overlying that depth, i.e., $q = dz(\partial\theta/\partial t)$. The whole procedure is presented in tabular form in Table 7.1. We next use Fig. 7.10 (the time variation of matric suction at each depth) to obtain the hydraulic head profiles existing at the times for which the flux values were calculated (Table 7.1). We do

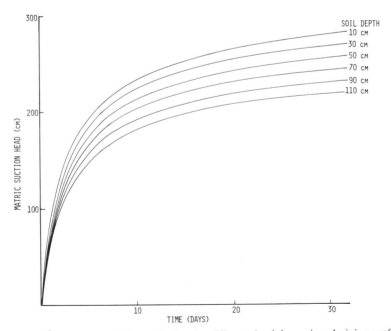

**Fig. 7.10.** Matric suction variation with time for different depth layers in a draining profile.]

this by adding the depth of each tensiometer to its matric suction values. The resulting hydraulic head profiles are plotted in Fig. 7.11. We can now divide each of the fluxes listed in Table 7.1 by its corresponding hydraulic gradient (namely, the slopes of the lines plotted in Fig. 7.11). We also note the volumetric wetness values which prevailed at each depth and time for which the hydraulic conductivity has been computed. This procedure is shown in Table 7.2. Our last step is to plot the hydraulic conductivity values against volumetric wetness and draw a best-fit curve through the plotted points (Fig. 7.12). In the example given, all the points seem to lie, more or less, on a straight line when plotted on a semilog coordinate system (log $K$ versus $\bar{\theta}$). This is not merely fortuitous: it is, admittedly, contrived. We chose a simple and well-behaved set of data for the sake of illustration. The real world is seldom so convenient. A more realistic example, pertaining to a texturally layered soil, was presented (in similarly excruciating detail) by Hillel *et al.* (1972).

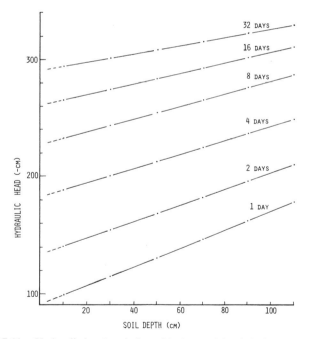

**Fig. 7.11.** Hydraulic head variation with time and depth during drainage.

**Table 7.1**

CALCULATION OF SOIL MOISTURE FLUX

| $t$ (day) | $z$ (cm) | $-\partial\theta/\partial t$ (day$^{-1}$) | $-dz(\partial\theta/\partial t)$ (cm/day) | $q = dz(\partial\theta/\partial t)$ (cm/day) |
|---|---|---|---|---|
| | 0–20 | 0.085 | 1.7 | 1.7 |
| | 20–40 | 0.065 | 1.3 | 3.0 |
| 1 | 40–60 | 0.045 | 0.9 | 3.9 |
| | 60–80 | 0.033 | 0.66 | 4.56 |
| | 80–100 | 0.030 | 0.60 | 5.16 |
| | 0–20 | 0.05 | 1.0 | 1.0 |
| | 20–40 | 0.04 | 0.8 | 1.8 |
| 2 | 40–60 | 0.035 | 0.7 | 2.5 |
| | 60–80 | 0.02 | 0.4 | 2.9 |
| | 80–100 | 0.016 | 0.32 | 3.22 |
| | 0–20 | 0.02 | 0.4 | 0.4 |
| | 20–40 | 0.02 | 0.4 | 0.8 |
| 4 | 40–60 | 0.02 | 0.4 | 1.2 |
| | 60–80 | 0.016 | 0.32 | 1.52 |
| | 80–100 | 0.012 | 0.24 | 1.76 |
| | 0–20 | 0.01 | 0.2 | 0.2 |
| | 20–40 | 0.01 | 0.2 | 0.4 |
| 8 | 40–60 | 0.008 | 0.16 | 0.56 |
| | 60–80 | 0.008 | 0.16 | 0.72 |
| | 80–100 | 0.006 | 0.12 | 0.84 |
| | 0–20 | 0.006 | 0.12 | 0.12 |
| | 20–40 | 0.005 | 0.10 | 0.22 |
| 16 | 40–60 | 0.005 | 0.10 | 0.32 |
| | 60–80 | 0.003 | 0.06 | 0.38 |
| | 80–100 | 0.003 | 0.06 | 0.44 |
| | 0–20 | 0.004 | 0.08 | 0.08 |
| | 20–40 | 0.003 | 0.06 | 0.14 |
| 32 | 40–60 | 0.003 | 0.06 | 0.20 |
| | 60–80 | 0.0025 | 0.05 | 0.25 |
| | 80–100 | 0.0025 | 0.05 | 0.30 |

**Table 7.2**

CALCULATION OF HYDRAULIC CONDUCTIVITY

| $z$ (cm) | $q^a$ (cm/day) | $\partial H/\partial z^b$ (cm/cm) | $K^c$ (cm/day) | $\bar{\theta}$ (%) |
|---|---|---|---|---|
| | 1.7 | 0.8 | 2.25 | 41.0 |
| | 1.0 | 0.68 | 1.47 | 38.2 |
| 20 | 0.4 | 0.6 | 0.67 | 31.5 |
| | 0.2 | 0.52 | 0.38 | 27.0 |
| | 0.12 | 0.44 | 0.27 | 24.0 |
| | 0.08 | 0.36 | 0.22 | 22.0 |
| | 3.0 | 0.8 | 3.75 | 44.5 |
| | 1.8 | 0.68 | 2.65 | 42.0 |
| 40 | 0.8 | 0.6 | 1.33 | 37.0 |
| | 0.4 | 0.52 | 0.77 | 31.0 |
| | 0.22 | 0.44 | 0.50 | 28.0 |
| | 0.14 | 0.36 | 0.39 | 26.0 |
| | 3.9 | 0.8 | 4.88 | 46.2 |
| | 2.5 | 0.68 | 3.68 | 42.5 |
| | 1.2 | 0.6 | 2.00 | 39.0 |
| 60 | 0.56 | 0.52 | 1.08 | 35.0 |
| | 0.32 | 0.44 | 0.73 | 31.0 |
| | 0.20 | 0.36 | 0.56 | 30.0 |
| | 4.56 | 0.8 | 5.70 | 46.6 |
| | 2.9 | 0.68 | 4.26 | 44.5 |
| 80 | 1.52 | 0.6 | 2.53 | 41.5 |
| | 0.72 | 0.52 | 1.38 | 38.0 |
| | 0.38 | 0.44 | 0.86 | 34.5 |
| | 0.25 | 0.36 | 0.69 | 33.0 |
| | 5.16 | 0.8 | 6.45 | 47.0 |
| | 3.22 | 0.68 | 4.74 | 45.0 |
| 100 | 1.76 | 0.6 | 2.93 | 42.0 |
| | 0.84 | 0.52 | 1.62 | 39.0 |
| | 0.44 | 0.44 | 1.00 | 37.0 |
| | 0.30 | 0.36 | 0.83 | 35.0 |

[a] Copied from Table 7.1.

[b] Obtained by measuring the slopes of the curves in Fig. 7.11 at the appropriate depths. In reality, these slopes are not necessarily constant.

[c] Obtained by dividing each flux $q$ by its corresponding hydraulic gradient $\partial H/\partial z$.

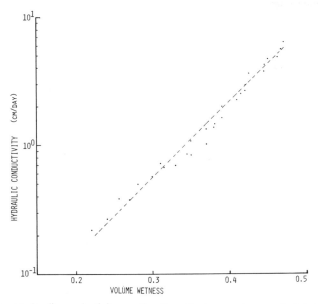

**Fig. 7.12.**   Hydraulic conductivity as a function of volume wetness, calculated from the data of Problem 2.

# Part IV    THE GASEOUS PHASE

# 8　Soil Air and Aeration

## A. Introduction

The process of *soil aeration* is one of the most important determinants of soil productivity. Plant roots adsorb oxygen and release carbon dioxide in the process of *respiration*. In most terrestrial plants (excepting such specialized plants as rice), the internal transfer of oxygen from the parts above the ground (leaves and stems) to those below the ground surface (roots) cannot take place at a rate sufficient to supply the oxygen requirements of the roots. Adequate root respiration requires that the soil itself be *aerated*, that is to say, that gaseous exchange take place between soil air and the atmosphere at such a rate as to prevent a deficiency of oxygen and an excess of carbon dioxide from developing in the root zone (Fig. 8.1). Soil microorganisms also respire, and, under conditions of restricted aeration, might compete with the roots of higher plants (Stotzky, 1965).

Gases can move either in the air phase (that is, in the pores which are drained of water, provided they are interconnected and open to the atmosphere) or in dissolved form through the water phase. The rate of diffusion of gases in the air phase is generally greater than in the water phase, hence soil aeration is dependent largely upon the volume fraction of air-filled pores.

Impeded aeration resulting from poor drainage and waterlogging, or from mechanical compaction of the soil, can strongly inhibit crop growth. In particular, the problem of soil compaction seems to have worsened in recent decades, along with the growing trend to use larger and heavier machinery and the tendency to tread over the field repeatedly for such purposes as

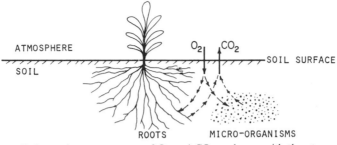

**Fig. 8.1.** Soil aeration as a process of $O_2$ and $CO_2$ exchange with the atmosphere (schematic). Among other gases involved in the soil atmosphere exchange are various volatile forms of nitrogen (e.g., $N_2$, $NH_3$, $NO$, $NO_2$), sulfur (e.g., $H_2S$, $SO_2$) and hydrocarbons (e.g., $CH_4$).

fertilization and pest control. Moreover, with greater use of fertilizers and irrigation, shortages of nutrients and water have been obviated in many places so that more and more it is aeration which has become a major limiting factor to the attainment of maximal productivity. It seems likely that root systems are commonly restricted in extent by the progressive decrease of aeration in the deeper regions of the soil profile. Poor aeration can decrease the uptake of water and induce early wilting (Stolzy *et al.*, 1963). According to Kramer (1956), restricted aeration also causes a decrease in the permeability of roots to water.

Anaerobic conditions in the soil induce a series of reduction reactions, both chemical and biochemical. Included among these reactions are *denitrification* (the processes by which nitrate is reduced to nitrite, thence to nitrous oxide and eventually to elemental nitrogen: $NO_3^- \rightarrow NO_2^- \rightarrow N_2O \rightarrow N_2$; *manganese reduction* from the manganic to the manganous form); *iron reduction* from the ferric to the ferrous form; and *sulfate reduction* to form hydrogen sulfide. Some of the numerous products of these anaerobic processes are toxic to plants (e.g., ferrous sulfide, ethylene, as well as acetic, butyric, and phenolic acids).

The subject of soil aeration has been reviewed repeatedly over the years and the evolving state of our knowledge is reflected in the successive publications by Russell (1952), Stolzy and Letey (1964), Grable (1966), Letey *et al.* (1967), Wesseling (1974), Currie (1975), and Cannell (1977).

## B. Volume Fraction of Soil Air

In most natural soils, the volume ratios of the three constituent phases, namely, the solids, water, and air, are continually changing as the soil as a whole undergoes wetting or drying, swelling or shrinkage, tillage or compaction, aggregation or dispersion, etc. Specifically, since the twin fluids—water and air—compete for the same pore space, their volume fractions are so related that an increase of the one generally entails a decrease of the other. From Eq. (2.8), we get

$$f_a = f - \theta \tag{8.1}$$

wherein $f_a$ is the volume fraction of air, $f$ the total porosity (the fractional volume of soil not occupied by solids), and $\theta$ the volume fraction of water (volume wetness).

Since the volume fraction of air is a transient property, if we wish to use it as an index of soil aeration we must obviously specify some characteristic and reproducible wetness value. The wetness value generally chosen for this purpose is the so-called "field capacity," discussed in detail in our Chapter 13. Although the term defies exact physical definition, suffice it to say at present that it is an approximation of the amount of water retained by a soil after the initially rapid stage of internal drainage. Objections to this concept notwithstanding for the moment, we might define an analogous soil aeration index, which we might call "*field air-capacity*," definable as the fractional volume of air in a soil at the field-capacity water content. Though under different names (e.g., noncapillary porosity, air-filled porosity, air content at field capacity, etc.), this index, or something very similar to it, has been investigated for many years in relation to soil aeration and plant response. It has been found to depend on numerous factors, not all of which are amenable to human control.

In the first place, air capacity depends on soil texture. In sandy soils, it is of the order of 25% or more, in loamy soils it is generally between 15 and 20%, and in clayey soils, which tend to retain the most water, it is likely to fall below 10% of total soil volume. In fine-textured soils, however, soil structure, too, has much to do with determining the air capacity. Strongly aggregated soils, with macroaggregates of the order of 5 mm or more in diameter, generally have a considerable volume of *macroscopic* (interaggregate) pores which drain very quickly and remain air filled practically all of the time. Hence such soils exhibit an air capacity of 20–30%. As the aggregates are dispersed, or broken down by mechanical forces, the macroscopic pores tend to disappear so that a strongly compacted soil may contain less than 5% air by volume at its characteristic field capacity value of soil moisture.

Investigators have long attempted to establish a value of the field air capacity, or air-filled porosity, at which soil aeration is likely to become limiting to root respiration and hence to plant growth (Vomocil and Flocker, 1961). The results, however, have been rather baffling. The reported limiting values have ranged between 5 and 20%, averaging around 10%.

There are, of course, fundamental limitations to the air capacity index as a means of characterizing soil aeration. First, the value is difficult to determine with a satisfactory degree of accuracy, since it generally depends on prior determination of two highly variable parameters which are themselves rather cumbersome to measure—namely, the field capacity and the total porosity. Both determinations depend on the method of sampling and their results can be grossly inaccurate. An even more serious objection to the use of air capacity as an index of soil aeration is that in principle the *rate of exchange* of soil air rather than simply the *content* of soil air constitutes the decisive factor. At high wetness values, soils often contain isolated pockets of occluded air which, though forming a part of the air-filled volume, do not contribute to active gas exchange. At times, even a thin surface crust, if highly compact or saturated, can form a bottleneck limiting aeration to the entire soil profile, regardless of the air-filled porosity beneath. We are thus led to the necessity of characterizing aeration in more dynamic terms.

## C. Composition of Soil Air

In a well-aerated soil, the composition of soil air is close to that of the external ("open") atmosphere, as the oxygen consumed in the soil is readily replaced from the atmosphere. Not so in a poorly aerated soil. Analyses of the actual composition of soil air in the field reveal it to be much more variable than the external atmosphere. Depending on such factors as time of year, temperature, soil moisture, depth below the soil surface, root growth, microbial activity, pH, and—above all—the rate of exchange of gases through the soil surface, soil air can differ to a greater or lesser degree from the composition of the external atmosphere. The greatest difference is in concentration of carbon dioxide ($CO_2$), which is the principal product of aerobic respiration by the roots of higher plants and by numerous macro- and microorganisms in the soil. The $CO_2$ concentration of the atmosphere is about 0.03%. In the soil, however, it frequently reaches levels which are ten or even one-hundred times greater.

As $CO_2$ is produced in the soil by oxidation of carbonaceous (organic) matter, an increase of $CO_2$ concentration is generally associated with a decrease in elemental oxygen ($O_2$) concentration (though not necessarily to an exactly commensurate degree, since additional sources of oxygen may exist in dissolved form and in easily reducible compounds). Since the $O_2$ concentration of air is normally about 20%, it would seem that even a hundredfold increase of $CO_2$ concentration, from 0.03 to 3%, can only diminish the $O_2$ concentration to about 17%. However, even before they begin to suffer from lack of oxygen per se, some plants may suffer from excessive concentrations of $CO_2$ and other gases in both the gaseous and aqueous phases. In more extreme cases of aeration restriction, the $O_2$ concentration can fall to near zero and prolonged anaerobic conditions can result in the development of a chemical environment characterized by reduction reactions such as denitrification, the evolution of such gases as hydrogen sulfide ($H_2S$), methane

(CH₄), and ethylene, and the reduction of mineral oxides (such as those of iron and manganese).

Another interesting difference between the atmosphere and soil air is that the latter is characterized by a high relative humidity, which nearly always approaches 100% except at the soil surface during prolonged dry spells.

## D. Convective Flow of Soil Air

The exchange of air between the soil and the atmosphere can occur by means of two different mechanisms: *convection* and *diffusion*. Each of these processes can be formulated in terms of a linear rate law, stating that the flux is proportional to the moving force. In the case of convection, also called mass flow, the moving force consists of a gradient of *total gas pressure*, and it results in the entire mass of air streaming from a zone of higher pressure to one of lower pressure. In the case of diffusion, on the other hand, the moving force is a gradient of *partial pressure* (or concentration) of any constituent member of the variable gas mixture which we call air, and it causes the molecules of the unevenly distributed constituent to migrate from a zone of higher to lower concentration even while the gas as a whole may remain isobaric and stationary. We propose to discuss convective flow of soil air in this section, and diffusion in the next.

A number of phenomena can cause pressure differences between soil air and the external atmosphere, thereby inducing convective flow into or out of the soil. Among these phenomena are barometric pressure changes in the atmosphere, temperature gradients, and wind gusts over the soil surface. Additional phenomena affecting the pressure of soil air are the penetration of water during infiltration, causing displacement of antecedent soil air, the fluctuation of a shallow water table pushing air upward or drawing air downward, and the extraction of soil water by plant roots. Short-term changes in soil air pressure can also occur during tillage or compaction by machinery.

The degree to which air pressure fluctuations and the resulting convective flow can contribute to the exchange of gases between the soil and the atmosphere has long been a subject of debate among soil physicists. The majority have tended to support the hypothesis that diffusion rather than convection is the more important mechanism of soil aeration (Keen, 1931; Penman, 1940; Russell, 1952). Recent evidence suggests, however, that previous analyses have been incomplete and that convection can in certain circumstances contribute significantly to soil aeration, particularly at shallow depths and in soils with large pores (Vomocil and Flocker, 1960; Grable, 1966; Farrel *et al.*, 1966; Scotter and Raats, 1968, 1969; Kimball and Lemon, 1971, 1972). For example, Vomocil and Flocker (1961), assuming that the $O_2$ consumption rate is 0.2 ml/hr gm of root tissue (fresh weight), that rooting density is 0.1 $gm/cm^3$, that the water extraction rate is 7.5 mm/day, and that soil bulk density remains constant, calculated that the process of root water extraction by itself can draw into the soil as much as 70% of the oxygen required for respiration of crop roots.

The convective flow of air in the soil is similar in some ways to the flow of water, and different in other ways. The similarity is in the fact that the flow of both fluids is impelled by, and is proportional to, a pressure gradient. The dissimilarity results from the relative incompressibility of water in comparison with air, which is highly compressible so that its density and viscosity are strongly dependent on pressure (as well as temperature). Quite another difference is that water has the greater affinity to the surfaces of mineral particles (i.e., it is the *wetting fluid*) and is thus drawn into narrow necks and pores, forming capillary films and wedges (menisci). In a three-phase system, therefore, air tends to occupy the larger pores. The two fluids—water and air—coexist in the soil by occupying different portions of the pore space having different geometric configurations. For this reason, the soil exhibits toward the two fluids different conductivity or permeability functions, as these relate to the different effective diameters and tortuosities of the pore sets occupied by each fluid. Only when the soil is completely permeated by one or the other, be it water or air, should either flowing fluid encounter the same transmission coefficient of the medium. (See the discussion of *intrinsic permeability* in Chapter 6.)

Notwithstanding the differences between water flow and air flow, it is possible to formulate the convective flow of air in the soil as an equation analogous to Darcy's law for water flow, as follows:

$$\mathbf{q}_v = -(k/\eta)\,\nabla P \tag{8.2}$$

where $\mathbf{q}_v$ is the *volume convective flux* of air (volume flowing through a unit cross-sectional area per unit time), $k$ is permeability of the air-filled pore space, $\eta$ is viscosity of soil air, and $\nabla P$ is the three-dimensional gradient of soil air pressure. In one dimension, this equation takes the form

$$q_v = -(k/\eta)(dP/dx) \tag{8.3}$$

If the flux is expressed in terms of mass (rather than volume) per unit area and per unit time, then the equation is

$$q_m = -(\rho k/\eta)(dP/dx) \tag{8.4}$$

wherein $q_m$ is the mass convective flux and $\rho$ the density of soil air.

Recalling that the density of a gas depends on its pressure and temperature, we shall now assume that soil air is an *ideal gas*, in which the relation of mass, volume, and temperature is given by the equation

$$PV = nRT \tag{8.5}$$

where $P$ is pressure, $V$ volume, $n$ number of moles of gas, $R$ the universal gas constant per mole, and $T$ absolute temperature. Since the density $\rho = M/V$, and the mass $M$ is equal to the number of moles $n$ times the molecular weight $m$, we have

$$\rho = (m/RT)P \tag{8.6}$$

We now introduce the *continuity equation* (conservation of mass) for a compressible fluid:

$$\partial \rho / \partial t = -\partial q_m / \partial x \tag{8.7}$$

Substituting the expression for $\rho$ from Eq. (8.6) and the expression for $q_m$ from Eq. (8.4) into Eq. (8.7), we obtain

$$\frac{m}{RT} \frac{\partial P}{\partial t} = \frac{\partial}{\partial x}\left( \frac{\rho k}{\eta} \frac{\partial P}{\partial x} \right) \tag{8.8}$$

If $\rho k / \eta$ are more or less constant (i.e., the pressure differences are small), we can write

$$\partial P / \partial t = \alpha \, \partial^2 P / \partial x^2 \tag{8.9}$$

wherein $\alpha = RTk/m$, a composite constant. This is an approximate equation for the transient state convective flow of air in soil. An assumption underlying the use of this equation is that flow is laminar, which it was indeed shown to be for small pressure differences (Muskat, 1946).

Quite a different mechanism of convective movement of gases in the soil is the transfer of dissolved gases by rain or irrigation water infiltrating into and percolating through soils. Some investigators have speculated that dissolved $O_2$ in percolating water may allow growth, or at least survival, of crop plants in soils covered by flowing water (Russell, 1952) or even permit roots to thrive below the water table (Robinson, 1964). However, oxygen-saturated air at atmospheric pressure contains only 6 ml of $O_2$ per liter, which is only enough to provide for the respiration of 1 gm (dry weight) of active roots. Except in the case of sandy soils with unusually high infiltration rates, therefore, it seems unlikely that oxygen supply by infiltrating water can have anything but a very temporary or marginal effect. However, a sudden heavy rain or flood irrigation can sometimes trap large volumes of air between the penetrating surface water and the water table and this air can serve as an oxygen reservoir during a flooding period lasting several days (Kemper and Amemiya, 1957).

### E.  Diffusion of Soil Air

The diffusive transport of gases such as $O_2$ and $CO_2$ in the soil occurs partly in the gaseous phase and partly in the liquid phase. Diffusion through the air-filled pores maintains the exchange of gases between the atmosphere and the soil, whereas diffusion through water films of various thickness maintains the supply of oxygen to, and disposal of $CO_2$ from, live tissues, which are typically hydrated. For both portions of the pathway, the diffusion process can be described by Fick's law:

$$q_d = -D \, dc/dx \qquad (8.10)$$

wherein $q_d$ is the diffusive flux (mass diffusing across a unit area per unit time), $D$ is the diffusion coefficient (generally having the dimensions of area per time), $c$ is concentration (mass of diffusing substance per volume), $x$ is distance, and $dc/dx$ is the concentration gradient. If partial pressure $p$ is used instead of concentration of the diffusing component, we get

$$q_d = -(D/\beta)(dp/dx) \qquad (8.11)$$

where $\beta$ is the ratio of the partial pressure to the concentration.

Considering first the diffusive path in the air phase, we note that the diffusion coefficient in the soil $D_s$ must be smaller than that in bulk air $D_0$ owing to the limited fraction of the total volume occupied by continuous air-filled pores and also to the tortuous nature of these pores. Hence we can expect $D_s$ to be some function of the air-filled porosity, $f_a$. Different workers have over the years found different relations between $D_s$ and $f_a$ for various soils. For instance, Buckingham (1904) reported the following nonlinear relation:

$$D_s/D_0 = \kappa f_a^2 \qquad (8.12)$$

On the other hand, Penman (1940) found a linear relation:

$$D_s/D_0 = 0.66 f_a \qquad (8.13)$$

wherein 0.66 is a tortuosity coefficient, suggesting that the apparent path is about two-thirds the length of the real average path of diffusion in the soil. Tortuosity very probably depends on the fractional volume of air-filled pores (i.e., it stands to reason that the tortuous path length should increase as the air-filled pore volume decreases); hence we can expect Penman's constant coefficient to hold for only a limited range of variation of air-filled porosity or of volume wetness.

In fact, other investigators have reported different values for different soils and ranges of air and water contents. To mention just a few:

1.  Blake and Page (1948) found that the ratio between $(D_s/D_0)$ and air-filled porosity $f_a$ varied between 0.62 and about 0.8, and that in an

aggregated soil $D_s$ approached zero when air content fell below about 10% of porosity, a phenomenon which they attributed to the discontinuity of blocked air-filled pores inside the aggregates.

2. Van Bavel (1952) found $D_s/D_0 = 0.61 f_a$.

3. Marshall (1959) took account of the size distribution of soil pores affecting diffusion and proposed the relationship $D_s/D_0 = f_a^{3/2}$, which accords with Penman's coefficient at high air-filled porosity values but falls below it at lower ones.

4. Millington (1959) proposed $D_s/D_0 = (f_a/f)^2 f_a^{4/3}$, where $f$ is total porosity.

5. Wesseling (1962) proposed $D_s/D_0 = 0.9 f_a - 0.1$, which suggests that $D_s$ becomes zero at an air-filled porosity value of $0.1/0.9 = 11\%$. The confusion was explained by de Vries (1952), who showed on the basis of theory that the relation sought between the effective diffusion coefficient and air-filled porosity should be curvilinear and dependent on pore geometry and hence is not expected to be the same for different soils and water versus air contents. Consequently, Grable and Siemer (1968) found that as air-filled porosity fell to around 10% the ratio $D_s/D_0$ fell to about 0.02. At even lower $f_a$ values of 4–5%, Lemon and Erickson found the $D_s/D_0$ ratio to be as low as 0.005.

Having established the variable nature of $D_s$, we now return to the mathematical formulation of diffusion processes. To account for transient conditions, we must once again introduce the continuity principle:

$$\partial c/\partial t = -\partial q_d/\partial x \qquad (8.14)$$

which states that the rate of change of concentration with time must equal the rate of change of diffusive flux with distance. The foregoing assumes that the diffusing substance is conserved throughout. As $O_2$ and $CO_2$ diffuse through the soil, however, $O_2$ is taken up and $CO_2$ is generated by aerobic biological activity along the diffusive path. To take account of the amount of a diffusing substance added to or subtracted from the system per unit time we add a $\pm S$ term to the right-hand side of Eq. (8.14). Note that a positive sign represents an increment rate (source) and a negative sign represents a decrement rate (sink) for the substance considered. Accordingly,

$$\partial c/\partial t = -(\partial q_d/\partial x) \pm S(x,t) \qquad (8.15)$$

The designation $S(x,t)$ implies that the source–sink term is a function of (i.e., varies with) both space and time.

We next substitute Eq. (8.10) into (8.15) and consider only the vertical direction $z$ (depth) to obtain

$$\frac{\partial c}{\partial t} = \frac{\partial}{\partial z}\left(D_s \frac{\partial c}{\partial z}\right) \pm S(z,t) \qquad (8.16)$$

Note that we use $D_s$, for which one may wish to substitute an expression such as the one by Penman ($D_s = 0.66D_0 f_a$) or any other empirical or theoretically based function.

In the event that $D_s$ is constant, Eq. (8.16) simplifies to

$$\frac{\partial c}{\partial t} = D_s \frac{\partial^2 c}{\partial z^2} \pm S(z,t) \tag{8.17}$$

In aggregated soils, gaseous diffusion can be expected to take place rapidly in the interaggregate macropores, which quickly drain out after a rain or irrigation and form a network of continuous air-filled voids. On the other hand, the intraaggregate micropores can remain nearly saturated for extended periods and thus restrict the internal aeration of aggregates. It is common observation that plant roots are generally confined to the larger pores between aggregates and scarcely penetrate the aggregates themselves, whether because the small internal pores of the aggregates and their mechanical rigidity do not permit penetration or because of aeration restriction per se. However, microorganisms do penetrate aggregates and, by their demand for oxygen, affect soil aeration as a whole. The centers of large, dense crumbs can be anaerobic even while the larger pores all around indicate good aeration. Thus we can have pockets of anaerobiosis in the midst of a seemingly well-aerated soil.

There is, incidentally, a fundamental difference between the functional dependence of a soil's permeability (or conductivity) upon pore geometry and the corresponding function for the diffusion coefficient. As the permeability pertains to pressure-induced convective-viscous flow, it obeys Poiseuille's law which states that flow varies as the fourth power of the pore radius. Hence permeability is strongly dependent upon *pore-size distribution*. Diffusion, on the other hand, depends primarily on the total volume and tortuosity of continuous pores available for diffusion. The reason that diffusion does not depend on pore-size distribution is that the *mean free path* of molecules in thermal motion (i.e., the distance an "average" molecule in random motion travels before it collides with another) is of the order of 0.0001 to 0.0005 mm and thus much smaller than the radii of the pores which generally account for most of a soil's air-filled porosity.

### F.  Soil Respiration and Aeration Requirements

The overall rate of respiration due to all biological activity in the soil—i.e., the amount of oxygen consumed and the amount of carbon dioxide produced by the entire profile—determines the *aeration requirement* of the soil. It is important to acquire quantitative knowledge of the aeration requirements of different crops and soils in varying circumstances if we are to devise means to

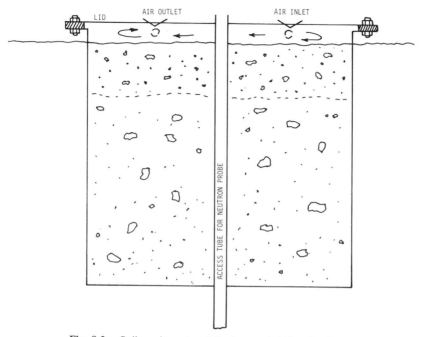

**Fig. 8.2.** Soil respirometer at Rothamsted. (After Currie, 1975.)

ensure that these requirements are indeed met. However, the information is difficult to obtain, as the rate and spatial distribution of soil respiration, as well as its temporal variation, depend on numerous factors, included among which are temperature, soil wetness, organic matter, and the time-variable respiratory activity of both macro- and microorganisms (Alexander, 1961).

*Anaerobiosis*, or oxygen stress, will occur in the soil whenever the rate of suppiy falls below the demand. This condition can develop quite quickly, since the storage of oxygen in the soil is generally rather low in relation to the quantity required for soil respiration. To illustrate, let us consider a soil with an effective root zone depth of 60 cm and 15% air-filled porosity, containing 90 liter of air under each square meter of soil surface. With an initial oxygen concentration of 20% in the gaseous phase, the storage of oxygen can be calculated to be 18 liters, equivalent to about 25 g.[1] If the oxygen requirement for soil respiration is of the order of 10 gm/day, per square meter of ground, the initial oxygen reserve in the soil would last only $2\frac{1}{2}$ day. Oxygen

---

[1] An ideal gas at standard temperature and pressure occupies a volume of 22.4 liter/mole. The mass of a mole of oxygen (atomic weight = 16) is 32 gm. The mass contained in 18 liter is therefore $(18/22.4) \times 32 = 25.7$ gm.

stress symptoms would probably begin even earlier. However, these figures are given only to provide an order of magnitude, since in actual conditions the aeration rate probably varies between wide limits. Plant growth probably depends more upon the occurrence and duration of periods of oxygen deficiency than upon average conditions (Erickson and van Doren, 1960).

Soil respiration values reported by various investigators have varied widely. Papendick and Runkles (1965), working under laboratory conditions, found respiration rates of $1.7 \times 10^{-11}$ mole/cm$^3$ sec, equivalent to 0.75 mole of oxygen per cubic meter per 12-hr day, or 24 gm/m$^3$ day. At the other extreme, Grable and Siemer (1968) reported values 10,000 times higher. Under field conditions, respiration values are likely to be smaller than those found in growth chambers. Greenwood (1971), working in England, measured average oxygen consumption rates of $1.3 \times 10^{-7}$ ml oxygen per second per milliliter of a soil carrying a mature crop, and maximum rates three times as high. The average value is equivalent to 8 gm of oxygen per cubic meter of soil per 12-hr day. Dasberg and Bakker (1970) found very similar values in cropped soils in Holland.

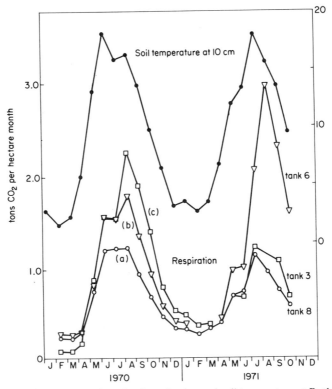

**Fig. 8.3.** Seasonal variation of soil respiration and soil temperature at Rothamsted. (After Currie, 1975.)

Comprehensive data on soil respiration rates have been obtained in England from *field respirometers*, first built at Wrest Park and later resited at the Rothamsted Experiment Station (Fig. 8.2). These installations have recently been described by Currie (1975). They consist of large containers (91 × 91 × 91 cm) filled with soil and sealed on top, with provisions to maintain normal atmospheric composition by continuously adding or removing measured amounts of oxygen and carbon dioxide as required. When the soil is cropped, the plants are grown through holes in the lid, using cold-setting silicone rubber to seal the gap around the stem. It is thus possible to measure daily carbon dioxide output and hourly oxygen uptake for soils with and without plants.

The results available to date demonstrate the primary importance of soil temperature. This is illustrated in terms of the seasonal variation of soil temperature and soil respiration in Fig. 8.3. Respiration rates in summer can be more than ten times as great as in winter. However, other factors can also have a strong influence. For example, respiration rates at a given temperature tend to be greater during spring than during autumn. The cause seems to be the more vigorous population of microorganisms and the greater availability of undecomposed organic residues in spring than in autumn. Soil respiration is obviously much greater in cropped than in fallow land, owing both to root respiration and to the enhanced microbial activity resulting from root exudates and root decay. The diurnal variation of soil respiration also follows the pattern of soil temperature, with oxygen uptake rates registering a more than twofold increase from early morning to mid afternoon. Thus, the respiration rates in soil vary from season to season, from day to day, and from hour to hour, and are related to crop growth stage and to microbial activity. The effect of waterlogging the soil is slight in winter, for although air diffusion is constricted, the oxygen requirements are slight. The same degree of waterlogging in summer, however, could be severely damaging to a crop at its most active growth stage.

## G. Measurement of Soil Aeration

The complex group of processes which we lump together in the term soil aeration is altogether too elusive and incompletely understood to be definable unequivocally by any single measurement. We are reduced, therefore, to measuring soil attributes and component processes which we consider to be pertinent to the problem at hand and to be *indicative* of soil aeration as a whole.

An early approach to the problem of measuring aeration was to determine the fractional air space, or *air-filled porosity*, at some standardized value of soil wetness. This could be done by taking an "undisturbed" sample from a soil presumed to be at its field capacity (e.g., two days after a deep wetting) or by saturating a soil sample in the laboratory and then subjecting it to some arbitrarily specified water tension. For example, the term *aeration porosity* has been defined (Kohnke, 1958) as the pore space filled with air when the soil sample is placed on a porous plate and equilibrated with a 50 cm hanging column of water. At a tension head of 50 cm, all pores with an effective diameter wider than 0.06 mm are drained of water.[2] The air space as a fraction of total porosity can then be measured directly with the aid of an *air pycnometer* (Page, 1948; Vomocil, 1965) or by taking the difference between total porosity (obtained from the measurement of bulk density) and volumetric wetness. By either method, the determination of fractional air space is fraught with uncertainties and gives no indication of aeration dynamics.

Another traditional approach is to measure the composition of soil air. Although still a static measurement, this appears to be a better diagnostic tool than measurement of air volume alone, for it can reveal more directly when a problem might exist—i.e., when the oxygen content of soil air falls significantly below that of the atmosphere owing to restricted gas exchange. The difficulty here is how to extract an air sample at once large enough to provide a reliable measurement, and yet small enough to represent the sampled point and to avoid disturbance and mixing of soil air, or even contamination from the atmosphere. The gas chromatograph technique, using a syringe to extract small samples (only 0.5 ml in volume) helps to make the measurement more reliable (Yamaguchi *et al.*, 1962). An alternative method permitting repeated monitoring of $O_2$ concentrations in soil air without extraction of samples is based on the use of membrane-covered electrodes such as described by Willey and Tanner (1963) and by McIntyre and Philip (1964). Electrodes have also been developed for measuring $CO_2$ concentrations (Letey and Stolzy, 1964). Such electrodes must be kept dry to measure oxygen concentration. Hence they are usually sealed in a hollow tube which is inserted into the soil. The volume of air surrounding an electrode must be equilibrated with soil air prior to the measurement. If the soil is very wet, this equilibration may require many hours or even several days.

A quite different approach to characterizing soil aeration is to measure the *air permeability*, i.e., the coefficient governing convective transmission of air through the soil in response to a total pressure gradient. This measurement can provide useful information on the effective sizes and the continuity of air-filled pores. The techniques which have been proposed include constant-pressure and falling-pressure devices (Fig. 8.4). The method has been

---

[2] Students may wish to verify this statement independently. Hint: use the capillary tension equation $h = 2\gamma/\rho g R$ (Chapter 7).

applied in the field (Grover, 1956) and found useful for assessing the "openness" of the surface layer to the entry of air, as affected by such cultural practices as tractor traffic and tillage.

Numerous techniques have been proposed for measuring diffusion processes in the soil, both in the gas phase and in the liquid phase. One of the earliest (Raney, 1950) was to fill a cavity in the soil with nitrogen gas, then sample the cavity periodically to determine the rate of diffusion of gases, particularly oxygen, from the surrounding soil. An alternative technique is to introduce some volatile substance into the cavity and measure the rate of its dissipation by diffusion into the surrounding soil's air phase. However, the measurement of diffusion rates through the air-filled pores alone fails to provide any indication of the possible impedance of the liquid envelope surrounding a root. As pointed out by Kristensen and Lemon (1964), it is possible to have high $O_2$ concentration in the network of large pores open to the atmosphere and yet have an inadequate supply to the root if the root is thickly hydrated.

An interesting method for measuring oxygen diffusion to a rootlike probe was introduced by Lemon and Erickson (1952). The process measured is the chemical reduction of elemental oxygen by a thin platinum electrode maintained at a constant potential. The resulting current measures the flux of oxygen to the moisture-covered electrode which acts as a sink, thus sim-

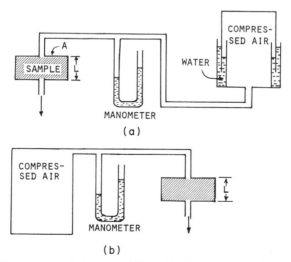

**Fig. 8.4.** Measurement of air permeability with (a) constant-pressure (variable volume) permeameter ($k = (L\eta/Af)(\Delta V/t)$), and (b) a falling-pressure permeameter (constant volume; $k = (2.3L\eta V/AP_a)[\log(p_1/p_2)/\Delta t]$). *Note:* $k$ is air permeability (cm²), $L$ sample length (cm), $A$ sample cross-sectional area (cm²), $V$ volume of the air cell (cm³), $P_a$ barometric pressure, $p$ cell air pressure; $t$ time, and $p_1$ and $p_2$ cell pressure at start and end of time step $\Delta t$.

ulating the action of a respiring root. In practice, one inserts the probe into a moist soil and waits until the reading becomes steady, at which time the flux of oxygen to the probe is taken to represent the oxygen-supplying power of the soil, and is commonly called ODR (for $O_2$ diffusion rate). The technique fails in relatively dry soils which, however, are unlikely to present aeration problems. Descriptions of the technique and the results obtainable by it have been published by Birkle *et al.* (1964), Letey and Stolzy (1964), and McIntyre (1970). Numerous investigations have shown correlations between ODR and plant response in soils of restricted aeration. According to Stolzy and Letey (1964) the roots of many plant species will not grow in an environment with an ODR of less than 0.2 mgm/cm² min. Note that the method is not based on solution of the diffusion equation for definable boundary conditions, and the results do not provide a value for the effective diffusion coefficient for oxygen in either soil air or soil water. Rather, what is measured is a flux, which, of course, depends on the size and shape of the electrode and the location of its insertion, as well as on the diffusion coefficients of the surrounding porous medium, the temperature, etc.

Additional methods for measuring soil aeration include determination of redox (reduction–oxidation) potential (Grable and Siemer, 1968; Dasberg and Bakker, 1970) and the already mentioned use of soil respirometers.


**Sample Problems**

**1.** Consider a cropped field with an effective root zone depth of 80 cm, a daily transpiration rate of 6 mm, and a daily soil respiration rate of 10 gm $O_2/m^2$. Calculate what fraction of the oxygen requirement is supplied by convection if air is drawn from the atmosphere in immediate response to the pressure deficit created in the soil by the extraction of soil moisture.

*Calculation:* Volume of water extracted per square meter: $1 \ m^2 \times 0.006 \ m = 0.006 \ m^3 = 6$ liter per day.
Volume of air drawn from the atmosphere = volume of water withdrawn from soil = 6 liter.
Volume of oxygen drawn from the atmosphere = $6 \times 21\% = 1.26$ liter.
Mass of oxygen drawn from the atmosphere = $(1.26/22.4) \times 32 = 1.8$ gm.
Percentage of daily oxygen requirement supplied by convection = $(1.8/10) \times 100 = 18\%$.
*Note:* 32 is the molecular weight of oxygen ($O_2$), and 22.4 liter is the volume of 1 mole of gas at standard temperature and pressure.

**2.** Consider a soil profile in which the air-phase oxygen concentration diminishes linearly from 21% at the soil surface to half of that at 100 cm depth. If the total porosity is a uniform 45% and the volume wetness 35%, calculate the diffusion rate using Penman's coefficient for the effective diffusion coefficient of oxygen in the soil ($D_s$). Assume steady-state diffusion. Use a value of $1.89 \times 10^{-1}$ cm$^2$/sec for the bulk-air diffusion coefficient.

Our first step is to estimate the effective diffusion coefficient $D_s$ using Penman's linear relation between $D_s$ and the air-filled porosity $f$:

$$D_s = 0.66 f_a D_0$$

where $f_a$ is air-filled porosity, $D_0$ is the diffusion coefficient in bulk air, and 0.66 is the tortuosity factor (assumed by Penman to be constant). Substituting the given values, we have

$$D_s = 0.66 \times (0.45 - 0.35) \times 0.189 \text{ cm}^2/\text{sec} = 0.0126 \text{ cm}^2/\text{sec}$$

We now use Fick's first law to calculate the steady state one-dimensional diffusive flux $q_d$ of oxygen through the soil profile from the external atmosphere to a plane at which the oxygen concentration is 50% of the atmosphere's at a depth of 100 cm:

$$q_d = D_s \, \Delta c / \Delta x$$

Recall that the concentration of oxygen in the external atmosphere is about $0.21 \times 32$ gm per 22.4 liter, i.e., $3 \times 10^{-4}$ gm/cm$^3$. Therefore

$$q_d = 1.26 \times 10^{-2} \times (3 \times 10^{-4} - 1.5 \times 10^{-4})/100$$
$$= 1.89 \times 10^{-8} \text{ gm/cm}^2 \text{ sec}$$

This quantity can be multiplied by $10^4$ cm$^2$/m$^2$ and by $8.64 \times 10^4$ sec/day to obtain the flux in grams per square meter of soil surface per day. Thus,

$$(q_d)_{\text{daily}} = 1.89 \times 10^{-8} \times 10^4 \times 8.64 \times 10^4 = 16.33 \text{ gm/m}^2 \text{ day}$$

Part V    COMPOSITE
          PROPERTIES AND
          BEHAVIOR

# 9    *Soil Temperature and Heat Flow*

## A. Introduction

Soil temperature, its value at any moment and the manner with which it varies in time and space, is a factor of primary importance in determining the rates and directions of soil physical processes and of energy and mass exchange with the atmosphere—including evaporation and aeration. Temperature also governs the types and rates of chemical reactions which take place in the soil. Finally, soil temperature strongly influences biological processes, such as seed germination, seedling emergence and growth, root development, and microbial activity.

Soil temperature varies in response to changes in the radiant, thermal, and latent energy exchange processes which take place primarily through the soil surface. The effects of these phenomena are propagated into the soil profile by a complex series of transport processes, the rates of which are affected by time-variable and space-variable soil properties. Hence the quantitative formulation and prediction of the soil thermal regime can be a formidable task. Even beyond passive prediction, the possibility of actively controlling or modifying the thermal regime requires a thorough knowledge of the processes at play and of the environmental and soil parameters which govern their rates. The pertinent soil parameters include the specific heat capacity, thermal conductivity, and thermal diffusivity (all of which are strongly affected by bulk density and wetness), as well as the internal sources and sinks of heat operating at any time.

Present day theory can provide at least a semiquantitative interpretation of observed influences of soil surface conditions, including the presence of mulching materials and various tillage treatments, on the soil's thermal regime. Moreover, available theory can help to explain why the annual temperature variation penetrates into the soil much more deeply than the diurnal variation; it can account for the obvious difference in temperature distribution among soils of differing constitution, such as sand, clay, or peat. It can also explain why the surface of a dry soil exhibits high maxima and low minima temperatures, and suggest how these extremes may be moderated when the soil moisture content is changed. Theories are now being developed to deal with freezing and thawing phenomena in cold-region soils. Finally, the important interaction of heat flow and water flow is being examined in an effort to understand how the transports of matter and energy occur simultaneously and interdependently in the soil.

Reviews of soil temperature and heat flow have been published over the years by Kersten (1949), Hagan (1952), van Rooyen and Winterkorn (1959), Smith *et al.* (1964), Taylor and Jackson (1965), van Wijk (1963), Chudnovskii (1966), van Bavel (1972), and de Vries (1975).


## B. Modes of Energy Transfer

We begin with some basic physics. In general, there are three principal modes of energy transfer: radiation, convection, and conduction.

By *radiation*, we refer to the emission of energy in the form of electromagnetic waves from all bodies above $0°K$. According to the *Stephan–Boltzmann* law, the total energy emitted by a body $J_t$ integrated over all wavelengths, is proportional to the fourth power of the absolute temperature $T$ of the body's surface. This law is usually formulated

$$J_t = \varepsilon\sigma T^4 \tag{9.1}$$

where $\sigma$ is a constant and $\varepsilon$ is the *emissivity coefficient* which equals unity for a perfect emitter (generally called a *black body*). The absolute temperature also determines the wavelength distribution of the emitted energy. *Wien's law* states that the wavelength of maximal radiation intensity $\lambda_m$ is inversely proportional to the absolute temperature:

$$\lambda_m = 2900/T \tag{9.2}$$

where $\lambda_m$ is in microns. The actual intensity distribution as a function of wavelength and temperature is given by *Planck's law:*

$$E_\lambda = C_1/\lambda^5[\exp(C_2/\lambda T) - 1] \tag{9.3}$$

where $E_\lambda$ is the energy flux emitted in a particular wavelength range, and

$C_1$, $C_2$ are constants. Since the temperature of the soil surface is generally of the order of 300°K (though it can range, of course, from below 273°K, the freezing point, to 330°K or even higher), the radiation emitted by the soil surface has its peak intensity at a wavelength of about 10 $\mu$m and its wavelength distribution over the range of 3–50 $\mu$m. This is in the realm of *infrared*, or *heat, radiation*. A very different spectrum is emitted by the sun, which acts as a black body at an effective surface temperature of about 6000°K. The sun's radiation includes the visible light range of 0.3–0.7 $\mu$m, as well as some infrared radiation of greater wavelength (up to about 3 $\mu$m) and some ultraviolet radiation ($\lambda < 0.3$ $\mu$m). Since there is very little overlap between the two spectra, it is customary to distinguish between them by calling the incoming solar spectrum *short-wave* radiation, and the spectrum emitted by the earth *long-wave* radiation.

The second mode of energy transfer, called *convection*, involves the movement of a heat-carrying mass, as in the case of ocean currents or atmospheric winds. An example more pertinent to soil physics would be the infiltration of hot waste water (from, say, a power plant) into an initially cold soil.

*Conduction*, the third mode of energy transfer, is the propagation of heat within a body by internal molecular motion. Since temperature is an expression of the kinetic energy of a body's molecules, the existence of a temperature difference within a body will normally cause the transfer of kinetic energy by the numerous collisions of rapidly moving molecules from the warmer region of the body with their neighbors in the colder region. The process of heat conduction is thus analogous to the process of diffusion, and in the same way that diffusion tends in time to equilibrate a mixture's composition throughout, heat conduction tends to equilibrate a body's internal distribution of molecular kinetic energy—that is to say, its temperature.

In addition to the three modes of energy transfer described, there is a composite phenomenon which one may recognize as a fourth mode, namely the *latent heat transfer*. A prime example is the process of *distillation*, which includes the heat-absorbing stage of evaporation, followed by the convective or diffusive movement of the vapor, and ending with the heat-releasing stage of condensation. A similar catenary process can also occur in transition back and forth from ice to liquid water.

The transfer of heat through the soil surface may occur by any or all of the above mechanisms. Within the soil, however, heat transfers by radiation, convection, and distillation are generally of secondary importance, and the primary process of heat transport is by molecular conduction.

## C.  Conduction of Heat in Soil

The conduction of heat in solids was analyzed as long ago as 1822 by Fourier, whose name is associated with the linear transport equations which have been used ever since to describe heat conduction. These equations are mathematically analogous to the diffusion equations (Fick's laws) as well as to Darcy's law for the conduction of fluids in porous media. An analogy can also be drawn between these laws and Ohm's law for the conduction of electricity (See Chapter 6). A definitive text on the mathematics of heat conduction was published by Carslaw and Jaeger (1959).

The first law of heat conduction, known as *Fourier's law*, states that the flux of heat in a homogeneous body is in the direction of and proportional to the temperature gradient:

$$\mathbf{q_h} = -\kappa \, \nabla T \tag{9.4}$$

where $\mathbf{q_h}$ is the *thermal flux* (i.e., the amount of heat conducted across a unit cross-sectional area in unit time), $\kappa$ is *thermal conductivity*, and $\nabla T$ the spatial gradient of temperature $T$. In one-dimensional form, this law is written

$$q_h = -\kappa_x \, dT/dx \qquad \text{or} \qquad q_h = -\kappa_z \, dT/dz \tag{9.5}$$

Here $dT/dx$ is the temperature gradient in any arbitrary direction designated $x$, and $dT/dz$ is, specifically, the vertical direction representing soil depth ($z = 0$ being the soil surface). The subscripts attached to the thermal conductivity term are meant to account for the possibility that this parameter may have different values in different directions. The negative sign in these equations is due to the fact that heat flows from a higher to a lower temperature (i.e., in the direction of a negative temperature gradient).

If $q_h$ is expressed in calories per square centimeter per second and the temperature gradient in degrees Kelvin per centimeter, $\kappa$ has the units of calories per centimeter-degree-second. If, on the other hand, the thermal flux is given in watts per meter and the gradient in degrees per meter, the thermal conductivity assumes the units of watts per meter-degree.

Equation (9.4) is sufficient to describe heat conduction under steady-state conditions, that is to say where the temperature at each point in the conducting medium and the flux remain constant in time. To account for nonsteady or transient conditions, we need a second law analogous to Fick's second law of diffusion as embodied in Eq. (7.15). To obtain the second law of heat conduction, we invoke the principle of *energy conservation* in the form of the *continuity equation*, which states that, in the absence of any sources or sinks of heat, the time rate of change in heat content of a volume element of the conducting medium (in our case, soil) must equal the change of flux with distance:

$$\rho c_m \, \partial T / \partial t = -\mathbf{V} \cdot \mathbf{q_h} \tag{9.6}$$

where $\rho$ is mass density and $c_m$ *specific heat capacity* per unit mass, (called simply specific heat and defined as the change in heat content of a unit mass of the body per unit change in temperature). The product $\rho c_m$ (often designated $C$) is the specific heat capacity per unit volume, and $\partial T / \partial t$ is the time rate of temperature change. Note that the symbol $\rho$ represents the total mass per unit volume, including the mass of water in the case of a moist soil. The symbol $\mathbf{V}$ (del) is the shorthand representation of the three-dimensional gradient. An equivalent form of Eq. (9.6) is

$$\rho c_m \, \partial T / \partial t = -(\partial q_x / \partial x + \partial q_y / \partial y + \partial q_z / \partial z)$$

where $x$, $y$, $z$ are the orthogonal direction coordinates.

Combining Eqs. (9.4) and (9.6), we obtain the desired *second law of heat conduction:*

$$\rho c_m \, \partial T / \partial t = \mathbf{V} \cdot (\kappa \mathbf{V} T) \tag{9.7}$$

which, in one-dimensional form, is

$$\rho c_m \frac{\partial T}{\partial t} = \frac{\partial}{\partial x}\left( \kappa \frac{\partial T}{\partial x} \right) \tag{9.8}$$

Sometimes we may need to account for the possible occurrence of heat sources or sinks in the realm where heat flow takes place. Heat sources include such phenomena as organic matter decomposition, wetting of initially dry soil material, and condensation of water vapor. Heat sinks are generally associated with evaporation. Lumping all these sources and sinks into a single term $S$, we can rewrite the last equation in the form

$$\rho c_m \frac{\partial T}{\partial t} = \frac{\partial}{\partial x}\left( \kappa \frac{\partial T}{\partial x} \right) \pm S(x,t) \tag{9.9}$$

in which the source–sink term is shown as a function of both space and time.

The ratio of the thermal conductivity $\kappa$ to the volumetric heat capacity $C(= \rho c_m)$ is called the *thermal diffusivity*, designated $D_T$. Thus,

$$D_T = \kappa / C \tag{9.10}$$

Substituting $D_T$ for $\kappa$, we can rewrite Eqs. (9.5) and (9.8):

$$q_h = -D_T C \, dT / dx \tag{9.11}$$

and

$$\frac{\partial T}{\partial t} = \frac{\partial}{\partial x}\left( D_T \frac{\partial T}{\partial x} \right) \tag{9.12}$$

In the special case where $D_T$ can be considered constant, i.e., not a function of distance $x$, we can write

$$\partial T/\partial t = D_T \, \partial^2 T/\partial x^2 \tag{9.13}$$

To solve the foregoing equations so as to obtain a description of how temperature varies in both space and time, we need to know, by means of measurement or calculation, the pertinent values of the three parameters just defined, namely, the volumetric heat capacity $C$, thermal conductivity $\kappa$, and thermal diffusivity $D_T$. Together, they are called the *thermal properties* of soils.

### D. Volumetric Heat Capacity of Soils

The *volumetric heat capacity* $C$ of a soil is defined as the change in heat content of a unit bulk volume of soil per unit change in temperature. Its units are calories per cubic centimeter per degree (Kelvin), or joules per cubic meter per degree. As such, $C$ depends on the composition of the soil's solid phase (the mineral and organic constituents present), bulk density, and the soil's wetness (see Table 9.1).

The value of $C$ can be calculated by addition of the heat capacities of the various constituents, weighted according to their volume fractions. As given by de Vries (1975),

$$C = \sum f_{si}C_{si} + f_w C_w + f_a C_a \tag{9.14}$$

Here, $f$ denotes the volume fraction of each phase: solid (subscripted s), water (w), and air (a). The solid phase includes a number of components, subscripted $i$, such as various minerals and organic matter, and the symbol $\sum$ indicates the summation of the products of their respective volume fractions and heat capacities. The $C$ value for water, air, and each component of the solid phase is the product of the particular density and specific heat per unit mass (i.e., $C_w = \rho_w c_{mw}$, $C_a = \rho_a c_{ma}$, $C_{si} = \rho_{si} c_{mi}$).

Most of the minerals composing soils have nearly the same values of density (about 2.65 gm/cm$^3$ or $2.65 \times 10^3$ kg/m$^3$) and of heat capacity (0.48 cal/cm$^3$ °K or $2.0 \times 10^6$ J/m$^3$ °K). Since it is difficult to separate the different kinds of organic matter present in soils, it is tempting to lump them all into a single constituent (with an average density of about 1.3 gm/cm$^3$ or $1.3 \times 10^3$ kg/m$^3$, and an average heat capacity of about 0.6 cal/cm$^3$ °K or $2.5 \times 10^6$ J/m$^3$ °K). The density of water is less than half that of mineral matter (1 gm/cm$^3$ or $1.0 \times 10^3$ kg/m$^3$) but the specific heat of water is more than twice as large (1 cal/cm$^3$ °K or $4.2 \times 10^6$ J/m$^3$ °K). Finally, since the density of air is only about 1/1000 that of water, its contribution to the specific heat of the composite soil can generally be neglected.

Thus, Eq. (9.14) can be simplified as follows:

$$C = f_m C_m + f_o C_o + f_w C_w \qquad (9.15)$$

**Table 9.1**

DENSITIES AND VOLUMETRIC HEAT CAPACITIES OF SOIL CONSTITUENTS (AT $10°C$)
AND OF ICE (AT $0°C$)

| Constituent | Density $\rho$ | | Heat capacity $C$ | |
|---|---|---|---|---|
| | (gm/cm$^3$) | (kg/m$^3$) | (cal/cm$^3$ $°$K) | (W/m$^3$ $°$K) |
| Quartz | 2.66 | $2.66 \times 10^3$ | 0.48 | $2.0 \times 10^6$ |
| Other minerals (average) | 2.65 | $2.65 \times 10^3$ | 0.48 | $2.0 \times 10^6$ |
| Organic matter | 1.3 | $1.3 \times 10^3$ | 0.6 | $2.5 \times 10^6$ |
| Water (liquid) | 1.0 | $1.0 \times 10^3$ | 1.0 | $4.2 \times 10^6$ |
| Ice | 0.92 | $0.92 \times 10^3$ | 0.45 | $1.9 \times 10^6$ |
| Air | 0.00125 | 1.25 | 0.003 | $1.25 \times 10^3$ |

where subscripts m, o, w refer to mineral matter, organic matter, and water, respectively. Note that $f_m + f_o + f_w = 1 - f_a$, and the total porosity $f = f_a + f_w$. The reader will recall that in our preceding chapters we designated the volume fraction of water $f_w$ as $\theta$. Knowing the approximate average values of $C_m$, $C_o$, and $C_w$ (0.46, 0.60, and 1.0 cal/gm, respectively), we can further simplify Eq. (9.15) to give

$$C = 0.48 \, f_m + 0.60 \, f_o + f_w \qquad (9.16)$$

The use of Eq. (9.15) must be qualified in the case of frozen or partially frozen soils, since the properties of ice differ somewhat from those of liquid water ($\rho = 0.92$ gm/cm$^3$ or $0.92 \times 10^3$ kg/m$^3$, and $C = 0.45$ cal/cm$^3$ $°$K or $1.9 \times 10^6$ J/m$^3$ $°$K).

In typical mineral soils, the volume fraction of solids is in the range of 0.45–0.65, and $C$ values range from less than 0.25 cal/cm$^3$ $°$K (about 1 MJ/m$^3$ $°$K) in the dry state to about 0.75 cal/cm$^3$ $°$K ($\approx$ 3MJ/m$^3$ $°$K) in the water-saturated state.

Apart from the method described for calculating a soil's volumetric heat capacity, it is, of course, also possible to measure it, using calorimetric techniques (Taylor and Jackson, 1965).

### E. Thermal Conductivity of Soils

Thermal conductivity, designated $\kappa$, is defined as the amount of heat transferred through a unit area in unit time under a unit temperature gradient. As shown in Table 9.2, the thermal conductivities of specific soil

**Table 9.2**

THERMAL CONDUCTIVITIES OF SOIL CONSTITUENTS (AT $10°$C)
AND OF ICE (AT $0°$C)

| Constituent | mcal/cm sec $°$K | W/m $°$K |
|---|---|---|
| Quartz | 21 | 8.8 |
| Other minerals (average) | 7 | 2.9 |
| Organic matter | 0.6 | 0.25 |
| Water (liquid) | 1.37 | 0.57 |
| Ice | 5.2 | 2.2 |
| Air | 0.06 | 0.025 |

constituents differ very markedly (see also Table 9.3). Hence the space-average (macroscopic) thermal conductivity of a soil depends upon its mineral composition and organic matter content, as well as on the volume fractions of water and air. Since the thermal conductivity of air is very much smaller than that of water or solid matter, a high air content (or low water content) corresponds to a low thermal conductivity. Moreover, since the proportions of water and air vary continuously, $\kappa$ is also time variable. Soil composition is seldom uniform in depth, hence $\kappa$ is generally a function of depth as well as of time. It also varies with temperature, but under normal

**Table 9.3**

AVERAGE THERMAL PROPERTIES OF SOILS AND SNOW[a]

| Soil type | Porosity $f$ | Volumetric wetness $\theta$ | Thermal conductivity ($10^{-3}$ cal/cm sec $°$C) | Volumetric heat capacity $C_v$ (cal/cm$^3$ $°$C) | Damping depth (diurnal) $d$ (cm) |
|---|---|---|---|---|---|
| Sand | 0.4 | 0.0 | 0.7 | 0.3 | 8.0 |
|  | 0.4 | 0.2 | 4.2 | 0.5 | 15.2 |
|  | 0.4 | 0.4 | 5.2 | 0.7 | 14.3 |
| Clay | 0.4 | 0.0 | 0.6 | 0.3 | 7.4 |
|  | 0.4 | 0.2 | 2.8 | 0.5 | 12.4 |
|  | 0.4 | 0.4 | 3.8 | 0.7 | 12.2 |
| Peat | 0.8 | 0.0 | 0.14 | 0.35 | 3.3 |
|  | 0.8 | 0.4 | 0.7 | 0.75 | 5.1 |
|  | 0.8 | 0.8 | 1.2 | 1.15 | 5.4 |
| Snow | 0.95 | 0.05 | 0.15 | 0.05 | 9.1 |
|  | 0.8 | 0.2 | 0.32 | 0.2 | 6.6 |
|  | 0.5 | 0.5 | 1.7 | 0.5 | 9.7 |

[a] After van Wijk and de Vries (1963).

conditions this variation is ignored. The factors which affect thermal conductivity $\kappa$ are the same as those which affect the volumetric heat capacity $C$, but the measure of their effect is different so that the variation in $\kappa$ is much greater than of $C$. In the normal range of soil wetness experienced in the field, $C$ may undergo a threefold or fourfold change, whereas the corresponding change in $\kappa$ may be hundredfold or more. One complicating factor is that, unlike heat capacity, thermal conductivity is sensitive not merely to the volume composition of a soil but also to the sizes, shapes, and spatial arrangements of the soil particles.

The problem of expressing the overall thermal conductivity of a soil as a function of the specific conductivities and volume fractions of the soil's constituents is very intricate, as it involves the internal geometry of soil structure and the transmission of heat from particle to particle and from phase to phase.

Two relatively simple alternative cases can be envisaged: a dry soil or a water-saturated soil with the same particle configuration. In either case, we have a two-phase system in which the particles are dispersed in a continuous medium of fluid (air or water) with a volume fraction $f_0$ and thermal conductivity $\kappa_0$. The particles then occupy a volume fraction $f_1 = 1 - f_0$ and have a thermal conductivity $\kappa_1$. A composite thermal conductivity for the medium as a whole can be defined as follows: Consider a representative cube of soil with side $l$, large in comparison with the diameters of the particles and pores. Assume that the upper face is at a temperature $T_1$ and the bottom face at a lower temperature $T_2$. A constant heat flux $q_h$ will then pass through the cube, proportional to the overall temperature gradient, with $\kappa_c$ as the factor of proportionality for the composite medium:

$$q_h = -\kappa_c\, dT/dx = \kappa_c(T_1 - T_2)/l$$

Since the cube is a mixture of two phases, the composite thermal conductivity $\kappa_c$ will be intermediate between $\kappa_0$ and $\kappa_1$. According to Burger (1915),

$$\kappa_c = (f_0\kappa_0 + kf_1\kappa_1)/(f_0 + kf_1) \tag{9.17}$$

wherein the factor $k$ is the ratio of the average temperature gradient in the particles to the corresponding gradient in the continuous fluid:

$$k = \frac{(dT/dz)_1}{(dT/dz)_0}$$

According to de Vries (1975) the value of $k$ depends not only on the ratio $\kappa_1/\kappa_0$, but also on the particle sizes, shapes, and mode of packing. These variables are difficult to characterize quantitatively in the case of particles of irregular shapes, distribution of sizes, and packing arrangements.

If there are several types of particles with different shapes or conductivities, Eq. (9.17) can be generalized:

$$\kappa_c = \sum_{i=1}^{n} k_i f_i \kappa_i / \sum_{i=1}^{n} k_i f_i \tag{9.18}$$

Here $n$ is the number of particle classes within which all particles have about the same shape and conductivity.

As shown by de Vries (1975), the thermal conductivity of soils of widely differing compositions can be estimated by Eq. (9.18) with a fair degree of accuracy. The deviations between measured and estimated values of thermal conductivity were reported to be less than 10% for $\kappa_1/\kappa_0$ ratios smaller than 10 (i.e., for saturated soils) and about 25% for $\kappa_1/\kappa_0$ ratios of the order of 100 (i.e., dry soils). In moist but unsaturated soils, water can be considered as a continuous medium in which soil particles and air are dispersed. Since the ratio $\kappa_1/\kappa_0$ is greater than unity for mineral particles and less than unity for organic matter and air, the errors caused by these different constituents may compensate each other, at least partially. Thus, Eq. (9.18) may still provide a fair estimate of the composite thermal conductivity even of a three-phase soil. The de Vries model can also be applied in two steps to calculate the thermal conductivity of a soil with aggregated particles. In the first step the conductivity of the aggregates is calculated, in the second step that of the soil as a whole. There are insufficient data to verify this model, however.

The following form of Eq. (9.18) for an unsaturated soil was used by van Bavel and Hillel (1975, 1976):

$$\kappa_c = \frac{f_w \kappa_w + k_s f_s \kappa_s + k_a f_a \kappa_a}{f_w + k_s f_s + k_a f_a}$$

wherein $\kappa_w$, $\kappa_a$, and $\kappa_s$ are the specific thermal conductivities of each of the soil constituents (water, air, and an average value for the solids, respectively). The factor $k_s$ represents the ratio between the space average of the temperature gradient in the solid relative to the water phase. The factor $k_s$ depends on the array of grain shapes as well as on mineral composition and organic matter content. The $k_a$ factor represents the corresponding ratio for the thermal gradient in the air and water phases.

The dependence of thermal conductivity and diffusivity on soil wetness is illustrated in Fig. 9.1. The influence of latent heat transfer by water vapor in the air-filled pores is proportional to the temperature gradient in these pores. It can be taken into account (van Bavel and Hillel, 1976; Hillel, 1977) by adding to the thermal conductivity of air an apparent conductivity due to evaporation, transport, and condensation of water vapor (the so-called vapor enhancement factor). This value is strongly temperature dependent, and rises rapidly with increasing temperature.

The flux of sensible heat associated with liquid water movement in the soil is generally considered negligible.

Given the complexities involved in any attempt to predict a soil's thermal conductivity by calculation based on theory, one might be justified in asking, "Why bother? Why not simply take measurements?" Indeed, one should never depend on theory alone. Measurements are necessary, if only to validate (or invalidate) theory. However, the task of measuring thermal conductivity presents its own difficulties and complexities. Because of the fact that soil water potential depends on temperature, the development of a temperature gradient generally induces the movement of water as well as of heat. Hence techniques for measuring heat transfer through a soil sample based on steady-state heat flow between two planes maintained at a constant temperature differential involve the risk of changing the sample's internal moisture distribution, and therefore its thermal properties, during the process of measurement: the soil near the warmer plane becomes drier, while that near the cooler plane becomes wetter. Early attempts to measure thermal conductivity (e.g., Smith, 1932) failed to recognize this pitfall as they purported to maintain constant soil moisture conditions during prolonged steady-state heat flow. Hence their results can only be considered approximations at best. While steady-state methods may be sufficiently accurate for

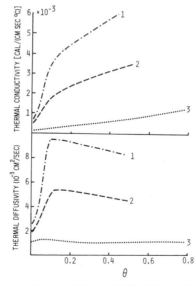

**Fig. 9.1.** Thermal conductivity and thermal diffusivity as functions of volume wetness (volume fraction of water) for (1) *sand* (bulk density 1.46 gm/cm$^3$; volume fraction of solids 0.55), (2) *loam* (bulk density 1.33 gm/cm$^3$; volume fraction of solids 0.5), and (3) *peat* (volume fraction of solids 0.2). (After de Vries, 1975.)

measuring thermal conductivity of dry soils, short-term transient heat-flow techniques are preferable, in principle, for moist soils.

The principal advantages of using transient-state methods in the measurement of thermal conductivity are that water movement in the soil volume of interest is obviated, or at least minimized, and that the long wait required for attainment of steady-state heat flow is avoided. In addition, whereas steady-state methods are confined practically entirely to the laboratory, transient-state methods are applicable both in the laboratory and in the field. At present one of the most practical of these methods for measuring thermal conductivity in situ is the cylindrical-probe heat source, which can be inserted into the soil to any depth in the field, and can be used in the laboratory as well (de Vries and Peck, 1958; Woodside, 1958). Its use is based on solution of the equation for heat conduction in the radial direction from a line source (Carslaw and Jaeger, 1959):

$$\frac{\partial T}{\partial t} = \kappa \frac{\partial^2 T}{\partial r^2} + \frac{1}{r} \frac{\partial T}{\partial r} \tag{9.19}$$

where $T$ is temperature, $t$ time, $r$ radial distance from the line source of heat, and $\kappa$, as before, thermal conductivity.

In practice, a cylindrical probe containing a heating wire is embedded in the soil, an electrical current is then supplied, and the rate of temperature rise is measured with a thermocouple or thermistor placed next to the wire. For a short distance from the line source, the rise in temperature $T - T_0$ is given by

$$T - T_0 = (q_h/4\pi\kappa)(c + \ln t) \tag{9.20}$$

wherein $T$ is the measured temperature, $T_0$ the initial temperature, $q_h$ the heat generated per unit time and unit length of heating wire, $\kappa$ the conductivity, $c$ a constant, and $t$ the time. A plot of temperature versus the logarithm of time permits a calculation of $\kappa$. A correction factor may be necessary to account for the dimensions of the probe (Jackson and Taylor, 1965).

The thermal diffusivity $D_h$, instead of the conductivity $\kappa$, is sometimes desired. It can be defined as the change in temperature produced in a unit volume by the quantity of heat flowing through the volume in unit time under a unit temperature gradient. An alternative definition, easier to perceive, is that the thermal diffusivity is the ratio of the conductivity to the product of the specific heat and density:

$$D_h = \kappa/c_s\rho = \kappa/C_v \tag{9.21}$$

where $C_v$ is the volumetric heat capacity. As shown in the preceding section, the specific heat and density of both solids and water must be considered when calculating the volumetric heat capacity:

$$C_v = \rho_s(c_s + c_w w) \tag{9.22}$$

where $\rho_s$ is the density of dry soil, $c_s$ the specific heat of dry soil, $c_w$ the specific heat of water, and $w$ the ratio of the mass of water to the mass of dry soil.

The thermal diffusivity can thus be calculated from prior measurements of thermal conductivity and volumetric heat capacity, or it can be measured directly as described by Jackson and Taylor (1965).

## F. Thermal Regime of Soil Profiles

In nature, soil temperature varies continuously in response to the ever-changing meteorological regime acting upon the soil–atmosphere interface. That regime is characterized by a regular periodic succession of days and nights, and of summers and winters. Yet the regular diurnal and annual cycles are perturbed by such irregular episodic phenomena as cloudiness, cold waves, warm waves, rainstorms or snowstorms, and periods of drought. Add to these external influences the soil's own changing properties (i.e., temporal changes in reflectivity, heat capacity, and thermal conductivity as the soil alternately wets and dries, and the variation of all these properties with depth), as well as the influences of geographic location and vegetation, and you can expect the thermal regime of soil profiles to be complex indeed. Yet not altogether unpredictable.

The simplest mathematical representation of nature's fluctuating thermal regime is to assume that at all depths in the soil the temperature oscillates as a pure harmonic (sinusoidal) function of time around an average value. Since nature's actual variations are not so orderly, this may be a rather crude approximation. Nevertheless, it is an instructive exercise in itself, and when used in conjunction with field data it can lead to a better understanding, and perhaps even provide a basis for the prediction, of a soil's thermal regime.

Now let us assume that, although soil temperature varies differently at different depths in the soil, the average temperature is the same for all depths. We next choose a starting time ($t = 0$) such that the surface is at the average temperature. The temperature at the surface can then be expressed as a function of time (Fig. 9.2):

$$T(0,t) = \overline{T} + A_0 \sin \omega t \qquad (9.23)$$

where $T(0,t)$ is the temperature at $z = 0$ (the soil surface) as a function of time $t$, $T$ is the average temperature of the surface (as well as of the profile), and $A_0$ is the amplitude of the surface temperature fluctuation [the range from maximum (or minimum) to average temperature]. Finally, $\omega$ is the radial frequency, which is $2\pi$ times the actual frequency. In the case of diurnal variation the period is 86,400 sec (24 hr) so $\omega = 2\pi/86,400 = 7.27 \times 10^{-5}/\text{sec}$. Note that the argument of the sine function is expressed in radians rather than in degrees.

The last equation is the boundary condition for $z = 0$. For the sake of convenience, let us assume that at infinite depth ($z = \infty$) the temperature is constant and equal to $\overline{T}$. Under these circumstances, temperature, at any depth $z$ and time $t$, is also a sine function of time, as shown in Fig. 9.3 (Lettau, 1962; van Wijk, 1963):

$$T(z,t) = \overline{T} + A_z \sin[\omega t + \varphi(z)] \tag{9.24}$$

in which $A_z$ is the amplitude at depth $z$. Both $A_z$ and $\varphi(z)$ are functions of $z$ but not of $t$. They can be determined by substituting the solution of Eq. (9.23) in the differential equation $\partial T/\partial t = D_h (\partial^2 T/\partial z^2)$. This leads to the solution

$$T(z,t) = \overline{T} + A_0[\sin(\omega t - z/d)]/e^{z/d} \tag{9.25}$$

The constant $d$ is a characteristic depth, called the *damping depth*, at which the temperature amplitude decreases to the fraction $1/e$ ($1/2.718 = 0.37$) of the amplitude at the soil surface $A_0$. The damping depth is related to the thermal properties of the soil and the frequency of the temperature fluctuation as follows:

$$d = (2\kappa/c\omega)^{1/2} = (2D_h/\omega)^{1/2} \tag{9.26}$$

It is seen that at any depth $z$ the amplitude of the temperature fluctuation $A_z$ is smaller than $A_0$ by a factor $e^{z/d}$ and that there is a phase shift (i.e., a time

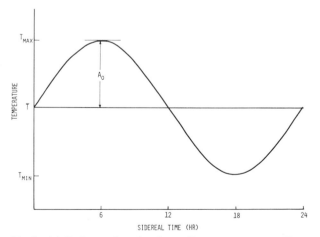

**Fig. 9.2.** Idealized daily fluctuation of surface soil temperature, according to the equation: $T = \overline{T} + A_0 \sin(\omega t/p)$, where $T$ is temperature, $\overline{T}$ average temperature, $A_0$ amplitude, $t$ time, and $p$ period of the oscillation (in this case, $p$ refers to the diurnal period of 24 hr).

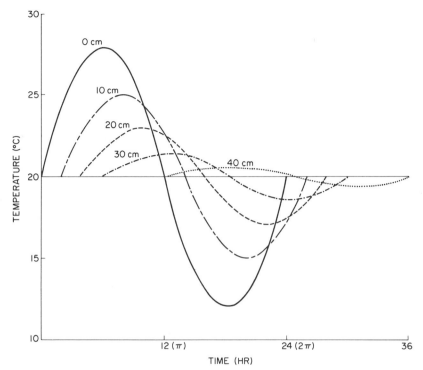

**Fig. 9.3.** Idealized variation of soil temperature with time for various depths. Note that at each succeeding depth the peak temperature is damped and shifted progressively in time. Thus, the peak at a depth of 40 cm lags about 12 hr behind the temperature peak at the surface and is only about $\frac{1}{16}$ of the latter. In this hypothetical case, a uniform soil was assumed, with a thermal conductivity of $4 \times 10^{-3}$ cal/cm sec deg and a volumetric heat capacity of 0.5 cal/ cm³ deg.

delay of the temperature peak) equal to $-z/d$. The decrease of amplitude and increase of phase lag with depth are typical phenomena in the propagation of a periodic temperature wave in the soil.

The physical explanation for the damping and retarding of the temperature waves with depth lies in the fact that a certain amount of heat is absorbed or released along the path of propagation when the temperature of the conducting soil increases or decreases, respectively. The damping depth is related inversely to the frequency, as can be seen from Eq. (9.26). Hence it depends directly on the period of the temperature fluctuation considered. The damping depth is $\sqrt{365} \approx 19$ times larger for the annual variation than for the diurnal variation in the same soil. For example, van Wijk and de Vries (1963) calculated the damping depth for a soil with $\kappa = 2.3 \times 10^{-3}$ cal/cm sec deg and obtained $d = 12$ cm for the diurnal temperature fluctuation and $d = 229$ cm for the annual fluctuation. Whereas at depth $z = d$ the amplitude is 0.37 as great as the amplitude at the surface, it is only about 0.05 of the surface amplitude at $z = 3d$ ($=36$ cm for the diurnal variation in the case of the soil used by these authors). When an arbitrary zero point $t_0$ is introduced into the time scale, Eq. (9.25) becomes

$$T(z,t) = \bar{T} + A_0[\sin(\omega t + \varphi_0 - z/d)]/e^{z/d} \qquad (9.27)$$

The constant $\varphi_0 = -\omega t_0$ is called the *phase constant*.

The annual variation of soil temperature down to considerable depth causes deviations from the simplistic assumption that the daily average temperature is the same for all depths in the profile. The combined effect of the annual and diurnal variation of soil temperature can be expressed by

$$\begin{aligned} T(z,t) = \; & \bar{T}_y + A_y[\sin(\omega_y t + \varphi_y - z/d_y)]/e^{z/d_y} \\ & + A_d[\sin(\omega_d t + \varphi_d - z/d_y)]/e^{z/d_d} \end{aligned} \qquad (9.28)$$

where the subscripted indices y and d refer to the yearly and daily temperature waves, respectively. Thus $\bar{T}_y$ is the annual mean temperature. The daily cycles are now seen to be short term perturbations superimposed upon the annual cycle. Vagaries of weather (e.g., spells of cloudiness or rain) can cause considerable deviations from simple harmonic fluctuations, particularly for the daily cycles. Longer term climatic irregularities can also affect the annual cycle, of course. Also, since the annual temperature wave penetrates much more deeply, the assumptions of soil homogeneity in depth and of the time constancy of soil thermal properties are clearly unrealistic.

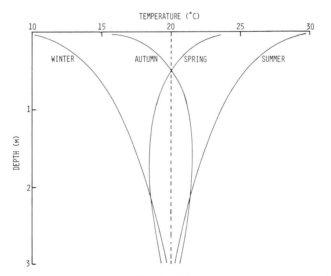

**Fig. 9.4.** The soil temperature profile as it might vary from season to season in a frost-free region.

An alternative theoretical approach is now possible, with fewer constraining assumptions. It is based on numerical, rather than analytical, methods for solving the differential equations of heat conduction. With the brute force of increasingly powerful digital computers, it is possible to construct and solve mathematical simulation models which allow soil thermal properties to vary in time and space (e.g., in response to periodic changes in soil wetness), so as to account for alternative surface saturation and desiccation and for profile layering, and which also allow boundary conditions (e.g., climatic inputs) to follow more realistic and irregular patterns. The surface amplitude of temperature need no longer be assumed to be an independent variable, but one which itself depends on the energy balance of the surface and thus is affected by both soil properties and above-

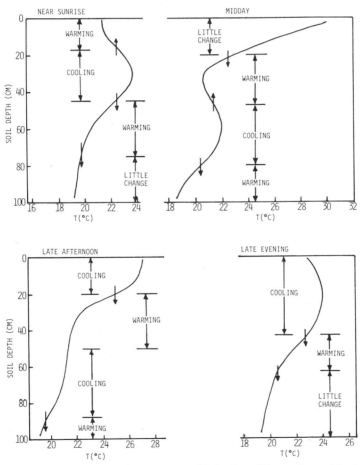

**Fig. 9.5.** Typical variation of temperature with depth at different times of day in summer. (From Sellers, 1965, based on data given by Carson, 1961.)

soil conditions. Examples of the numerical approach can be found in the published works of Wierenga and deWit (1970), van Bavel and Hillel (1975, 1976), and Hillel (1977).

Other recent innovations of practical importance involve the development of techniques for monitoring the soil thermal regime more accurately and precisely than previously available techniques. One such innovation is the infrared *radiation thermometer* for the scanning or remote sensing of surface temperature for both fallow and vegetated soils without disturbance of the measured surface. Knowledge of the surface temperature and its variation in time is important in assessing energy exchange between soil and atmosphere as well as in determining boundary conditions for in-soil heat transfer.

Another important technique is the use of *heat flux plates*. These are flat and narrow plates or disks of constant thermal conductivity which allow precise measurement of the temperature difference between their two sides so as to yield the heat flux through them. When embedded horizontally in the soil at regular depth intervals, a series of such heat flux plates can provide a continuous record of heat transfer throughout the profile. There are problems, however. The presence of heat flux plates can distort the flow of heat in the surrounding medium if their thermal conductivity is very different from that of the soil. The experimental error can be minimized by constructing plates of maximal thermal conductivity and minimal thickness, and by calibrating them in a medium with a thermal conductivity close to that of the soil in which they are to be placed. Another problem is that such plates do not allow vapor flow, which can be an important component of heat transfer. Studies demonstrating the use of heat flux plates were reported by Fuchs and Tanner (1968) and by Hadas and Fuchs (1973).

The soil temperature profile as it might vary from season to season in a frost-free region is illustrated in Fig. 9.4. The diurnal variation of temperature and the directions of heat flow within a soil profile are illustrated in Fig. 9.5.

### Sample Problems

**1.** Assuming steady-state conditions, calculate the one-dimensional thermal flux and total heat transfer through a 20-cm thick layer if the thermal conductivity is $3.6 \times 10^{-3}$ cal/cm sec deg and a temperature differential of $10°C$ is maintained across the sample for 1 hr.

Using Eq. (9.5) in discrete form, we can write

$$q_h = \kappa \, \Delta T / \Delta x = 3.6 \times 10^{-3} \text{ cal/cm sec deg} \times 10 \text{ deg}/20 \text{ cm}$$
$$= 1.8 \times 10^{-3} \text{ cal/cm}^2 \text{ sec}$$

Total heat transfer is

$$q_h t = 1.8 \times 10^{-3} \text{ cal/cm}^2 \text{ sec} \times 3600 \text{ sec} = 6.48 \text{ cal/cm}^2$$

**2.** A thermal flux of $10^{-3}$ cal/cm$^2$ sec is maintained into the upper surface of a 10-cm thick sample, the bottom of which is thermally insulated. Calculate the time rate of temperature change and the total temperature rise per hour if the bulk density is 1.2 gm/cm$^3$ and the specific heat capacity is 0.6 cal/gm deg.

For this case of heat flow, we use a discrete form of Eq. (9.6)

$$dT/dt = (\Delta q_h/\Delta x)(1/\rho_b c_m)$$

Using the data provided, the time rate of temperature change

$$dT/dt = (10^{-3} \text{ cal/cm}^2 \text{ sec}/10 \text{ cm})(1.2 \text{ gm/cm}^3 \times 0.6 \text{ cal/gm deg})^{-1}$$
$$= 1.39 \times 10^{-4}$$

Total temperature rise is $1.39 \times 10^{-4}$ deg/sec $\times$ 3600 sec/hr $= 0.5°C$/hr.

**3.** Calculate the volumetric heat capacity $C$ of a soil with a bulk density of 1.46 gm/cm$^3$ when completely dry, when completely saturated. Assume that the density of solids is 2.60 gm/cm$^3$ and that organic matter occupies 10% of the solid matter (by volume).

First calculate the volume fraction of pores (the porosity):

$$f = (\rho_s - \rho_b)/\rho_s = (2.60 - 1.46) \text{ gm/cm}^3/2.60 \text{ gm/cm}^3 = 0.44$$

Hence the volume fraction of solids is $1 - 0.44 = 0.56$. Since organic matter constitutes 10% of the soil's solid phase, the volume fraction of mineral matter is

$$f_m = 0.56 \times 0.9 = 0.504$$

The volume fraction of organic matter is

$$f_o = 0.56 \times 0.1 = 0.056$$

The volumetric heat capacity can be calculated using Eq. (9.15):

$$C = f_m C_m + f_o C_o + f_w C_w$$

where $f_m$, $f_o$, and $f_w$, are the volume fractions of mineral matter, organic matter, and water, respectively; and $C_m$, $C_o$, and $C_w$ refer to heat capacities of the same constituents (namely, 0.48 cal/cm$^3$ deg for mineral matter, 0.6 cal/cm$^3$ deg for organic matter, and 1 cal/cm$^3$ deg for water). Accordingly, when the soil is completely dry,

$$C = (0.48 \times 0.504) + (0.60 \times 0.05) = 0.24 + 0.03 = 0.27 \text{ cal/cm}^3 \text{ deg}$$

When the soil is saturated, its volume fraction of water equals the porosity. Thus

$$C = 0.27 \text{ cal/cm}^3 \text{ deg} + 0.44 \times 1 \text{ cal/cm}^3 \text{ deg} = 0.71$$

*Note:* We have completely neglected the heat capacity of air, since it is too small to make any significant difference (see Table 9.1).

**4.**   The daily maximum soil-surface temperature is 40°C and the minimum is 10°C. Assuming that the diurnal temperature wave is symmetrical, that the mean temperature is equal throughout the profile (with the surface temperature equal to the mean value at 6 A.M. and 6 P.M.), and that the "damping depth" is 10 cm, calculate the temperatures at noon and midnight for depths 0, 5, 10, and 20 cm.

Since the temperature range is 30°C and the mean ($\bar{T}$) 25°C, the amplitude at the surface $A_0$, the maximum value above the mean, is 15.

We use Eq. (9.25) to calculate the temperature $T$ at any depth $z$ and time $t$:

$$T(z,t) = \bar{T} + A_0[\sin(\omega t - z/d)]/e^{z/d}$$

where $\omega$ is the radial frequency ($2\pi/24$ hr) and $d$ is the "damping depth" at which the temperature amplitude is $1/e(= 0.37)$ of $A_0$. *Note:* The radial angle is expressed in radians rather than in degrees.

*At depth zero (the soil surface):* Noontime temperature (6 hr after $T = \bar{T}$):

$$T(0,6) = 25 + 15 \times [\sin(\pi/2 - 0)]/e^0 = 25 + 15 = 40°C$$

Midnight temperature (18 hr after $T = \bar{T}$)

$$T(0,18) = 25 + 15 \times [\sin(3\pi/2 - 0)]/e^0 = 25 - 15 = 10°C$$

*At depth 5 cm:* Noontime temperature:

$$
\begin{aligned}
T(5,6) &= 25 + 15 \times [\sin(\pi/2 - 5/10)]/e^{5/10} \\
&= 25 + 15 \times [\sin(1.57 - 0.5)]/e^{1/2} \\
&= 25 + 15 \times [\sin(1.07)]/1.65 \\
&= 25 + 15 \times (0.87720/1.65) = 32.97°C
\end{aligned}
$$

Midnight temperature:

$$
\begin{aligned}
T(5,18) &= 25 + 15 \times [\sin(3\pi/s - 5/10)]/1.65 \\
&= 25 + 15 \times [\sin(4.71 - 0.5)]/1.65 \\
&= 25 + 15 \times (-0.87720/1.65) = 17.03°C
\end{aligned}
$$

*At depth 10 cm (the damping depth):* Noontime temperature:

$$
\begin{aligned}
T(10,6) &= 25 + 15 \times [\sin(\pi/2 - 1)]/e^1 = 25 + 15 \times \sin(0.57)/e \\
&= 25 + 15 \times (0.53963/2.718) = 27.98°C
\end{aligned}
$$

Midnight temperature :

$$
\begin{aligned}
T(10,18) &= 25 + 15 \times [\sin(3\pi/2 - 1)]/e \\
&= 25 + 15 \times [\sin(4.71 - 1)]/2.718 \\
&= 25 + 15 \times (-0.53963/2.718) = 22.02°C
\end{aligned}
$$

*At depth 20 cm:* Noontime temperature:

$$T(20,6) = 25 + 15 \times [\sin(1.57 - 20/10)]/e^{20/10}$$
$$= 25 + 15 \times [\sin(-0.43)]/e^2$$
$$= 25 + 15 \times (-0.41687/7.39) = 25 - 0.85 = 24.15°C$$

Midnight temperature:

$$T(20,18) = 25 + 15 \times [\sin(4.71 - 2)]/7.39$$
$$= 25 + 15 \times (0.41687/7.39 = 25.85°C$$

*Note:* At a depth of 20 cm the phase shift is so pronounced that the temperature at midnight is actually higher than at noontime.

A useful exercise for students at this point is to calculate and to plot the sinsusoidal variation of temperature at each depth so as to observe how the phase shift (time lag of maximum and minimum values) increases and the amplitude decreases with depth.

# 10 Soil Compaction and Consolidation

## A. Introduction

When subjected to pressure, soil tends to compress; that is to say, it tends to increase its density. Hypothetically speaking, we can imagine several possible mechanisms by which soil compression, or densification, could take place. One rather unlikely mechanism is that the solid particles themselves might be compressed, or be crushed against each other. We can dismiss this possibility because the compressibility of most solid-phase components (soil minerals) is too low to provide appreciable compression, and their strength too great to allow crushing, in the range of pressures normally encountered or administered in agricultural and engineering practice. Another possible mechanism is the compression of the liquid phase, i.e., of soil water. Deep aquifers, under great overburden pressure, do indeed exhibit appreciable water compression. Not so, however, near the soil surface, where soil water is seldom confined (and hence is free to flow away when placed under pressure) and the pressures are seldom great enough to result in significant compression of water.

Still another mechanism, perhaps more likely to be of some importance, is the possible compression of confined air. In a very wet soil, air may be occluded (trapped) in isolated small pockets during the wetting phase, or it may effervesce out of solution to form bubbles when the temperature rises. Gases may also be released by biological or chemical reactions. Air is, of course, very much more compressible than either mineral solids or water. (The presence of compressible air bubbles has in fact been offered as an explanation for the "springiness" of wet soil so familiar to those who have had frequent occasion to walk in cultivated fields during or shortly after a rain.) However, the fractional volume of occluded air bubbles in a wet soil is generally small and, as the soil begins to drain, the air phase in the soil soon becomes continuous and open to the atmosphere so that air compression in an unsaturated soil is itself not likely to be significant except in special cases.

If the densities of none of the individual components of the soil changes materially, how then can the soil as a whole compress? What remains as the principal mechanism of soil compression is, obviously, the reduction of soil porosity through the partial expulsion of either or both of the permeating fluids, air and water, from the compressing soil body. Here we can consider two extreme cases and a whole range of intermediate conditions. The first case is that of a totally dry soil. Its compression under static pressure or by vibration causes the particles to reorient and to assume a closer packing arrangement, thereby reducing the fractional volume of air. In the opposite case of a water-saturated soil, any such decrease of porosity must necessarily take place at the expense of the fractional volume of water. The difference is, of course, that the viscosity of water is 50–100 times greater than that of air. Hence the expulsion of air is a rapid, almost instantaneous, phenomenon, whereas the expulsion of water is generally a very much slower process, particularly as progressive compression repeatedly closes the largest pores and requires subsequent flow to take place in narrower and more tortuous pores. Application of pressure to soil in any intermediate state between dryness and saturation, i.e., to a moist soil, will result first in expulsion of air and a gradual approach to saturation; only after essentially all of the air has been driven out and the soil in effect has become saturated will the further application of uncompensated compressive stress begin to result in removal of water.

To distinguish between the two processes or phases described, conventional wisdom requires that we designate them by different names. Traditionally, the term *compaction* has been applied to the compression of an unsaturated soil body resulting in reduction of the fractional air volume. The term *consolidation*, on the other hand, has long been used to signify the compression of a saturated soil by "squeezing out" water. (Note that all along we have been using the overall term compression to depict all processes of soil densification, including both compaction and consolidation.) We shall proceed to describe each of these processes in turn.

## B. Soil Compactibility in Relation to Wetness

Having as their goal the attainment of high soil bulk density so as to minimize future subsidence, increase shearing strength, and reduce percolation, soil engineers have long sought methods for achieving the maximal possible degree of compaction with the least expenditure of energy. An empirical approach to this problem was developed nearly 50 years ago by R. R. Proctor, whose testing procedure, known as the *Proctor test*, was designed to determine the "optimal" soil wetness at which compaction of a given soil can be achieved most effectively by a given compactive effort. Although modified repeatedly and given different names, Proctor's basic procedure has enjoyed universal acceptance and has been adopted as the standard criterion for soil compaction in engineering practice (ASTM, 1958).

In principle, compactive stresses can be administered in several different ways. Perhaps the simplest is to confine a soil sample in a rigid-walled container and apply a *static load* (e.g., a weight) by means of a piston resting on the sample's surface. An alternative way is to apply a *dynamic load*, i.e., a time-variable load, as, for example, by means of an impacting or hammering device. Still another way, found to be most effective in the case of dry granular materials, is to vibrate the sample. Finally, one could attempt to compact soil by applying a space-variable and time-variable set of compressive and shearing stresses in combination. Such action, which in effect causes a churning (or "puddling") of the soil, is commonly called *kneading compaction*. It is this latter method which forms the basis of the Proctor test, which is, in essence, an imitation of the common engineering practice of compacting soil in the field by means of spiked rollers. The standard test is carried out in a cylindrical mold, 10 cm in diameter and having the capacity of 100 cm$^3$. The soil material is compacted in three layers, each layer being worked by a 5-cm diameter impact-driven tamper in a standardized manner.

For any given amount of compactive effort, the resulting bulk density is a function of soil moisture (wetness). This functional dependence, illustrated in Fig. 10.1 indicates that, starting from a dry condition, the attainable bulk density at first increases with increasing soil wetness, then reaches a peak called maximal density at a wetness value called optimum moisture, beyond which the density decreases. This behavior is readily explainable, at least qualitatively. Typically, a dry soil resists compaction because of its stiff matrix and high degree of particle-to-particle bonding, interlocking, and/or frictional resistance to deformation. As soil wetness increases, the moisture films weaken the interparticle bonds, cause swelling, and seem to reduce internal friction by "lubricating" the particles, thus making the soil more workable and compactible. As soil wetness nears saturation, however, the fractional volume of expellable air is reduced and the soil can no longer be compacted by a given compactive effort to the same degree as before. Henceforth, any further increase in soil moisture reduces, rather than increases, soil compactibility. Finally, at saturation, no amount of kneading can cause any increase in soil bulk density (unless water is expelled, as in consolidation, which, however, is not the subject of this section). At high wetness values, water is seen to prevent closer packing of the soil matrix. Rather, the water

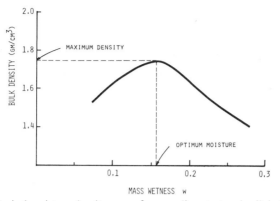

**Fig. 10.1.** Typical moisture-density curve for a medium-textured soil, indicating the maximum density obtainable with a particular compactive effort.

which hydrates the grains pushes them apart and causes swelling, thus reducing the attainable bulk density.

The function described, depicting the dependence of attainable bulk density upon soil moisture, does not constitute a single characteristic curve for a given soil but a family of curves, as shown in Fig. 10.2. For each level of compactive effort, there is a separate curve. With higher levels of compactive effort, the curve is shifted upward and leftward, indicating higher attainable bulk density at lower values of "optimal moisture."

Experience shows that the curved line connecting the peaks of all the bulk density versus wetness curves corresponds approximately to the 80% degree-of-saturation line, and that the descending portions of these curves tend to converge on a curved line representing a degree of saturation of about 85–90%. To the extent that the Proctor test portrays or simulates compaction in the field, it can serve as a criterion by which to guide earth construction works. A typical job specification might require a contractor to compact an earth fill to, say, 90% of attainable maximal density as defined by the Proctor test (or by any of the derivative tests such the "AASHO" or "Modified AASHO" test [ASTM, 1958b]). The contractor, if he knows anything at all about soil mechanics, will then examine the results of the test to determine the optimal wetness value (or range of values) at which to carry out the compacting operation. In practice, an earth fill is laid by depositing and rolling successive layers while controlling soil moisture through measured applications (sprinkling) of water to the deposited material. On the other hand, if the soil material is too wet, it must be dried by drainage and, weather permitting, by evaporation.

In our discussion thus far we have made the implicit assumption that our "typical" soil is a fine-textured or medium-textured one (e.g., a clay, clay–

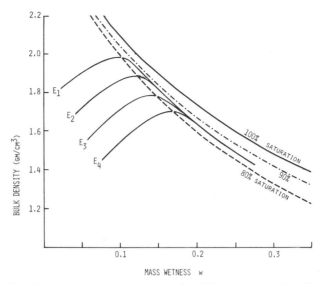

**Fig. 10.2.** Family of moisture-density curves for different compactive efforts ($E_1 > E_2 > E_3 > E_4$). Note that the 100% saturation line, representing zero air-filled porosity, was calculated assuming a particle density value of 2.65 gm/cm$^3$.

loam, silt–loam, or loam) with some appreciable cohesiveness. If, however, the material to be compacted is coarse-textured (granular), then the most effective method of compaction is not by rolling or kneading but by vibration. Various types of ramming or jolting devices have been tried for this purpose.

## C. Occurrence of Soil Compaction in Agricultural Fields

In the agronomic context, soils or soil layers are considered to be compacted when the total porosity, and particularly the air-filled porosity, are so low as to restrict aeration, as well as when the soil is so tight, and its pores so small, as to impede root penetration and drainage. Still another manifestation of soil compaction is the difficulty it creates in field management, particularly with respect to tillage.

Soils and soil layers may become compact naturally as a consequence of their textural composition, moisture regime, or the manner in which they were formed in place. *Surface crusts* can form over exposed soils under the beating and dispersing action of raindrops and the subsequent drying of a compacted layer of oriented particles. Naturally compact *subsurface* layers may consist of densely packed granular sediments, which may be partially cemented. Indurated layers, called *hardpans*, can be of variable texture and may, in extreme cases, exhibit rocklike properties (in which case they have been termed *fragipans*, or *ortstein*) and become almost totally impenetrable to roots, water, and air. Usually such hardpans are found at the junction of two distinctly different layers where penetration of water and/or dissolved or suspended materials is retarded by a clay layer, a water table, or bedrock (Lutz, 1952).

A *claypan* is a tight, restrictive subsoil layer of high clay content which tends to be plastic and relatively impermeable to water and air when wet. In humid climates, such layers may remain perpetually wet and give rise to perched water-table conditions above them, thus inducing anaerobic conditions within the root zone. Claypans may be depositional or may have developed in place. They occur at various depths in the profile. High clay content alone does not necessarily result in the formation of a claypan, as much depends on soil structure as well as on texture. Many good agricultural soils having clayey B horizons exhibit well-developed structure with large interped pores which permit the unobstructed passage of water and air. In claypans, however, structural development is poor and the clay may even be in a somewhat dispersed state. Claypans and hardpan conditions are difficult to rectify. Mechanical fragmentation by tillage alone may result in only temporary relief, as these layers tend to re-form spontaneously.

Another natural factor which can contribute to soil compaction is the tendency of clayey soil to shrink upon drying, particularly if it had been puddled in the wet state. Gill (1959) found that the bulk density of compressed samples increased from 1.54 gm/cm$^3$ at a mass wetness of 25% to 1.75 gm/cm$^3$ at a wetness of 20%. Well-structured soils, however, break into numerous small aggregates (crumbs) as they dry, so that even though individual aggregates shrink the layer as a whole becomes loose and porous.

Quite apart from the natural formation of compacted layers, soil compaction can, and all too often does, take place under the influence of man-induced mechanical forces applied to the soil surface. One such cause of soil compaction is trampling by livestock. In an example reported by Tanner and Mamaril (1959), grazing animals caused an increase in topsoil bulk density from 1.22 to 1.43 gm/cm$^3$, corresponding to a decrease of air-filled porosity from 17.3 to 7.2% and an increase in penetrometer resistance from 3.2 to 19.5 bar. However, by far the most common cause of soil compaction in modern agriculture is the effect of machinery, imposed on the soil by wheels, tracks, and soil-engaging tools.

## D. Pressures Caused by Machinery

The magnitudes of pressures exerted on the soil surface by wheeled and tracked vehicles depend in a combined way on characteristics of the soil surface zone and of the wheels or tracks involved. The manner in which these

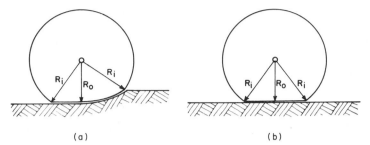

(a)                                                    (b)

**Fig. 10.3.** Deformation of a pneumatic tire in contact with the soil surface: (a) deformation in the tire and soil; (b) deformation in the tire only. (From Gill and Vanden Berg, 1967.)

pressures are distributed within the soil, and the deformations they cause, depend, in turn, on the pattern of surface pressure as well as on the mechanical or rheological characteristics of the soil in depth. This general topic was reviewed by Chancellor (1976). An invaluable store of data is available in the book by Gill and Vanden Berg (1967).

A general rule of thumb which applies to *pneumatic* (air-inflated) *tires* is that the pressure exerted upon the supporting surface is approximately equal to the inflation pressure. Since the total weight of a vehicle at rest should equal the sum of the products of the pressures exerted by the wheels and their respective contact areas, one can reason that an increase in weight should be compensated by a flattening of the tires and a commensurate increase in contact area without change in inflation pressure (Fig. 10.3). In reality, however, there are considerable deviations from this principle, which disregards such factors as the stiffness of the tire walls, the presence of lugs or ribs, the shearing and slippage which are affected by drawbar pull, and the properties of the soil underfoot. Gill and Vanden Berg (1967) showed that the stiff walls of the tire carcass cause the pressures at the edges of the tire-to-soil contact area to be greater than those near the center (Fig. 10.4). Moreover, the longitudinal distribution of the pressures exerted during forward travel depends on the softness of the soil and the extent of sinkage. When inflation pressure is very high and the soil very soft, a pneumatic tire may behave like a rigid wheel and tire–soil contact pressures may all be below inflation pressure (Chancellor, 1976). Lugs on tires have the effect of concentrating a high pressure on a small fraction of the normal tire–soil contact area. This uneven distribution of pressures is, however, dissipated within the upper 15–25 cm of soil so that at greater depth there remains little difference between the pressures introduced by lugged and those by smooth tires. However, within the zone of influence, the kneading effect of the lugs may be a factor in soil puddling.

In the case of *rigid wheels*, the pressures transmitted to the soil depend very much on the softness or hardness of the soil itself. If the surface is firm, the

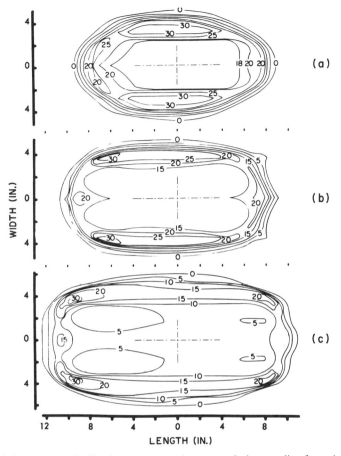

**Fig. 10.4.** Pressure distribution patterns under a smooth tire traveling from right to left on firm sand. Inflation pressures were (a) 14 lb/in.$^2$, (b) 10 lb/in.$^2$, and (c) 6 lb/in.$^2$. The centers of the patterns have pressures approximately equal to the inflation pressures, while the edge patterns show greater pressure due to tire carcass stiffness. Note that 14.7 lb/in.$^2$ ≈ 1 bar. (From Vanden Berg and Gill, 1962.)

contact area is reduced and the pressure is relatively high. If the soil is soft and easily deformable, the wheels sink into the ground so that the same load is distributed over a larger area. Hence soil pressures do not increase in proportion to the added load.

The ground pressure values for *crawler tracks* can be estimated by dividing the total weight of the vehicle by its contact area with the ground. There are, however, several factors which can cause deviations from this principle (Chancellor, 1976). During actual operation, tractors generally tilt backwards (Taylor and Vanden Berg, 1966), so that pressures on the rear side of the traveling track may be twice or thrice as great as the average pressure.

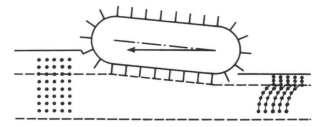

**Fig. 10.5.** Combined compaction and shearing deformation produced within the soil by a moving crawler tractor. (From Taylor and Vanden Berg, 1966.)

This tilting and pressure shift, accompanied by shearing, tend to increase as drawbar pull by the crawler tractor is increased. Figure 10.5 illustrates this effect. A similar pattern of soil shearing and displacement has been observed under moving wheels, as shown in Fig. 10.6.

Determination of the magnitudes and distributions of the pressures exerted on the soil surface is the first step in any attempt to estimate the distribution of stresses within the soil body (Reaves and Cooper, 1960; Onafenko and Reece, 1967). Procedures for making such calculations were summarized by Chancellor (1976) on the basis of foundation theory (Tschebotarioff, 1973). The governing equations in this theory are known as the *Boussinesq* equations, which, though originally intended only to describe pressure distribution in uniform elastic materials, have long been used in soil mechanics [see, for example, the detailed description given by Plummer and Dore (1940)] and have been reported (Söhne, 1958) to be applicable to agricultural soils

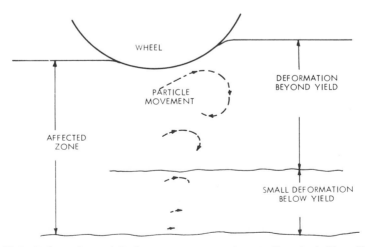

**Fig. 10.6.** Deformation and displacement patterns under a moving wheel. (From Yong and Osler, 1966.)

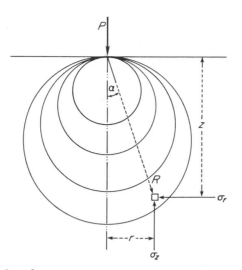

**Fig. 10.7.** Distribution of pressures under a concentrated vertical load applied to the soil surface, according to the Boussinesq theory. The circles represent lines of equal vertical stress $\sigma_z$.

also. The equations give the pressure at any point within a body of infinite extent due to a *concentrated load F* applied normal to the surface (Fig. 10.7). Disregarding the horizontal components of the stress, the equation for vertical pressure $\sigma_z$ as a function of depth $z$ and horizontal distance $r$ (from the axis of the load) is

$$\sigma_z = 3Fz^3/2\pi(r^2 + z^2)^{5/2} \qquad (10.1)$$

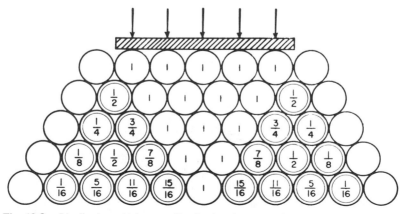

**Fig. 10.8.** Idealized vertical stress distribution in a granular material under partial-area load. (After Zelenin, 1950.)

Note that directly under load ($r = 0$) the pressure decreases as the depth squared (i.e., $\sigma_z = 3F/2\pi z^2$).

To apply this theory to predict the downward propagation of pressures applied over a *finite area* of soil surface, we can divide the finite area into small units and assume each to be subject to a concentrated load, the effect of which spreads downward at an angle of 30° from the vertical. To estimate the total pressure acting at any point on any horizontal plane below the surface, one must add up the pressure contributions due to the forces acting on each surface element, as illustrated in Figs. 10.8 and 10.9. The theory predicts that, on any such plane at any depth, the pressure is maximal directly under the center of the loaded surface area and decreases toward the spreading edges. The magnitude of this maximal pressure felt at any depth within the soil depends not only on the magnitude of the pressure at the surface but also on the width of the surface area over which it is applied. Thus, increasing the surface area subjected to a given pressure will have the effect of increasing the maximal pressures experienced at all depths within the profile under the center of the loaded area. Consequently reducing the surface pressure by distributing the load over a larger surface area will decrease maximal subsurface pressures but to a degree less than proportional to the decrease in surface pressure. A pressure distribution pattern under a rear tractor tire, computed by Söhne (1958) on the basis of the theory described, is illustrated in Fig. 10.10.

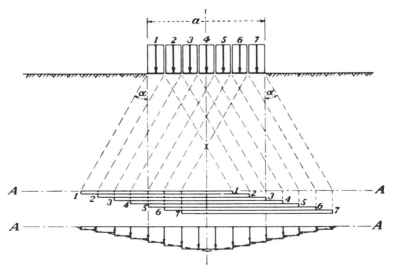

**Fig. 10.9.** Hypothetical distribution of vertical pressures within the soil, as under a crawler track, represented as a set of vertical blocks (1–7). Angle $\alpha$ is generally taken to be 30°. (From Tschebotarioff, 1973.)

Pressures are imparted to the soil not only by vehicles traveling on its surface, but also by tillage tools operating beneath the surface. As tools of various designs are thrust into and through the soil, several different effects may occur simultaneously, as the soil is cut, compressed, sheared, lifted, displaced, and mixed. Inevitably, some of the soil is pushed ahead by the moving tool against the resistance of the static soil body, and is thus compacted. Space does not permit an extended account of pressure distributions caused by different types of implements (such as moldboard plows, subsoilers, bulldozers, cultivators, rototillers, etc.), nor is the information available complete by any means. Reported pressures caused by tillage implements have been of the order of 1 bar for a bulldozer (Hettiaratchi *et al.*, 1966), 2–4 bar for plowshares and plowbottoms (Gill and McCreery, 1960; Mayanskas, 1959), and an estimated $5\frac{1}{2}$ bar for a subsoiler (Chancellor, 1976).

In the interest of simplification, we have thus far concentrated on the pressures applied by machinery to the soil. We must remember, however, that, in addition to pressures, machines quite inevitably impart shearing stresses as well. As shown by Taylor and Vanden Berg (1966), the very

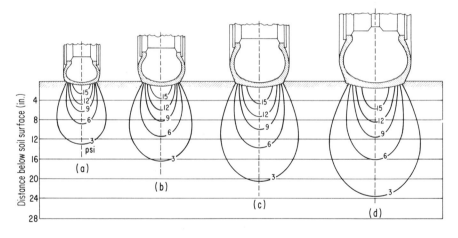

| Tire size | Load (lb) | Inflation pressure (psi) |
|---|---|---|
| (a) 7–24 | 660 | 12 |
| (b) 9–24 | 1100 | 12 |
| (c) 11–24 | 1650 | 12 |
| (d) 13–30 | 2200 | 12 |

**Fig. 10.10.** Vertical pressures under tractor tires, computed on the basis of the Boussinesq theory, assuming a concentration factor $v = 5$ and soil with normal density and water content. (From Söhne, 1958.)

mechanism by which crawler tracks and drive wheels develop the traction needed to propel a vehicle forward over a soil base involves imparting shear stresses to the soil. An attempt to analyze the distributions of both normal and shearing stresses in the soil as affected by variously shaped and operated machines can indeed be a formidable task. Suffice it so say in the present context that the simultaneous application of pressures and shearing stresses can contribute greatly to soil compaction, as we have already mentioned in a preceding section in connection with the kneading compaction of soil samples in the laboratory.

## E. Consequences of Soil Compaction

Instances of soil compaction are highly prevalent and becoming more so in modern agriculture. The trend to reduce labor on the farm by the use of larger and heavier machinery increases the likelihood and damage of untimely cultivation and the frequency of traffic over the soil. Expecially insidious is the common practice of plowing with the tractor wheel running over the bottom of the open furrow, where the soil is likely to be even more compactible than at the surface, and to greater depth, owing to higher moisture and lower organic matter content. Furthermore, compaction in depth is much more difficult to rectify and hence longer lasting than compaction at the surface.

Compaction is not limited to initial plowing (called *primary tillage*). As pointed out by Soane (1975), some 90% of the soil surface may be traversed by tractor wheels during the traditional preparation of seedbeds for close-growing crops such as cereal grains, followed by a further trampling of at least 25% during combine harvesting and as much as 60% where straw is baled and carted off. The compaction caused by all this traffic, particularly during seedbed operations, can increase bulk density to a depth of at least 30 cm and can remain throughout the life of the crop. Row crops such as cotton often sustain intensive wheel traffic, as repeated passes over the field are made for application of pesticides (Dumas *et al.*, 1973). Especially damaging is the practice of cultivating a clayey soil with heavy equipment when the soil is at a wet state and its strength is low. "Smearing" of the plow-layer bottom by plowshares creates a "plow sole" or "plow pan." The churning action by wheels, including the effects of wheel slip, can be more important than loading per se in bringing about degradation of soil structure and subsequent compaction (Davies *et al.*, 1973). Traffic associated with harvesting potatoes and sugar beets in wet conditions causes deep rutting, smearing, and compaction, which can also inhibit drainage (Swain, 1975).

As shown by Wiersum (1957), Taylor and Bruce (1968), and Cannell (1977), roots are unable to decrease in diameter to enter pores narrower than their root caps; thus, if they are to grow through compacted soil they must displace soil particles to widen the pores by exerting a pressure greater than the soil's mechanical strength. In addition to this mechanical constraint, soil compaction also impedes the movement of water and air through the soil by reducing the number of large pores. The resulting restriction of aeration and drainage thus exposes roots to several simultaneous stresses. Studies on the effect of mechanical impedance on root growth have been reported by, among others, Gill and Miller (1956), Barley (1962), Abdalla *et al.* (1969), Goss (1970), and Russell and Goss (1974). According to the latter, where barley roots must overcome externally imposed pressure of only 0.2 bar to enlarge pores, the rate of root elongation is reduced by 50%, and if the pressure in the soil is increased to just 0.5 bar, root extension is reduced by 80%. It is noteworthy, in this connection, that rigidly confined roots are capable of exerting pressures as great as 10 bar, a fact that was discovered as early as 1893 by Pfeffer [cited by Gill and Bolt (1955)] and since reconfirmed by several investigators (e.g., Stolzy and Barley, 1968; Taylor and Ratliff, 1969). This fact may explain the ability of roots to eventually penetrate compacted zones, as well as their oft-cited ability to penetrate, and enlarge, crevices in rocks, roads, and foundations. However, from the standpoint of promoting crop growth, we are generally interested in minimizing soil resistance to root elongation, rather than in mobilizing maximal root pressure.

The interactions among soil compaction, strength, moisture retention, aeration, and root growth have been studied by, among others, Taylor and Gardner (1963), Hopkins and Patrick (1969), Klute and Peters (1969), Greacen *et al.* (1969), and Kays *et al.* (1974). In many of the studies of root penetration in compact media, soil resistance has been characterized by means of penetrometers of various types. The usefulness and universality of penetrometer measurements is, however, still limited by the lack of a standard design and standardized procedures.

Where the soil is highly compact and rigid, root growth may be confined almost entirely to cracks and cleavage planes (Taylor *et al.*, 1966). In seeking out such cracks, plant roots are not merely passive agents, as they promote differential shrinkage through preferred extraction of moisture from zones of greater penetrability.

## F. Control of Soil Compaction

Control of soil compaction is a continual requirement in modern agriculture. The performance of field operations necessarily involves *some* compaction. Hence a major task of soil management is, first, to minimize soil compaction to the extent possible, and, second, to alleviate or rectify that unavoidable measure of compaction caused by traffic and tillage once it occurs.

The most obvious approach to the prevention of soil compaction is the avoidance of all but truly essential pressure-inducing operations. This calls for reducing the number of operations involved in primary and secondary tillage (plowing and subsequent cultivations, respectively), using the most efficient implement at the most appropriate time so as to effect the desirable soil condition in a single pass rather than by a repeated sequence of passes. It has long been known that excessive soil manipulation, beyond what is required to prepare a seedbed and check weeds, leads to a loss of yield and deterioration of soil structure (Keen and Russell, 1973). Indeed overly intensive cultivation can cause disastrous results, such as accelerated erosion by water and wind.

In recent years, increasing awareness of these hazards has resulted in the development of radically new field management systems, variously called "zero tillage," "minimum tillage," or "conservation tillage" (Wittmus et al., 1973, 1975; Lewis, 1973; McGregor et al., 1975). Such systems reduce the number of operations, avoid unnecessary inversion of topsoil, and generally retain crop residues as a protective mulch over the surface. In former times, the repeated and frequent cultivation of orchards and of row-cropped fields was required for weed control. This requirement was reduced to a considerable extent by the introduction of sprayable herbicides. The potential benefits of reduced tillage go beyond the avoidance of compaction and preservation of soil structure, as they include savings in time, labor, and energy, the latter having become important enough in itself. However, excessive reliance on phytotoxic chemicals can pose serious environmental problems.

Since random traffic over the field by heavy machinery is a major cause of soil compaction, cultural systems have been proposed to confine vehicular traffic to permanent, narrow lanes and to reduce the fractional area trampled by wheels to less than 10% of the land surface if possible (Dumas et al., 1973). In row crops seedbed preparation, involving the formation of a highly pulverized top layer vulnerable to structural breakdown, should be confined to the narrow strips where planting of row crops takes place rather than be carried out over the entire surface as was the practice in former times. (In too many places, "former times" persist to this day.) The interrow zone can now be left in an open, cloddy condition which promotes water and air penetration and reduces erosion by water and wind. The field is thus divided into three (if possible, permanent) zones: (1) narrow planting strips, precision tilled; (2) narrow traffic lanes, permanently compacted; and (3) interrow water-management beds, maintained in a rough and cloddy condition and covered with a mulch of plant residues.

An extremely important factor is the timing of field operations in relation to soil moisture. Moist soils can be highly vulnerable to compaction. Operations which impose high pressures should, if possible, be carried out on dry soil, which is much less compactible. These include traffic in general, and— more specifically—such tillage methods as subsoiling, which in fact may be more effective in shattering dry soils than wet ones.

Much can be done by equipment designers to distribute loads more evenly over the tracked surface rather than concentrate them under the rear wheels and to provide independent power to implements (via the power take-off or a separate engine) as, for instance, in the case of rotary tillers. Powered tillage implements can reduce dependence on the tractor's draw-bar pull, which increases tractor wheel slippage and hence soil puddling. Tractors must generally be twice as heavy as the pulling forces they are expected to generate, so that reduction of draw-bar pull requirements can permit reduction of tractor weight. This and other means (e.g., decrease of tire inflation pressure) to reduce the stresses transmitted to the soil should constitute important design criteria for the manufacture and selection of tractors and implements. A basic discussion of tillage is given in our next chapter.

On the other side of the issue, it should be pointed out that not all instances of soil compaction are necessarily harmful. The practice of adding packing wheels to seeding machines has long been known to enhance germination, by ensuring better seed-to-soil contact and increasing the unsaturated hydraulic conductivity of the otherwise excessively loose and porous seedbed (Dasberg *et al.*, 1966).

## G. Soil Consolidation

Thus far in this chapter we have confined our attention to the process of soil compaction, which, we recall, has been defined as the compression of unsaturated soil due to reduction of its air-filled pore space without change in mass wetness. As already pointed out, continued compression of an unsaturated soil will ultimately result in the practically complete expulsion of the air and, if compression is continued still, in expulsion of water as well. This is analogous to pressing a moist sponge or wringing a moist piece of cloth until water is eventually squeezed out. So, the process by which a body of soil, either initially saturated or compacted to the point of saturation, is compressed in a manner that results in reduction of pore volume by expulsion of water, is called *consolidation*. Because of the slower nature of consolidation relative to compaction (water being 50–100 times more viscous than air at ordinarily encountered temperatures), consolidation is not an immediate response of soil to transient pressures such as those caused by episodic traffic or tillage. Rather, consolidation is manifested in the gradual settlement, or subsidence, of soil under long term loading such as that due to a permanent structure (e.g., a single building, a housing development, or even an entire city). Hence the practical importance of consolidation lies not so much in agriculture as in engineering, where the amount and uniformity of foundation settlement are of vital concern to those interested in the stability and safety of structures.

To arrive at a conceptual understanding of consolidation, let us consider a sample of water-saturated soil confined in a rigid-walled cylinder and subjected to compressive load applied to a piston in such a manner as to allow no possible outlet for water. If the confined water is assumed to be incompressible, then the applied pressure can cause no compression of the soil matrix and must be borne entirely by the water phase. If we now bore a narrow hole through the piston, some of the pressured water would tend to squirt out, thus relieving the hydrostatic pressure inside the cell and gradually transmitting the load to the soil matrix, which would tend to compress at a rate commensurate with the volume of water extruded. The water contained in the voids of a saturated compressible layer of soil behaves under pressure in a manner similar to that of the water under the hypothetical piston just described. In this case, however, the water must find its way out through the exceedingly narrow and tortuous pores of the soil itself. Imagine all this taking place under an unevenly distributed load, resulting in different hydrostatic pressures at various points in the soil, and you will have an idea of the complex, space-variable, and time-variable nature of the consolidation process.

In general, granular soils such as dense sands are the least compressible soils, and such consolidation as does take place usually occurs relatively rapidly, thanks to these soils' high permeability. Clays and other fine-

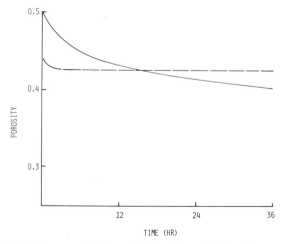

**Fig. 10.11.** Relative compressibilities and rates of consolidation for a sandy and a clayey soil (hypothetical). Solid line, clayey soil; dashed line, sandy soil.

textured soils, with higher porosity, generally have a much greater total compressibility, but the rate of their consolidation is likely to be slow, sometimes extending over a period of months or even years. This difference is illustrated in Fig. 10.11.

A classical mechanical model, shown in Fig. 10.12 has been found to be useful in illustrating the principles involved in the consolidation process. The model consists of a perforated or porous piston pressed into a water-filled cylinder fitted with springs. When the piston is first loaded and before any downward motion has yet taken place, the pressure applied is transmitted to the water only. Soon, however, water begins to ooze out of the piston, and as the piston moves down, the springs begin to contract. As the process continues, more and more of the water escapes and the springs contract to an increasing degree. If the piston is placed under a constant load, the downward movement will finally stop and static equilibrium will prevail when the resistance of the compressed springs just equals the downward force acting on the piston. At this point and henceforth, the entire load will be borne by the springs, the excess water pressure having been relieved by the outflow. Thus at the beginning of the process the water is stressed but not the springs, and at the end the situation is reversed. At any intermediate point during the process, the load is carried partly by the water and partly by the springs. The springs are quite obviously analogous to the soil matrix, and the water in the cylinder to the water permeating soil pores.

The concept described in so many words can be summed up tersely in the form of an equation, known as *Terzaghi's effective stress equation* (Terzaghi, 1953):

$$\sigma_{tot} = p + \sigma_{eff}$$

or

$$\sigma_{eff} = \sigma_{tot} - p \qquad (10.2)$$

Here $\sigma_{tot}$ is the *total stress*, $p$ is the hydrostatic pressure known as the *pore water pressure*, and $\sigma_{eff}$ is the so-called *effective stress* borne by the solid

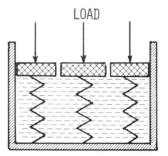

**Fig. 10.12.** Conceptual model illustrating the process of consolidation.

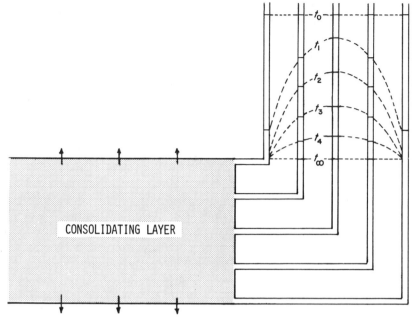

**Fig. 10.13.** A consolidating layer draining both upward and downward. The piezometers on the right indicate the isochrones at various times from the start (at $t_0$) to the eventual end (at $t_\infty$) of the process.

matrix and therefore often termed the *intergranular stress*. The term $p$ is a *neutral stress*, inasmuch as the liquid pressure in a saturated soil acts equally in all directions, hence it has no effect on compressing the matrix. When a total stress $\sigma_{tot}$ is suddenly applied to a soil and then maintained for an indefinite period, $p$ decreases and $\sigma_{eff}$ increases as shown hypothetically in Fig. 10.13.

The theory of consolidation, also due to Terzaghi, is based on several assumptions:

1. The soil is completely saturated.
2. The water and the soil particles are incompressible.
3. Water flow and hence consolidation occur in one dimension only (i.e., vertically).
4. Water movement in the consolidating layer obeys Darcy's law, and the hydraulic conductivity is constant.
5. Loading is instantaneous and the load constant, and the time lag of compression is caused entirely by the slow outflow of water from the compressing soil.
6. There is a linear relationship between the effective stress acting on the soil matrix and the volume change produced.

Since the soil volume decrement is due entirely to, and equal to, the change in fractional volume of voids $V_v$, which is, in turn, caused by the Darcian outflow of water, we can write an equation of continuity,

$$\partial V_v / \partial t = - K \, \partial^2 h / \partial z^2 \tag{10.3}$$

wherein $t$ is time, $K$ hydraulic conductivity, $h$ pressure head of soil water (actual pore pressure $P$ divided by liquid density $\rho$ and gravitational acceleration $g$: $h = P/\rho g$). The change in volume of voids can be related to the change in pore water pressure using the *modulus of volume change*, $m_v$, where

$$m_v = - \frac{\partial V_v / \partial t}{\partial P / \partial t} = - \frac{\partial V_v}{\partial P} \tag{10.4}$$

Hence

$$m_v = \frac{\partial P}{\partial t} = \frac{K}{\rho g} \frac{\partial^2 P / \partial z^2}{\partial P / \partial t}$$

or

$$\frac{\partial P}{\partial t} = \frac{K}{\rho g m_v} \frac{\partial^2 P}{\partial z^2} = C_c \frac{\partial^2 P}{\partial z^2} \tag{10.5}$$

in which $C_c \,(= K/\rho g m_v)$ is called the *coefficient of consolidation*.

To characterize soils with respect to their consolidation behavior, laboratory tests are generally carried out with small samples, preferably "undisturbed" cores, placed in an apparatus as shown in Fig. 10.14. Extraction of a sample from some depth in the field generally entails release of the original in situ, confining pressure, or overburden. The sample, however, retains what might be called a memory of its state of *preconsolidation* pressure. When reconsolidated, it will not compress significantly until its

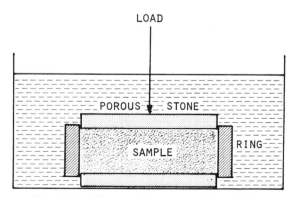

**Fig. 10.14.** The consolidation test (schematic).

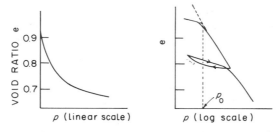

**Fig. 10.15.** Void ratio versus effective pressure curve for a compressible soil during consolidation. The semilogarithmic plot indicates the apparent preconsolidation pressure $p_0$ and a pressure-release ("rebound")–reconsolidation loop.

preconsolidation pressure is exceeded. Thereafter, its compression will tend to follow a straight line relation for void ratio versus the logarithm of pressure, at least until particle interlocking causes a significant increase of interparticle frictional resistance to further compression. Continued compression beyond this range is often called *secondary consolidation*. A technique for assessing a soil's preconsolidation pressure was developed by Casagrande and has been widely adopted (Casagrande and Fadum, 1940). The pattern of consolidation is illustrated in Fig. 10.16.

As long as the stress pattern does not exceed the yield strength of the soil, compression (volume strain) predominates over shear, and the latter can be ignored. If, however, the imposed shearing stresses exceed soil strength, shearing deformation may take place to an extent that can no longer be disregarded.

### Sample Problems

**1.** Calculate and plot the functional dependence of bulk density upon mass wetness and upon volume wetness for a saturated soil, and for 90 and 80% degrees of saturation.

To derive the appropriate relationship, we can begin with the familiar equations relating bulk density $\rho_b$ to porosity $f$ [see Eqs. (2.13) and (2.14)]:

$$f = 1 - (\rho_b/\rho_s) \tag{10.6}$$

or

$$\rho_b = \rho_s(1 - f) \tag{10.7}$$

By definition (Chapter 2) the degree of saturation $s$ equals the volume of water $V_w$ as a fraction of the volume of pores $V_f$:

$$s = V_w/V_f = \theta/f \tag{10.8}$$

where $\theta$ is volume wetness ($V_w/V_t$) and $f$ is porosity ($V_f/V_t$), $V_t$ being the total volume of a soil body.

Hence,

$$f = \theta/s \tag{10.9}$$

We incorporate the last expression in Eq. (10.7) to obtain a relationship between bulk density and volume wetness for any degree of saturation:

$$\rho_b = \rho_s[1 - (\theta/s)] \tag{10.10}$$

To obtain a relationship between bulk density and mass wetness $w$, we recall the well-known relation between volume wetness and mass wetness:

$$\theta = w\rho_b/\rho_w \tag{10.11}$$

and substitute this into Eq. (10.10):

$$\rho_b = \rho_s\left(1 - \frac{w}{s}\frac{\rho_b}{\rho_w}\right) \tag{10.12}$$

from which we can extract

$$\rho_b = \frac{\rho_w}{(\rho_w/\rho_s) + (w/s)} \tag{10.13}$$

Assuming that the density of solids $= 2.65$ gm/cm$^3$ and that of water 1 gm/cm$^3$, we can write

$$\rho_b = 1/[0.377 + (w/s)](gm/cm^3) \tag{10.14}$$

For complete saturation use $s = 1.0$ and plot $\rho_b$ against $w$. For the 90 and 80% degrees of saturation, plot $\rho_b$ against $w$ with $s$ set equal to 0.9 and 0.8, respectively. The curves should resemble those of Fig. 10.2.

**2.** A Protor compaction test was performed on medium-textured soil with the following results:

| Mass wetness $w$: | 6% | 9% | 12% | 15% | 18% | 21% |
|---|---|---|---|---|---|---|
| Bulk density $\rho_b$: | 1.80 | 1.91 | 1.94 | 1.86 | 1.75 | 1.65 gm/cm$^3$ |

Plot the data and determine the optimal moisture and maximum density values. Calculate the volume wetness and degree of saturation at optimal moisture content.

An examination of the data, confirmed by the plotted graph, shows that the optimal moisture is at or about a mass wetness $w$ value of 12%, and the maximal density $\rho_b$ is 1.94 gm/cm$^3$.

Accordingly, the volume wetness is

$$\theta = w\frac{\rho_b}{\rho_w} = 0.12 \times \frac{1.94 \text{ gm/cm}^3}{1 \text{ gm/cm}^3} = 0.23 = 23\%$$

and the degree of saturation is

$$s = \theta/f$$

where $f$, the porosity, can be obtained from the soil's bulk and particle densities:

$$f = 1 - (\rho_b/\rho_s)$$

Hence

$$s = \frac{\theta}{1 - (\rho_b/\rho_s)} = \frac{0.23}{1 - 1.94/2.65} = 0.85 = 85\%$$

3.   A concentrated load of 1000 kg [equivalent to a force of $10^4$ N (newton), or $10^9$ dyn] is applied to the soil surface. Estimate the vertical stress at depths of 5 and 10, directly under the load and at horizontal distances of 2 and 4 m.

We use the Boussinesq equation (10.1)

$$\sigma_z = \frac{3z^3F}{2\pi(r^2 + z^2)^{5/2}}$$

where $\sigma_z$ is the vertical stress, $F$ the downward force acting on the surface, $z$ the depth, and $r$ the radial distance from the axis of the vertical force. For the sake of convenience, this equation can be rearranged as

$$\sigma_z = F\frac{3z^3}{2\pi\{z^2[(r^2/z^2) + 1]\}^{5/2}} = F\frac{3z^3}{2\pi z^5[(r/z)^2 + 1]^{5/2}}$$

$$= \frac{F}{z^2}\frac{3}{2\pi[1 + (r/z)^2]^{5/2}} = \frac{F}{z^2}A$$

wherein $A$ is a composite function of $r/z$.

For $z = 5$ m and $r = 0$, $A = 0.478$. Thus,

$$\sigma_z = \frac{0.478 \times 10^9 \text{ dyn}}{(500 \text{ cm})^2} = \frac{4.78 \times 10^8 \text{ dyn}}{2.5 \times 10^5 \text{ cm}^2}$$
$$= 1.91 \times 10^3 \text{ dyn/cm}^2 = 1.91 \times 10^{-3} \text{ bar}$$

For $z = 5$ m and $r = 2$ m, $A = 0.329$.

$$\sigma_z = \frac{0.329 \times 10^9 \text{ dyn}}{(500 \text{ cm})^2} = 1.32 \times 10^3 \text{ dyn/cm}^2 = 1.32 \times 10^{-3} \text{ bar}$$

For $z = 5$ m and $r = 4$ m, $A = 0.139$.

$$\sigma_z = \frac{0.139 \times 10^9 \text{ dyn}}{(500 \text{ cm})^2} = 0.56 \times 10^3 \text{ dyn/cm}^2 = 5.6 \times 10^{-4} \text{ bar}$$

For $z = 10$ m and $r = 0$, $A = 0.478$.

$$\sigma_z = \frac{0.478 \times 10^9 \text{ dyn}}{(1000 \text{ cm})^2} = 0.478 \times 10^3 \text{ dyn/cm}^2 = 4.78 \times 10^{-4} \text{ bar}$$

For $z = 10$ m and $r = 2$ m, $A = 0.433$.

$$\sigma_z = \frac{0.433 \times 10^9 \text{ dyn}}{(1000 \text{ cm})^2} = 0.433 \times 10^3 \text{ dyn/cm}^2 = 4.33 \times 10^{-4} \text{ bar}$$

For $z = 10$ m and $r = 4$ m, $A = 0.329$.

$$\sigma_z = \frac{0.329 \times 10^9 \text{ dyn}}{(1000 \text{ cm})^2} = 0.329 \times 10^3 \text{ dyn/cm}^2 = 3.29 \times 10^{-4} \text{ bar}$$

*Note:* In these calculations, we took no account of the soil's own overburden pressure. In reality, the vertical stresses caused by the load applied at the surface are incremental to the soil's overburden pressure, which tends to increase with depth.

# 11 Tillage and Soil Structure Management

## A. Introduction

Tillage is usually defined as the mechanical manipulation of the soil aimed at improving soil conditions affecting crop production.

Three primary aims are generally attributed to tillage: control of weeds, incorporation of organic matter into the soil, and improvement of soil structure. An auxiliary function of tillage, still insufficiently well understood, is the conservation of soil moisture, where the processes of rain infiltration, runoff, and evaporation are involved (Hillel *et al.*, 1969).

Two decades ago, the advent of chemical herbicides seemed to reduce the importance of tillage as the primary method for eradication of weeds. However, more lately there have been growing objections to the application of more and more toxic chemicals in agriculture, owing to their residual damage to the larger environment and their increasing cost. Hence there is now once again increasing interest in the weed-controlling aspects of tillage. In the less developed countries, the unavailability as well as the lack of knowledge of appropriate herbicides may limit their use in any case.

The practice of inverting the topsoil in order to bury manures and crop residues has become a less important function of tillage in modern field management, where the use of animal and green manures is rather uncommon. Crop residues can, and in many cases should, be left over the surface as a *stubble mulch* to protect against evaporation and erosion. On the other hand, the use of agricultural land for disposal of waste products may once again reawaken interest in the soil-mixing aspect of tillage.

We come finally to the essential task of *soil structure management*, which has a bearing on planting, germination, water and air exchange, as well as erosion by rain and wind. Here we find that tillage practices suitable in one location may become harmful in another. Arid-zone soils with low organic matter contents and unstable aggregates are particularly vulnerable to compaction, crusting, and erosion. The precise effects of tillage on soil structure must be defined and optimized in each case if tillage is to be transformed from a hit or miss art to a scientifically based, dependable, and sustainable means of production. A basic discussion of soil compaction is given in our preceding chapter.

Tillage operations are especially consumptive of energy. The amount of earth-work involved in repeatedly loosening, pulverizing, inverting, and then recompacting the topsoil is indeed very considerable. In a typical small field of 1 h, the topsoil to a depth of only 30 cm weighs no less than 4000 tons. In an extensive farm of 1000 ha the mass of soil thus manipulated may exceed 4,000,000 tons.

In nonmechanized farming, a single tillage operation with a primitive plow requires a 40 km walk on rough land for each hectare!

The consumption of energy, as well as the wear and tear of tractors and implements, increase steeply as the depth of tillage increases. Thus, the cost of deep plowing (to a depth of 45–50 cm) is roughly double the cost of moderately deep plowing (to about 35 cm), quadruple the cost of normal plowing (about 25 cm deep) and tenfold greater than the cost of shallow plowing (approximately 15 cm deep). With the rising costs of fuel, the absolute and relative costs of tillage are almost certain to rise so that certain practices now common may become prohibitive, especially in the context of small-scale farming in the developing countries.

## B. Traditional and Modern Approaches to Tillage

The traditional and still widely accepted practice of tillage is based on a series of primary cultivations (aimed at breaking the soil mass into a loose system of clods of mixed sizes) followed by secondary cultivations (aimed at further pulverization, repacking, and smoothing of the soil surface). These practices, performed uniformly over the entire field, often involve a whole series of successive operations, each of which is necessary to correct or supplement the previous operation. In the process, energy is often wasted and natural soil structure may be destroyed (Voorhees and Hendrick, 1977).

The more modern approach to soil structure management conceives of a field typically planted to row crops as consisting of at least two distinctly different zones:

(1)   A *planting zone*, where conditions are to be optimal for sowing and conducive to rapid and complete germination and seedling establishment.

(2)   A *management zone* in the interrow areas, where soil structure is to be coarse and open, allowing maximal intake of water and air, and minimal erosion and weed infestation.

These two zones differ in function as well as in mode of preparation and management.

Ideally, the planting row should be finely pulverized and loose enough to allow germination, emergence, and early root growth. On the other hand, it should be dense enough to allow intimate contact between seed and soil so as to promote rapid supply of water and nutrients. Sprayable mulches have been developed to enhance moisture and temperature control in the seedbed, but such treatments are as yet too expensive for common practice.

The interrow management zone, which generally consists of the greater part of the surface area in the field, normally remains bare and exposed for prolonged periods. Its condition often determines the water and air economy of the growing crop, as well as the water and soil conservation of the field as a whole.

A third zone or function must be considered. Modern farming almost inevitably involves passage over the field by tractors and implements. Random traffic often causes compaction all over the field, an effect which can be minimized if travel is confined to special tracks and travel over the seedling and water-management zones is avoided (Cooper *et al.*, 1969; Dumas *et al.*, 1975; Trouse, 1978).

The three zones described can be established for a perennial cropping program. Once established, these zones can be maintained consistently over a period of years, in row-cropped fields as well as in orchards and nurseries.

Recent trends in tillage research have been aimed at minimizing tillage operations and travel (both to reduce costs and to avoid soil compaction) while tailoring each operation to its specific zone and objective (Phillips and Young, 1973). This approach, in numerous variations, underlies the methods variously termed "zero tillage," "minimum tillage," "plow–plant," "wheel-track planting," "precision tillage," "permanent bed," etc. Space will not permit a detailed elucidation of each of these methods, some of which overlap in several respects. Suffice it to say in the present context that methods developed in one location may not be suitable for another location, where conditions can differ greatly. Hence there are no universal prescriptions for what constitutes efficient tillage. Some but not all soils have suitable tilth quite naturally and require little if any tillage to serve as favorable media for crop growth. Others, however, exhibit pans or barriers which inhibit root penetration and hence can be improved by appropriate tillage (Gill, 1974).

## C. Problems of Tillage Research

The study of soil structure management has two aspects: defining the optimal soil physical state for any given purpose, and determining the most feasible means to achieve such an optimal state. The former task is difficult enough; the latter is *very* difficult. Any attempt to define soil–crop–tillage interactions in a fundamental way involves a complex array of factors relating to the mechanics of implement design and mode of operation, as well as to the dynamics of soil deformation and failure and of soil aggregation. The problem is most often approached empirically from two partial and alternative points of view, the engineering and the agronomic. The engineering approach aims at improving operational efficiency, while the agronomic approach aims directly at improving crop yields. Needed, and often lacking, is a unified approach. Part of the difficulty encountered so far has been the lack of measurable criteria of universal significance.

Various methods of tillage evaluation have been suggested by Nichols (1929), Nichols and Reed (1934), Browning (1950), Lyles and Woodruff (1961), Söhne (1956), Byers and Webber (1957), Payne (1956), Fountaine and Payne (1952), Pereira and Jones (1954), Hawkins (1959), Larson and Gill (1973), and many others. In most cases, pertinent soil physical properties are measured before and after mechanically defined tillage operations, and correlations are sought between soil and tillage parameters. To date, the most comprehensive review and analysis of tillage research is contained in the book by Gill and VandenBerg (1967).

Because of the many factors (mechanical, soil physical, climatic, and agronomic) and the complex interactions encompassed, tillage investigations must necessarily be long-term undertakings. Short-term experiments so often attempted under constraints of time, means, and personnel seldom yield conclusive results even for the location in which they are carried out. Much less are they likely to yield results of universal applicability. Though the tillage methods compared may differ widely in cost, performance, and measurable effect on the soil, final crop yields may indicate no consistent or significant differences. This is probably due in large part to faulty selections of measurable criteria, inexact measurements, and soil variability, as well as to the fact that crop response to tillage often tends to be masked by numerous unpredictable, yet decisive, variables acting in the field (e.g., limitation or excess of water due to vagaries of climate, fertility, pests, diseases). It often seems that the greatest chance for increasing the benefit-to-cost ratio in the field lies not so much in the expectation of obtaining greater yields from new and perhaps more elaborate tillage methods, but more in the possibility of maintaining the same high yields while simplifying and economizing tillage operations. Traditional field practice based on excessive manipulation of the soil (as in repeatedly and alternately loosening and recompacting it, as well as in preparing a seedbed over an entire field when such treatment is only required in the narrow planting rows) is now giving way to management systems based on minimal and precision methods of tillage (Reicosky et al., 1977).

## D. Operation of Tillage Tools

Plows (moldboard and disk types) are used for *primary tillage* to break out, crumble, and invert furrows of soil (Kepner et al., 1972). Subsoilers and chisels, also used for primary tillage, break and loosen the soil without inverting it. Disk harrows, spike harrows, sweeps, drags, cultipackers, and other implements are used to refine coarse soil conditions during *secondary tillage*. Rotary hoes, as well as various harrows, are often used for light tillage, for the control of weeds, and for disrupting soil crusts. A series of special-purpose tools, including hillers, ridgers, subsoilers, and listers, are also used to shape the soil surface zone for planting on the ridge or in the furrow, or for irrigation.

As pointed out by Söhne (1966) the moldboard plow has long been and remains the most important tillage implement for the following reasons:

(1) It is a classically simple implement, consisting of a share which cuts the soil, a sole and landside which maintain the necessary depth and side stability, and an arched plate (the moldboard) which turns, loosens, and throws the soil sideways.

(2) The shape of the moldboard can be varied for different soil and climatic conditions, and it can be adapted to different speeds (up to about 8 km/hr at present).

(3) Finally, the segmented parts of the moldboard plow, particularly the rapidly wearing parts, can be quickly and inexpensively replaced.

The actual operation of the moldboard plow can be described as follows:

A continuous slice of soil is cut by the coulter and share and is forced upward over the curved moldboard. Unless the soil is in a plastic state (i.e., if the soil is friable, as it typically is when plowed) it cannot be deformed and moved over the moldboard without developing shear planes and breaking into fragments or clods of various sizes. Nichols (1929) was the first to describe the pattern by which a furrow slice breaks into primary and secondary shearing planes. As it is lifted, turned, and finally cast off the plow, the furrow slice is accelerated so that when it finally strikes the ground it is generally further shattered. As a result, the soil is deposited as a loose and highly porous assemblage of variously sized clods. The manner in which the furrow slice is deposited and mixed is illustrated in Fig. 11.1. Universal shapes for plow bodies designed to operate at two speeds are illustrated in Fig. 11.2. Note that, because of the greater acceleration imparted to the soil, high-speed plowing generally encounters greater resistance and requires more power than low-speed plowing.

Disk plows and disc harrows consist of sharp-edged convex discs which roll obliquely to the direction of movement, thus cutting the soil in a manner similar to that of a moldboard plow. Each disk plow is mounted on its own axis and can be inclined at an angle of up to 20°, whereas all disks of a disk harrow assembly are mounted on the same axis. Vertical force is needed to cause the disks to penetrate into the soil, and at times high loads are neces-sary. Despite the reduction of friction at the rolling disc, it has been found in practice that drawbar pull (and hence energy consumption) for disk plows is at least as great as for moldboard plows of comparable size. Disk plows are particularly suitable for hard and abrasive soils, but are not widely used in most agricultural areas. Disk harrows, however, have won wide acceptance and are especially useful in the tillage of stubble fields.

As mentioned in Section C, energy losses are sustained whenever engine power is transmitted to the tillage tool by tractor wheels because of the rolling resistance and slip occurring between wheels and soil. According to Söhne (1966), the efficiency of transmission between tires and soil averages about 60–65% and in unfavorable conditions (soft soil) may fall below 40%. How-ever, attempts to transmit power directly to rotating tillage implements (e.g., using the tractor's power take-off or by means of auxiliary engines) have met with only partial success. Figure 11.3 shows several types of rotating

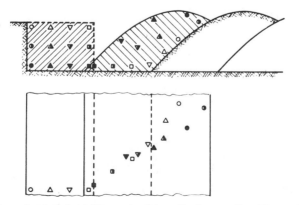

**Fig. 11.1.** The pattern of deposition and mixing of a furrow slice. The upper part of the figure shows a forward view, and the lower part gives a top view. (After Söhne, 1966.)

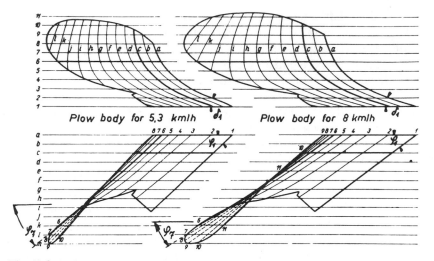

**Fig. 11.2.** Universal shapes of moldboard plows for low- and high-speed plowing. (After Söhne, 1966.)

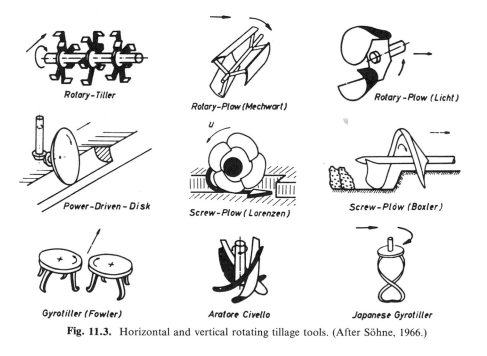

**Fig. 11.3.** Horizontal and vertical rotating tillage tools. (After Söhne, 1966.)

implements, none of which has so far even begun to replace the moldboard plow as a universal implement. Rotary tillers have been found useful on a limited scale in horticultural practice. The thorough pulverization of the soil by such implements can be advantageous, as it allows a seedbed to be prepared in a single operation. As often as not, however, it is disadvantageous, since a rotary-tilled soil can lose its structure if the aggregates are not very stable. Moreover, because of the high cutting velocity and the resulting high acceleration imparted to soil fragments, and because of the much larger cutting surface compared to moldboard plows, rotary tillers require on light-textured soil about 2.5 times and on heavy-textured soils about 3.5 times the power required by a plow for similar width and depth of tillage.

# Part VI    THE FIELD–WATER CYCLE AND ITS MANAGEMENT

The movement of water in the field can be characterized as a continuous, cyclic, repetitive sequence of processes, without beginning or end. However, we can conceive of the cycle as if it begins with the entry of water into the soil by the process of infiltration, continues with the temporary storage of water in the soil, and ends with its removal from the soil by drainage, evaporation, or plant uptake. Several fairly distinct stages of the cycle can be recognized, and, although these stages are interdependent and may at times be simultaneous, we shall attempt, for the sake of clarity, to describe them separately in the following several chapters.

# 12    *Infiltration and Surface Runoff*

## A. Introduction

When water is supplied to the soil surface, whether by precipitation or irrigation, some of the arriving water penetrates the surface and is absorbed into the soil, while some may fail to penetrate but instead accrue at the surface or flow over it. The water which does penetrate is itself later partitioned between that amount which returns to the atmosphere by evapotranspiration and that which seeps downward, with some of the latter reemerging as streamflow while the remainder recharges the groundwater reservoir.

*Infiltration* is the term applied to the process of water entry into the soil, generally by downward flow through all or part of the soil surface.[1] The rate of this process, relative to the rate of water supply, determines how much water will enter the root zone, and how much, if any, will run off. Hence the rate of infiltration affects not only the water economy of plant communities, but also the amount of surface runoff and its attendant danger of soil erosion. Where the rate of infiltration is restricted, plants may be denied sufficient moisture while the amount of erosion increases. Knowledge of the infiltration process as it is affected by the soil's properties and transient conditions, and by the mode of water supply, is therefore a prerequisite for efficient soil and water management.

---

[1] Water may enter the soil through the entire surface uniformly, as under ponding or rain, or it may enter the soil through furrows or crevices. It may also move up into the soil from a source below (e.g., a high water table).

Comprehensive reviews of the principles governing the infiltration process have been published by Philip (1969a) and by Swartzendruber and Hillel (1973).

## B. "Infiltration Capacity" or Infiltrability

If we sprinkle water over the soil surface at a steadily increasing rate, sooner or later the rising supply rate will exceed the soil's limited rate of absorption, and the excess will accrue over the soil surface or run off it (Fig. 12.1). The *infiltration rate* is defined as the volume flux of water flowing into the profile per unit of soil surface area. This flux, with units of velocity, has also been referred to as "infiltration velocity." For the special condition wherein the rainfall rate exceeds the ability of the soil to absorb water, infiltration proceeds at a maximal rate, which Horton (1940) called the soil's "infiltration capacity." This term was not an apt choice, as pointed out by Richards (1952), since it implies an *extensive* aspect (e.g., one speaks of the *capacity* of a reservoir, when referring to its *total volume*) rather than an *intensive* aspect (e.g., a flow rate in terms of volume per units of area and time), as is more appropriate for a flux. Richards then proposed "infiltration rate" instead of "infiltration capacity," with "infiltration velocity" instead of "infiltration rate," but this suggestion has not been widely adopted.

More recently, Hillel (1971) has coined the term *infiltrability* to designate the infiltration flux resulting when water at *atmospheric pressure* is made *freely available* at the soil surface. This single-word replacement avoids the extensity–intensity contradiction in the term infiltration capacity and allows the use of the term infiltration rate in the ordinary literal sense to represent the surface flux under any set of circumstances, whatever the rate or pressure at which the water is supplied to the soil. For example, the infiltration rate can be expected to exceed infiltrability whenever water is ponded over the soil to a depth sufficient to cause the pressure at the surface to be significantly

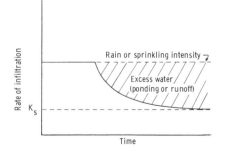

**Fig. 12.1.** Dependence of the infiltration rate upon time, under an irrigation of constant intensity lower than the initial value, but higher than the final value, of soil infiltrability.

greater than atmospheric pressure. On the other hand, if water is applied slowly or at a subatmospheric pressure, the infiltration rate may well be smaller than the infiltrability. In other words, as long as the rate of water delivery to the surface is smaller than the soil's infiltrability, water infiltrates as fast as it arrives and the supply rate determines the infiltration rate, i.e., the process is *supply controlled* (or *flux controlled*). However, once the delivery rate exceeds the soil's infiltrability, it is the latter which determines the actual infiltration rate, and thus the process becomes *surface controlled* or *profile controlled*. Horton, considered by many to be the father of modern physical hydrology, hypothesized that the soil surface zone determines what he called the infiltration-capacity, more or less apart from conditions within the soil profile. On the other hand, Childs (1969) regarded the infiltration process as a consequence of both the hydraulic conductivity and the hydraulic gradient prevailing in the soil's surface zone, allowing for the possibility that the gradient might be affected by the conditions existing throughout the profile.

If a shallow layer of water is instantaneously applied, and thereafter maintained, over the surface of an initially unsaturated soil, the full measure of soil infiltrability comes into play from the start. Many trials of infiltration under shallow ponding have shown infiltrability to vary, and generally to decrease, in time. Thus, the cumulative infiltration, being the time integral of the infiltration rate, has a curvilinear time dependence, with a gradually decreasing slope (Fig. 12.2).

Soil infiltrability and its variation with time are known to depend upon the initial wetness and suction, as well as on the texture, structure, and uniformity (or layering sequence) of the profile. In general, soil infiltrability is high in the early stages of infiltration, particularly where the soil is initially quite dry, but tends to decrease monotonically and eventually to approach asymptotically a constant rate, which is often termed the *final infiltration capacity*[2] but which we prefer to call the *steady-state infiltrability*.

The decrease of infiltrability from an initially high rate can in some cases result from gradual deterioration of the soil structure and the consequent partial sealing of the profile by the formation of a dense surface crust, from the detachment and migration of pore-blocking particles, from swelling of clay, or from entrapment of air bubbles or the bulk compression of soil air if it is prevented from escaping as it is displaced by incoming water. Primarily, however, the decrease in infiltrability results from the inevitable decrease in the *matric suction gradient* (constituting one of the forces drawing water into the soil) which occurs as infiltration proceeds.

---

[2] The adjective "final" in this context does not signify the end of the process (since infiltration can persist practically indefinitely if profile conditions permit), but it does indicate that soil infiltrability has finally attained a constant value from which it appears to decrease no more.

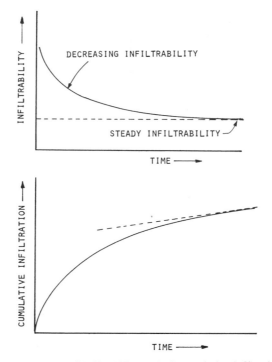

**Fig. 12.2.** Time dependence of infiltrability and of cumulative infiltration under shallow ponding.

If the surface of an initially dry soil is suddenly saturated, as, for instance, if the surface is ponded, the matric suction gradient acting in the surface layer is at first very steep. As the wetted zone deepens, however, this gradient is reduced, and, as the wetted part of the profile becomes thicker and thicker, the suction gradient tends eventually to become vanishingly small. In a horizontal column, the infiltration rate eventually tends to zero. However, in downward flow into a vertical column under continuous ponding, the infiltration rate (equal to the infiltrability) can be expected to settle down to a steady gravity-induced, rate which, as we shall later show, is practically equal to the saturated hydraulic conductivity if the profile is homogeneous and structurally stable (Fig. 12.2). In summary, soil infiltrability depends on the following factors:

(1)   Time from the onset of rain or irrigation. The infiltration rate is apt to be relatively high at first, then to decrease, and eventually to approach a constant rate that is characteristic for the soil profile.

(2)   Initial water content. The wetter the soil is initially, the lower will

be the initial infiltrability (owing to smaller suction gradients) and the quicker will be the attainment of the final (constant) rate, which itself is generally independent of the initial water content.

(3) Hydraulic conductivity. The higher the saturated hydraulic conductivity of the soil is, the higher its infiltrability tends to be.

(4) Soil surface conditions. When the soil surface is highly porous and of "open" structure the initial infiltrability is greater than that of a uniform soil, but the final infiltrability remains unchanged, as it is limited by the lower conductivity of the transmission zone beneath. On the other hand, when the soil surface is compacted and the profile covered by a surface crust of lower conductivity the infiltration rate is lower than that of the uncrusted (uniform) soil. The surface crust acts as an hydraulic barrier, or bottleneck, impeding infiltration. This effect, which becomes more pronounced with a thicker and denser crust, reduces both the initial infiltrability and the eventually attained steady infiltrability. A soil of unstable structure tends to form such a crust during infiltration, especially as the result of the slaking action of beating raindrops. In such a soil, a plant cover or a surface mulch of plant residues can serve to intercept and break the impact of the raindrops and thus help to prevent surface sealing.

(5) The presence of impeding layers inside the profile. Layers which differ in texture or structure from the overlying soil may retard water movement during infiltration. Perhaps surprisingly, clay layers and sand layers can have a similar effect, although for opposite reasons. The clay layer impedes flow owing to its lower *saturated* conductivity, while a sand layer retards the wetting front (where unsaturated conditions prevail) owing to the lower *unsaturated* conductivity of the sand at equal matric suction. Flow into a dry sand layer can take place only after the pressure head has built up sufficiently for water to move into and fill the large pores of the sand.

Methods to measure soil infiltrability were reviewed by Bertrand (1965). These methods include flooding, applying artificial rainfall, and analysis of watershed hydrographs.

## C. Profile Moisture Distribution during Infiltration

If we examine a homogeneous profile at any moment during infiltration under ponding, we shall find that the surface of the soil is saturated, perhaps to a depth of several millimeters or centimeters, and that beneath this zone of complete saturation is a less than saturated, lengthening zone of apparently uniform wetness, known as the *transmission zone*. Beyond this zone there is a *wetting zone*, in which soil wetness decreases with depth at a steepening

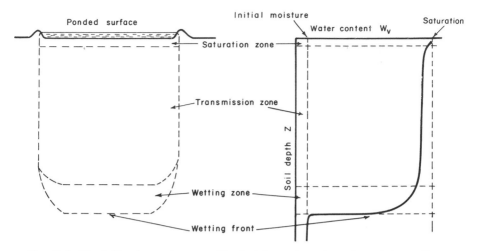

**Fig. 12.3.** The infiltration moisture profile. At left, a schematic section of the profile; at right, the water content versus depth curve. The common occurrence of a saturation zone as distinct from the transmission zone may result from the structural instability of the surface zone soil.

gradient down to a *wetting front*, where the moisture gradient is so steep[3] that there appears to be a sharp boundary between the moistened soil above and the relatively dry soil beneath.

The typical moisture profile during infiltration, first described by Bodman and Coleman (1944) by and Coleman and Bodman (1945), is shown schematically in Fig. 12.3. Later investigations have cast some doubt as to whether a saturation zone distinct from the transmission zone necessarily exists, or whether it is merely an experimental artifact or anomaly resulting from the looseness, structural instability, slaking, or swelling of the soil at the surface. The saturation zone might also result from the air-entry value, or from air entrapment. The surface soil, being unconfined and subject to the disruptive and slaking action of raindrops and turbulent water, often experiences aggregate breakdown and colloidal dispersion, resulting in the formation of an impeding crust, which, in turn, affects the moisture profile below.

However, if we continue periodically to examine the moisture profile of a stable soil during infiltration, we shall find that the nearly saturated transmission zone lengthens (deepens) continuously, and that the wetting zone and wetting front move downward continuously (though at a diminishing

---

[3] The reason for the steepening gradient is that as the water content decreases the hydraulic conductivity generally decreases exponentially. Since the flux is the product of the gradient and the conductivity, it follows that to get a certain flux moving in the soil the gradient must increase as the conductivity decreases.

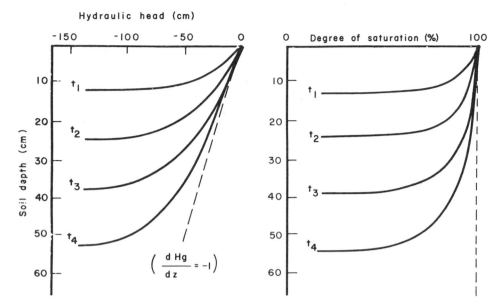

**Fig. 12.4.** Water-content profiles (at right) and hydraulic-head profiles (at left) at different times ($t_1$, $t_2$, $t_3$, $t_4$) during infiltration into a uniform soil ponded at the surface. $dH_g/dZ$ is the gravitational head gradient. In this figure the possible existence of a saturation zone distinct from the transmission zone is disregarded.

rate), with the latter becoming less steep as it moves deeper into the profile. Typical families of successive moisture and hydraulic-head profiles are shown in Fig. 12.4.

With the foregoing qualitative description of infiltration as a background, we can now proceed to consider some of the quantitative aspects of the process as it occurs under various conditions.

### D.  Infiltrability Equations

Numerous formulations, some entirely empirical and others theoretically based, have been proposed over the years in repeated attempts to express infiltrability as a function of time or of the total quantity of water infiltrated into the soil. Following the comprehensive critique of these equations given by Swartzendruber and Hillel (1973), we shall now present some of the more widely applied equations in their historical order of appearance. We use the symbol $I$ to represent the cumulative volume of water infiltrated in time $t$ per unit area of soil surface and the symbol $i$ for the infiltrability as a volume flux (i.e., the volume of water entering a unit soil surface area per unit time). Thus $i = dI/dt$ and $I = \int_0^t i \, dt$.

The earliest equation was introduced by Green and Ampt (1911),

$$i = i_c + b/I \qquad (12.1)$$

Here $i_c$ and $b$ are the characterizing constants, with $i_c$ being the asymptotic steady infiltration flux reached when $t$ (and hence $I$) become large. Not that at $t = 0$ $I$ is also zero, so that Eq. (12.1) predicts $i$ to be infinite initially and then to decrease gradually to its eventual value $i_c$. A more detailed elucidation of the Green and Ampt approach will be given in the next section.

The next equation is that of Kostiakov (1932),

$$i = Bt^{-n} \qquad (12.2)$$

where $B$ and $n$ are constants. This strictly empirical formulation also provides an infinite initial $i$ but implies that $i$ approaches zero as $t$ increases, rather than a constant nonzero $i_c$. This could have relevance for purely horizontal water absorption (in the absence of a gravity gradient) but is clearly deficient for downward infiltration.

The third equation is due to Horton (1940),

$$i = i_c + (i_0 - i_c)e^{-kt} \qquad (12.3)$$

where $i_c$, $i_0$, and $k$ are the characterizing constants. At $t = 0$ the infiltrability is not infinite but takes on the finite value $i_0$. The constant $k$ determines how quickly $i$ will decrease from $i_0$ to $i_c$. This form is also integrable and provides $I$ as an explicit function of $t$. However, this equation is cumbersome in practice, since it contains three constants which must be evaluated experimentally.

The fourth equation is that of Philip (1957c),

$$i = i_c + s/2t^{1/2} \qquad (12.4)$$

where $i_c$ and $s$ are the characterizing constants. This equation is a truncated form of the series Eq. (12.9) to be presented in Section E. Once again, the infiltrability is represented as infinite at zero time, and, as in Eq. (12.1), only two constants are required. Equation (12.4) will integrate to provide either $I$ as an explicit function of $t$ or $t$ as an explicit function of $I$.

The fifth equation was proposed by Holtan (1961):

$$i = i_c + a(M - I)^n \qquad (12.5)$$

where $i_c$, $a$, $M$, and $n$ are constants. Holtan further specified $M$ as the water-storage capacity of the soil above the first impeding stratum (total porosity minus the antecedent soil water, expressed in units of equivalent depth), but the meaning of $M$ for a soil without an impeding stratum was not made clear.

Furthermore, what is usually not stated explicitly about Eq. (12.5) is that it can only be construed as holding for the range $0 \leq I \leq M$, since $i = i_c$ can only occur at the single point $I = M$. When $I$ exceeds $M$, then the quantity $(M - I)^n$ becomes either positive and increasing, negative and decreasing, or imaginary, depending upon whether $n$ is even, odd, or fractional, respectively. Thus, in addition to needing the condition $0 \leq I \leq M$ on Eq. (12.5), to be complete one must also state

$$i = i_c \qquad \text{for} \quad I > M \tag{12.6}$$

since there is no reason whatever to suppose that infiltration should cease once $I = M$. Hence, instead of a single equation good for all $I > 0$, the Holtan expression must be recognized as the two-form mathematical specification represented by Eq. (12.5) and (12.6). Equation (12.6) is integrable, but only provides $t$ as an explicit function of $I$ and not vice versa.

The Green–Ampt and Philip equations both arise out of mathematical solutions to well-defined physically based theories of infiltration. The Horton and Holtan equations, on the other hand, are essentially empirical expressions selected to have the correct qualitative shape. For this reason, therefore, they are not quite so inherently restrictive as to the mode of water application, since they do not imply surface ponding from time zero on, as do the Green–Ampt and Philip equations. In fact, for a flux-controlled type of beginning infiltration, the finite initial infiltration rate of the Horton and the Holtan equations is obviously the more realistic. The larger number of characterizing constants in these two equations, however, can at times hinder their usefulness, while at other times it might help to provide a better description of the phenomenon. Both of these aspects have been recognized and demonstrated (Skaggs *et al.*, 1969).

## E. Basic Infiltration Theory

We begin by combining the Darcy equation with the continuity equation to obtain the general flow equation for water in soil. For one-dimensional flow the appropriate form of the flow equation is

$$\frac{\partial \theta}{\partial t} = \frac{\partial}{\partial x} \left( K \frac{\partial H}{\partial x} \right) \tag{12.7}$$

where $\theta$ is the volumetric wetness, $t$ is time, $x$ is distance along the direction of flow, and $H$ is the hydraulic head. The simplest application of this equation is in the description of the infiltration (termed absorption, Philip, 1969a) of water in the horizontal direction. In this case, the gravity force is zero, and water is drawn into the soil by matric suction gradients only. If the soil is homogeneous, the diffusivity equation can be applied directly:

$$\frac{\partial \theta}{\partial t} = \frac{\partial}{\partial x}\left[ D(\theta)\frac{\partial \theta}{\partial x}\right] \tag{12.8}$$

Downward infiltration into an initially unsaturated soil generally occurs under the combined influence of suction and gravity gradients. As the water penetrates deeper and the wetted part of the profile lengthens, the average suction gradient decreases, since the overall difference in pressure head (between the saturated soil surface and the unwetted soil inside the profile) divides itself along an ever-increasing distance. This trend continues until eventually the suction gradient in the upper part of the profile becomes negligible, leaving the constant gravitational gradient in effect as the only remaining force moving water downward. Since the gravitational head gradient has the value of unity (the gravitational head decreasing at the rate of 1 cm with each centimeter of vertical depth below the surface), it follows that the flux tends to approach the hydraulic conductivity as a limiting value. In a uniform soil (without crust) under prolonged ponding, the water content of the wetted zone approaches saturation.[4]

Darcy's equation for vertical flow is

$$q = -K\frac{dH}{dz} = -K\frac{d}{dz}(H_{\mathrm{p}} - z) \tag{12.9}$$

where $q$ is the flux, $H$ the total hydraulic head,[6] $H_{\mathrm{p}}$ the pressure head, $z$ the vertical distance from the soil surface downward (i.e., the depth), and $K$ the hydraulic conductivity. At the soil surface, $q = i$, the infiltration rate. In an unsaturated soil, $H_{\mathrm{p}}$ is negative and can be expressed as a suction head $\psi$. Hence,

$$q = K\frac{d\psi}{dz} + K \tag{12.10}$$

Combining these formulations of Darcy's equation, (12.9) and (12.10) with the continuity equation $\partial \theta / \partial t = -\partial q / \partial z$ gives the general flow equation

$$\frac{\partial \theta}{\partial t} = \frac{\partial}{\partial z}\left( K\frac{\partial H}{\partial z}\right) = -\frac{\partial}{\partial z}\left( K\frac{\partial \psi}{\partial z}\right) - \frac{\partial K}{\partial z} \tag{12.11}$$

If soil wetness $\theta$ and suction head $\psi$ are uniquely related, then the left-

---

[4] In practice, because of air entrapment, the soil-water content may not attain total saturation but some maximal value lower than saturation which has been called "satiation." Total saturation is assured only when a soil sample is wetted under vacuum.

hand side of Eq. (12.11) can be written $\partial\theta / \partial t = (\partial\theta / \partial\psi)(\partial\psi / \partial t)$, which transforms Eq. (12.11) into

$$C \frac{\partial\psi}{\partial t} = \frac{\partial}{\partial z}\left(K \frac{\partial\psi}{\partial z}\right) + \frac{\partial K}{\partial z} \qquad (12.12)$$

where $C\,(= -\partial\theta/\partial\psi)$ is defined as the *specific* (or differential) *water capacity* (i.e., the change in water content in a unit volume of soil per unit change in matric potential).

Alternatively, we can transform the right-hand side of Eq. (12.11) once again using the chain rule to render $-\partial\psi/\partial z = -(\partial\psi/\partial\theta)(\partial\theta/\partial z) = (1/C)(\partial\theta/\partial z)$. *We thus obtain*

$$\frac{\partial\theta}{\partial t} = \frac{\partial}{\partial z}\left(\frac{K}{C} \frac{\partial\theta}{\partial z}\right) - \frac{\partial K}{\partial z} \qquad \text{or} \qquad \frac{\partial\theta}{\partial t} = \frac{\partial}{\partial z}\left(D \frac{\partial\theta}{\partial z}\right) - \frac{\partial K}{\partial z} \quad (12.13)$$

where $D$, once again, is the soil-moisture diffusivity, which we propose calling the *hydraulic diffusivity* (see Chapter 6). Equations (12.11), (12.12), and (12.13) can all be considered as forms of the *Richards equation* (Swartzendruber, 1969).

Note that the above three equations contain two terms on their right-hand sides, the first term expressing the contribution of the suction, (or wetness) gradient, and the second term expressing the contribution of gravity. Whether the one or the other term predominates depends on the initial and boundary conditions and on the stage of the process considered. For instance, when infiltration takes place into an initially dry soil, the suction gradients at first can be much greater than the gravitational gradient, and the initial infiltration rate into a horizontal column tends to approximate the infiltration rate into a vertical column. Water from a furrow will therfore tend at first to infiltrate laterally almost to the same extent as vertically (Fig. 12.5). On the other hand, when infiltration takes place into an initially wet soil, the suction gradients are small from the start and become negligible much sooner (Fig. 12.6).

The first mathematically rigorous solution of the flow equation applied to vertical infiltration was given by Philip (1957a). A more recent comprehensive review of infiltration theory was provided by the same author (Philip, 1969a). His original solution pertained to the case of an infinitely deep uniform soil of constant initial wetness $\theta_i$, assumed at time zero to become submerged under a thin layer of water that instantaneously increases soil wetness at the surface from its initial value to a new value $\theta_0$ (near saturation) that is thereafter maintained constant. Mathematically, these conditions are stated as

$$
\begin{array}{lll}
t = 0, & z > 0, & \theta = \theta_i \\
t \geq 0, & z = 0, & \theta = \theta_0
\end{array}
\qquad (12.14)
$$

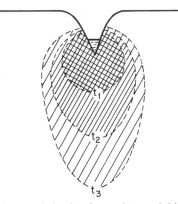

**Fig. 12.5.** Infiltration from an irrigation furrow into an initially dry soil. The wetting front is shown after different periods of time ($t_1 < t_2 < t_3$). At first the strong suction gradients cause infiltration to be nearly uniform in all directions; eventually the suction gradients decrease and the gravitational gradient predominates.

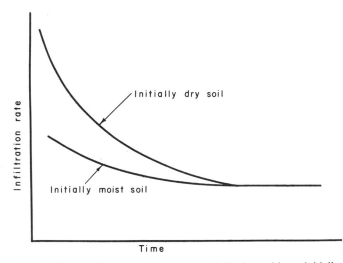

**Fig. 12.6.** Infiltrability as a function of time in an initially dry and in an initially moist soil.

His solution took the form of a power series:

$$z(\theta,t) = \sum_{n=1}^{\infty} f_n(\theta) t^{n/2}$$
$$= f_1(\theta) t^{1/2} + f_2(\theta) t + f_3(\theta) t^{3/2} + f_4(\theta) t^2 + \cdots \quad (12.15)$$

where $z$ is the depth to any particular value of wetness $\theta$ and the coefficients $f_n(\theta)$ are calculated successively from the diffusivity and conductivity functions. This solution indicates that at small times the advance of any $\theta$ value proceeds as $\sqrt{t}$ (just as in horizontal infiltration), while at larger times the downward advance of soil wetness approaches a constant rate $(K_0 - K_i)/(\theta_0 - \theta_i)$, where $K_0$ and $K_i$ are the conductivities at the wetness values of $\theta_0$ (wetted surface) and $\theta_i$ (initial soil wetness), respectively.

Philip's solution also describes the time dependence of cumulative infiltration $I$ in terms of a power series:

$$I(t) = \sum_{n=1}^{\infty} j_n(\theta)t^{n/2}$$

$$= st^{1/2} + (A_2 + K_0)t + A_3 t^{3/2} + A_4 t^2 + \cdots + A_n t^{n/2} \quad (12.16)$$

in which the coefficients $j_n(\theta)$ are, again, calculated from $K(\theta)$ and $D(\theta)$, and the coefficient $s$ is called the *sorptivity*. Differentiating Eq. (12.16) with respect to $t$ we obtain the series for the infiltration rate $i(t)$:

$$i(t) = \tfrac{1}{2}st^{-1/2} + (A_2 + K_0)$$
$$+ \tfrac{3}{2}A_3 t^{1/2} + 2A_4 t + \cdots + \tfrac{n}{2}A_n t^{n/2 - 1} \quad (12.17)$$

In practice, it is generally sufficient for an approximate description of infiltration to replace Eq. (12.16) and (12.17) by two-parameter equations of the type

$$I(t) = st^{1/2} + At, \qquad i(t) = \tfrac{1}{2}st^{-1/2} + A \quad (12.18)$$

where $t$ is not too large. In the limit as $t$ approaches infinity, the infiltration rate decreases monotonically to its final asymptotic value $i(\infty)$. Philip (1969a) pointed out that this does not imply that $A = K_0$, particularly not at small or intermediate times. However at *large times* (for which the infinite series does not converge) it is possible to represent Eq. (12.18) as

$$I = st^{1/2} + Kt, \qquad i = \tfrac{1}{2}st^{-1/2} + K \quad (12.19)$$

where $K$ is the hydraulic conductivity of the soil's upper layer (the transmission zone), which in a uniform soil under ponding is nearly equal to the saturated conductivity $K_s$. For very large values of time Philip's analysis yields

$$i = K \quad (12.20)$$

This theoretical analysis was used to interpret the earlier experimental findings of Bodman and Coleman (1944).

Recall that the sorptivity has already been defined (according to Philip, 1969a) in terms of the horizontal infiltration equation:

$$s = I/t^{1/2} \quad (12.21)$$

As such, it embodies in a single parameter the influence of the matric suction and conductivity on the transient flow process that follows a step-function change in surface wetness or suction. Strictly speaking, one should write $s(\theta_0, \theta_i)$ or $s(\psi_0, \psi_i)$, since $s$ has meaning only in relation to an initial state of the medium and an imposed boundary condition. Philip also defined an "intrinsic sorptivity"—a parameter which takes into account the viscosity and surface tension of the fluid.

It should be obvious from the foregoing that the effects of ponding depth and initial wetness (Fig. 12.6) can be significant during early stages of infiltration, but decrease in time and eventually tend to vanish in a very deeply wetted profile. Typical values of the "final" (steady) infiltration rate are shown in Table 12.1. These values merely give an order of magnitude, while in actual cases the infiltration rate can be considerably higher (particularly during the initial stages of the process and in well-aggregated or cracked soils) or lower (as in the presence of a surface crust).

**Table 12.1**

| Soil type | Steady infiltration rate (mm/hr) |
|---|---|
| Sands | $>20$ |
| Sandy and silty soils | $10-20$ |
| Loams | $5-10$ |
| Clayey soils | $1-5$ |
| Sodic clayey soils | $<1$ |

## F. Infiltration into Layered Profiles

Notwithstanding the instructive value of the solutions described thus far to the flow equation, pertaining to infiltration into uniform profiles, it is obvious that in nature infiltration seldom, if ever, takes place under such ideal conditions. More typically, infiltration processes occur in soils which are uniform neither in texture nor in initial wetness. The solution of the flow equation for such conditions is apparently impossible by analytical means, and hence progress in the development of more realistic theories of infiltration was frustrated for a long time. Experimental studies, however, have shown that the process of infiltration can be greatly affected by soil profile heterogeneity. For example, Miller and Gardner (1962), who conducted experiments on the effects of thin layers of different texture sandwiched into otherwise uniform profiles, showed that although in any conducting soil the matric suction and hydraulic head must be continuous throughout the profile regardless of layering sequence, the wetness and conductivity may exhibit abrupt discontinuities at interlayer boundaries. The steady-state downflow of water through a two-layer profile had earlier been analyzed by Takagi (1960), who showed that where the upper layer is less pervious than the lower negative pressures (suctions) develop in the lower layer, and these can remain constant throughout a considerable depth range.

The advent of high-speed computers and of computer-based numerical methods for solving nonlinear differential equations subject to complex boundary conditions made possible a new approach to infiltration theory. No longer is the theory shackled to the restrictive assumptions which had been necessary in order to formulate the process mathematically in closed form, as required for the attainment of analytical solutions. One by one these restrictive conditions have been relaxed as increasingly sophisticated numerical methods were developed so that gradually more complex, time-variable, space-variable, and multidimensional processes could be handled.

An early demonstration that the computer can be used effectively to obtain a mathematical description of infiltration in heterogeneous soil profiles with strata of differing properties and initial wetness values was given by Hanks and Bowers (1962). In their model the initial wetness $\theta_i$ was allowed to vary with depth $z$, and different $K(\psi)$ and $\theta(\psi)$ functions were assigned to soil layers of different thicknesses. Later Wang and Lakshminarayana (1968) modified this formulation to make more explicit the expression of heterogeneity, i.e., the dependence of $K$ and $\psi$ on $z$ and $\theta$ in the form $K = K(\theta, z)$ and $\psi = \psi(\theta, z)$. Both models were still based on the surface-ponding boundary condition. More recent investigations have provided even more comprehensive descriptions of infiltration into variously constituted soil profiles (e.g., Hillel and Talpaz, 1977; Hillel, 1977). Space limitations preclude a detailed presentation of the numerous models now available. Hence the following discussion will provide only a qualitative description of some of the features characterizing infiltration into layered profiles.

One typical situation is that of a coarse layer of higher saturated hydraulic conductivity overlying a less conductive finer-textured layer. In such a case the infiltration rate is at first controlled by the coarse layer, but when the wetting front reaches and penetrates into the finer-textured layer, the infiltration rate can be expected to drop and tend to that of the finer soil alone. Thus, in the long run, it is the layer of lesser conductivity which controls the process. If infiltration continues for long, then positive pressure heads (a "perched water table") can develop in the coarse soil, just above its boundary with the impeding finer layer.

In the opposite case of infiltration into a profile with a fine-textured layer over a coarse-textured one, the initial infiltration rate is again determined by the upper layer. As water reaches the interface with the coarse lower layer, however, the infiltration rate may decrease. Water at the wetting front is normally under suction, and this suction may be too high to permit entry into the relatively large pores of the coarse layer. This explains the observation (Miller and Gardner, 1962) that the wetting-front advance stops for a time (though infiltration at the surface does not stop) until the pressure head at the interface builds up sufficiently for water to penetrate into the coarse material. Thus, under unsaturated flow conditions a layer of sand or gravel in a medium- or fine-textured soil, instead of enhancing water movement in the profile, may actually impede it. The lower layer, in any case, cannot become saturated, since the restricted rate of flow through the less permeable upper layer cannot sustain flow at the saturated hydraulic conductivity of the coarse lower layer (except when the externally applied pressure, i.e., the ponding depth, is large or where both the layers are under the water table).

## G. Infiltration into Crust-Topped Soils

A very important special case of a layered soil is that of a profile which develops a thin crust, or seal, at the surface. Such a seal can develop under the beating action of raindrops (Ekern, 1950; McIntyre, 1958; Tackett and Pearson, 1965), or as a result of the spontaneous slaking and breakdown of soil aggregates during wetting (Hillel, 1960).

The action of a raindrop striking at an exposed soil surface is believed to be related to its kinetic energy $E_k$ :

$$E_k = \tfrac{1}{2}mv^2 \tag{12.22a}$$

Here $m$ is raindrop mass and $v$ velocity. The total action of a rainstorm might be expected therefore to be a function of the sum of the kinetic energies of all the drops:

$$E_{k,tot} = \tfrac{1}{2}\sum m_i v_i^2 \tag{12.22b}$$

wherein $m_i$, $v_i$ are the masses and velocities of raindrops of successive size groups. As a raindrop of any given size falls earthward through the atmosphere, it is accelerated by gravity but encounters the viscous resistance of the atmosphere, which increases with velocity. When the resistance equals the downward force of gravity ($mg$), acceleration stops and the drop continues to fall at a constant, "terminal" velocity which depends on its mass. The spectrum of drop sizes, and hence also of terminal velocities, varies from storm to storm and form place to place. The impact of raindrops on the soil can be mitigated by the presence of a protective cover of vegetation or mulch (which can intercept the drops before they strike ground) and is also influenced by surface roughness, slope, incident angle, and standing water on the surface.

Another important factor affecting the formation of a surface seal (crust) is the chemical composition of the infiltrating solution. A high sodium adsorption ratio (SAR), coupled with a low overall concentration of salts, can induce dispersion and swelling of the clay present in the soil's surface layer, which in turn can have a strong effect on soil infiltrability. The dispersed aggregates collapse and close the interaggregate cavities, and migrating clay particles tend to lodge in soil pores beneath the surface. Evidence of these phenomena was provided by Chen and Banin (1975), by Frenkel *et al.* (1978), and by Oster and Schroer (1979). The effect of sodium in the added water is particularly pronounced at the soil surface because of the mechanical action resulting from the kinetic energy of the applied water and because of the absence of a confining matrix. Extreme clay dispersion and clogging can occur at SAR values as low as 10 if solution concentration is lower than 2 mole/m$^3$ (Oster and Schroer, 1979). Particularly destructive to surface soil structure is the alternation of sodic irrigation water with rainfall. Where such conditions occur drastic reduction of soil infiltrability can result.

Surface crusts are characterized by greater density, finer pores, and lower saturated conductivity than the underlying soil. Once formed a surface crust can greatly impede water intake by the soil (Fig. 12.7), even if the crust is quite thin (say, not more than several millimeters in thickness) and the soil is otherwise highly permeable. Failure to account for the formation of a crust can result in gross overestimation of infiltration.

An analysis of the effect of a developing surface crust upon infiltration was carried out by Edwards and Larson (1970), who adapted the Hanks and Bowers (1962) numerical solution to this problem. Hillel (1964), and Hillel and Gardner (1969, 1970) used a quasianalytical approach to calculate fluxes during steady and transient infiltration into crust-capped profiles from

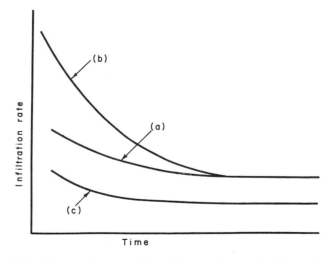

**Fig. 12.7.** Infiltrability as a function of time (a) in a uniform soil, (b) in a soil with a more permeable upper layer, and (c) in a soil covered by a surface crust.

knowledge of the basic hydraulic properties of the crust and of the underlying soil.

## H.  Rain Infiltration

When rain or sprinkling intensity exceeds soil infiltrability, in principle the infiltration process is similar to the case of shallow ponding. If rain intensity is less than the initial infiltrability value of the soil but greater than the final value, then at first the soil will absorb at less than its potential rate and the flow of water in the soil will occur under unsaturated conditions; however, if the rain is continued at the same intensity, and as soil infiltrability decreases, the soil surface will eventually become saturated and henceforth the process will continue as in the case of ponding infiltration. Finally, if rain intensity is at all times lower than soil infiltrability (i.e., lower than the effective saturated hydraulic conductivity), the soil will continue to absorb the water as fast as it is applied without ever reaching saturation. After a long time, as the suction gradients become negligible, the wetted profile will attain a wetness for which the conductivity is equal to the water supply rate, and the lower this rate, the lower the degree of saturation of the infiltrating profile. This effect is illustrated in Fig. 12.8.

The process of infiltration under rain or sprinkler irrigation was studied by Youngs (1964) and by Rubin and Steinhardt (1963, 1964), Rubin *et al.* (1964),

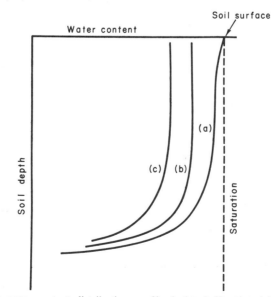

**Fig. 12.8.**  The water-content distribution profile during infiltration (a) under ponding, (b) under sprinkling at relatively high intensity, and (c) under sprinkling at a very low intensity.

and Rubin (1966). The latter author, who used a numerical solution of the flow equation for conditions pertinent to this problem, recognized three modes of infiltration due to rainfall: (1) *nonponding infiltration*, involving rain not intense enough to produce ponding, (2) *preponding infiltration*, due to rain that can produce ponding but that has not yet done so, and (3) *rainpond infiltration*, characterized by the presence of ponded water. Rainpond infiltration is usually preceded by preponding infiltration, the transition between the two being called *incipient ponding*. Thus, nonponding and preponding infiltration rates are dictated by rain intensity, and are therefore *supply controlled* (or *flux controlled*), whereas rainpond infiltration rate is determined by the pressure (or depth) of water above the soil surface as well as by the suction conditions and conductivity relations of the soil. Where the pressure at the surface is small, rainpond infiltration, like ponding infiltration in general, is *profile controlled.*

In the analysis of rainpond or ponding infiltration, the surface boundary condition generally assumed is that of a constant pressure at the surface, whereas in the analysis of nonponding and preponding infiltration, the water flux through the surface is considered to be equal either to the rainfall rate or to the soil's infiltrability, whichever is the lesser. In actual field conditions, rain intensity might increase and decrease alternately, at times exceeding the soil's saturated conductivity (and its infiltrability) and at other times dropping below it. However, since periods of decreasing rain intensity involve complicated hysteresis phenomena (Hillel, 1980), the analysis of variable-intensity rainstorms is rather difficult.

Figure 12.9 describes infiltration rates into a sandy soil during preponding and rainpond infiltration under three rain intensities. The horizontal parts of the curves correspond to preponding infiltration, and the descending parts to rainpond infiltration periods. As pointed out by Rubin (1966), the rainpond infiltration curves are of the same general shape and approach the same limiting infiltration rate, but they do not constitute horizontally displaced

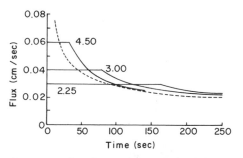

**Fig. 12.9.** Relation between surface flux and time during infiltration into Rehovot sand due to rainfall (solid lines) and flooding (dashed line). The numbers labeling the curves indicate the magnitude of the relative rain intensity. (After Rubin, 1966.)

parts of a single curve and do not coincide with the infiltration rate under flooding, which is shown as a broken line in the same graph.

A rainstorm of any considerable duration typically consists of spurts of high-intensity rain punctuated by periods of low-intensity rain. During such respite periods, surface soil moisture tends to decrease because of internal drainage, thus reestablishing a somewhat higher infiltrability. The next spurt of rainfall is therefore absorbed more readily at first, but soil infiltrability quickly falls back to, or even below, the value it had at the end of the last spurt of rain. The recovery of infiltrability under an intermittent (as opposed to a continuous) supply of water was demonstrated by Busscher (1979). A complete description would, of course, necessitate taking account of the hysteresis phenomenon in the alternately wetting-and-draining surface zone.

A combined theoretical and experimental study of rain infiltration was reported by Swartzendruber and Hillel (1975) who used a sprinkling infiltrometer (Rawitz *et al.*, 1972) to test a set of equations designed to predict the time of appearance and the quantity of *surface water excess* (i.e., the excess of rainfall over infiltration) for different rain intensities. Still more comprehensive theoretical studies of rain infiltration and subsequent soil moisture storage for different soil textures and variously composed soil profiles were carried out by Hillel and van Bavel (1976) and by Hillel and Talpaz (1977).

## I. Surface Runoff

Whenever the rate of water supply to the soil surface (whether it be by rain, irrigation, or snow melt) exceeds the rate of infiltration, free water, called *surface water excess*, tends to accumulate over the soil surface. This water collects in depressions, often called *pockets*, thus forming puddles, the total volume of which, per unit area, is called the *surface storage capacity*. It depends upon the geometric irregularities of the surface as well as upon the overall slope of the land (Fig. 12.10). Only when surface storage is filled and the puddles begin to overflow can actual runoff begin. The term *surface runoff* thus represents the portion of the water supply to the surface which neither is absorbed by the soil nor accumulates on the surface, but which runs downslope and eventually collects in channels variously called *rills* or *gullies*. These channels generally form a treelike pattern of connecting or converging branches with numerous confluences leading to larger and larger streams.

As pointed out by Eagleson (1970), there exists a whole spectrum of channel geometries and flow types. On the one extreme is the thin sheetlike runoff called *overland flow*. It is likely to be the primary type in surface runoff from small natural areas or fields having little topographic relief. The

**Fig. 2.10.** Effect of surface roughness and slope on pocket storage of rainfall excess. The water thus accumulated eventually infiltrates after cessation of the rain.

next distinctive type is found in the smallest stream channels, which gather the overland flow in a continuous fashion along their length to form the lowest order of *stream flow*. As these smallest streams merge with one another, they form streams of "higher order" which have concentrated tributaries as well as continuous lateral inflows.

### Sample Problems

**1.** The infiltration rate under shallow ponding was monitored as a function of cumulative rainfall and found to be 20 mm/hr when a total of 100 mm had infiltrated. If the eventual steady rate of infiltration is 5 mm/hr, estimate the infiltration rate at a cumulative infiltration of 200 and 400 cm. Use the Green and Ampt theory.

We refer to Eq. (12.1):

$$i = i_c + b/I$$

where $i$ is the infiltration rate, $i_c$ the eventual steady rate, $I$ the cumulative infiltration, and $b$ a constant. Substituting the values given above, and solving for $b$,

$$20 \text{ mm/hr} = 5 \text{ mm/hr} + (b \text{ mm}^2/\text{hr})/100 \text{ mm},$$
$$b = 100 (20 - 5) = 1500 \text{ mm}^2/\text{hr}$$

We can now use the calculated value of $b$ to estimate the infiltration rate at any value of cumulative infiltration.

*At I = 200 mm*

$$i = 5 + 1500/200 = 12.5 \text{ mm/hr}$$

*At I = 400 mm*

$$i = 5 + 1500/400 = 8.75 \text{ mm/hr}$$

**2.** Given the soil conditions described in the preceding problem, how much water can be delivered to the root zone of a crop without exceeding the soil's infiltrability if the sprinkling irrigation rate is 15 or 25 mm/hr? What is the highest steady sprinkling rate we can use if we wish to provide an irrigation of 250 mm in the shortest possible time?

At a sprinkling rate of 15 mm/hr,

$$15 \text{ min/hr} = 5 \text{ mm/hr} + (1500 \text{ mm}^2/\text{hr})/I \text{ mm},$$
$$I = 1500/(15 - 5) = 150 \text{ mm}$$

At a sprinkling rate of 25 mm/hr,

$$I = 1500/(25 - 5) = 75 \text{ mm}$$

Note that the higher the sprinkling rate, the smaller the total amount of water which can infiltrate without exceeding the soil's infiltrability (and thus causing flooding or runoff).

To infiltrate 250 mm by steady sprinkling without exceeding infiltrability, we can apply

$$i = 5 + \frac{1500}{250} = 11 \text{ mm/hr}$$

**3.**   A horizontal infiltration trial was conducted with a soil-filled tube having a cross-sectional area of 50 cm². After 15 min, cumulative infiltration totaled 150 cm³. What is the expectable cumulative infiltration and infiltration rate at the end of 1, 4, and 16 hr?

We use Philip's horizontal infiltration ("absorption") theory to calculate the sorptivity $s$ according to Eq. (12.21):

$$s = I/t^{1/2}$$

where $t$ is time and $I$ is cumulative infiltration.

$$s = \frac{150 \text{ cm}^3/50 \text{ cm}^2}{(15 \text{ min} \times 60 \text{ sec/min})^{1/2}} = \frac{3.0 \text{ cm}}{30 \text{ sec}^{1/2}} = 0.1 \text{ cm/sec}^{1/2}$$

We can now apply this value of $s$ to estimate the cumulative infiltration at various times, again using Eq. (12.21):

$$I = st^{1/2}$$

At the end of 1 hr

$$I = 0.1 \text{ cm/sec}^{1/2} \times (3600 \text{ sec})^{1/2} = 6.0 \text{ cm}$$

At the end of 4 hr

$$I = 0.1 \text{ cm/sec}^{1/2} \times (14400 \text{ sec})^{1/2} = 12.0 \text{ cm}$$

At the end of 16 hr:

$$I = 0.1 \text{ cm/sec}^{1/2} \times (57600 \text{ sec})^{1/2} = 24.0 \text{ cm}$$

The corresponding infiltration rates are obtainable by differentiating the above equation with respect to time $t$:

$$i = dI/dt = s/2t^{1/2}$$

At 1 hr

$i = (0.1 \text{ cm/sec}^{1/2})/2(3600 \text{ sec})^{1/2} = 8.33 \times 10^{-4} \text{ cm/sec} \cong 30 \text{ mm/hr}$

At 4 hr

$i = (0.1 \text{ cm/sec}^{1/2})/2(14400 \text{ sec})^{1\ 2} = 4.166 \times 10^{-4} \text{ cm/sec} \cong 15 \text{ mm/hr}$

At 16 hr

$i = (0.1 \text{ cm/sec}^{1/2})/2(57600 \text{ sec})^{1/2} = 2.0833 \times 10^{-4} \text{ cm/sec} \cong 7.5 \text{ mm/hr}$

**4.** If the effective saturated hydraulic conductivity of the soil in the preceding problem is $2 \times 10^{-4}$ cm/sec, estimate the cumulative infiltration and infiltration rate values for a vertical column at 1, 4, and 16 hr. If the initial wetness is (0.05 cm$^3$ H$_2$O/(cm$^3$ soil) and the saturation wetness is 0.45 cm$^3$/cm$^3$, estimate the depth of the wetting front at the same times.

To calculate cumulative infiltration into a vertical column, we use Eq. (12.19)

$$I = st^{1/2} + Kt$$

At 1 hr (3600 sec)

$I = 0.1 \text{ cm/sec}^{1/2} \times 60 \text{ sec}^{1/2} + 2 \times 10^{-4} \text{ cm/sec} \times 3600 \text{ sec}$
$= 6.0 \text{ cm} + 0.72 \text{ cm} = 6.72 \text{ cm}$

At 4 hr (14400 sec)

$I = 0.1 \text{ cm/sec}^{1/2} \times 120 \text{ sec}^{1/2} + 2 \times 10^{-4} \text{ cm/sec} \times 14400 \text{ sec}$
$= 12 \text{ cm} + 2.88 \text{ cm} = 14.88 \text{ cm}$

At 16 hr (57600 sec)

$I = 0.1 \text{ cm/sec}^{1/2} \times 240 \text{ sec}^{1/2} + 2 \times 10^{-4} \text{ cm/sec} \times 57600 \text{ sec}$
$= 24 \text{ cm} + 11.52 \text{ cm} = 35.52 \text{ cm}$

To calculate the infiltration rate into a vertical column, we differentiate Eq. (12.19) with respect to time:

$$i = dI/dt = s/2t^{1/2} + K$$

At 1 hr

$i = (0.1 \text{ cm/sec}^{1/2})/(2 \times 60 \text{ sec}^{1/2}) + 2 \times 10^{-4} \text{ cm/sec}$
$= 8.33 \times 10^{-4} \text{ cm/sec} + 2 \times 10^{-4} \text{ cm/sec} = 10.33 \times 10^{-4} \text{ cm/sec}$
$\cong 37.19 \text{ mm/hr}$

At 4 hr

$i = (0.1 \text{ cm/sec}^{1/2}(/(2 \times 120 \text{ sec}^{1/2}) + 2 \times 10^{-4} \text{ cm/sec}$
$= 4.166 \times 10^{-4} \text{ cm/sec} + 2 \times 10^{-4} \text{ cm/sec} = 6.166 \times 10^{-4} \text{ cm/sec}$
$\cong 22.22 \text{ mm/hr}$

At 16 hr

$$i = (0.1 \text{ cm/sec}^{1/2})/(2 \times 240 \text{ sec}^{1/2}) + 2 \times 10^{-4} \text{ cm/sec}$$
$$= 2.083 \times 10^{-4} \text{ cm/sec} + 2 \times 10^{-4} \text{ cm/sec} = 4.083 \times 10^{-4} \text{ cm/sec}$$
$$\cong 14.7 \text{ mm/hr}$$

To estimate the depth of the wetting front $L_f$ we divide the cumulative infiltration I by the change in soil wetness $\Delta\theta$ from its initial value to saturation:

$$L_f = I/\Delta\theta$$

At 1 hr

$$L_f = (6.72 \text{ cm})/(0.45 - 0.05) = 16.8 \text{ cm}$$

At 4 hr

$$L_f = (14.88 \text{ cm}/0.4 = 37.2 \text{ cm}$$

At 16 hr

$$L_f = (35.52 \text{ cm})/0.4 = 88.8 \text{ cm}$$

*Note:* The student is invited, for his own edification, to plot the calculated values of $I$, $i$, and $L_f$ against time and against the square root of time. The relationships involved will become much clearer in the process. A comparison between the corresponding data of Problems 3 and 4 illustrates the difference between horizontal and vertical infiltration.

# 13 *Internal Drainage and Redistribution Following Infiltration*

## A. Introduction

When rain or irrigation ceases and surface storage is depleted by evaporation or infiltration, the infiltration process comes to an end, since no more water enters into the soil. Downward water movement within the soil, however, does not cease immediately and may in fact persist for a long time as soil water redistributes within the profile. The soil layer wetted to near saturation during infiltration does not retain its full water content, as part of its water moves down into the lower layers under the influence of gravity and possibly also of suction gradients. In the presence of a high water table, or if the profile considered is initially saturated throughout, this postinfiltration movement is herein termed *internal drainage*. In the absence of groundwater, or where the water table is too deep to affect the relevant depth zone, and if the profile is not initially wetted to saturation throughout its depth, this movement is called *redistribution*, since its effect is to redistribute soil water by increasing the wetness of successively deeper layers at the expense of the infiltration-wetted layers of the soil profile.

In some cases, the rate of redistribution decreases rapidly, becoming imperceptible after several days so that thereafter the initially wetted part of the soil appears to retain its moisture, unless this moisture is evaporated or is taken up by plants. In other cases, redistribution may continue at an appreciable, though diminishing, rate for many days or weeks.

The importance of the redistribution process should be self-evident, as it determines the amount of water retained at various times by the different depth zones in the soil profile and hence can affect the water economy of plants. The rate and duration of downward flow during redistribution determine the effective *soil water storage*, a property that is vitally important, particularly in relatively dry regions, where water supply is infrequent and plants must rely for long periods on the unreplenished reservoir of water within the rooting zone. Even in relatively humid regions, where the water supply by precipitation would seem to be sufficient, inadequate soil moisture storage can deprive the crops to be grown of a major portion of the water supply and can cause crop failure. Finally, the redistribution process is important because it often determines how much water will flow through the root zone (rather than be detained, or temporarily stored, within it) and hence how much leaching of solutes will take place.

As we shall explain more fully in the sections to follow, soil-water storage is generally not a fixed quantity or a static property but a dynamic phenomenon, determined by the time-variable rates of soil-water inflow to, and outflow from, the relevant soil volume.

## B. Internal Drainage in Thoroughly Wetted Profiles

We have already made a distinction between the postinfiltration movement of soil water in cases where a groundwater table is present fairly close to the soil surface (i.e., at a depth not exceeding a few meters) and in cases where groundwater is either nonexistent or too deep to affect the state and movement of soil moisture in the root zone.

At the groundwater table, also called the *phreatic surface*, soil water is at atmospheric pressure. Beneath this level the hydrostatic pressure exceeds atmospheric pressure, while above this level soil water is under suction. Internal drainage in the presence of a water table tends to a state of equilibrium, in which the suction at each point corresponds to its height above the free water level. The downward drainage flux decreases as the hydraulic gradient decreases, both approaching zero in time (provided, of course, there is no further addition of water by another episode of infiltration, or abstraction of water by evapotranspiration—processes which would prevent the attainment of static equilibrium). At equilibrium, if attainable, soil wetness would increase in depth to a value of saturation just above the water table, a depth distribution which would mirror the soil moisture characteristic curve. Phenomena involving falling water tables and groundwater drainage will be treated in our next chapter.

In the present section, we wish to concentrate upon the internal drainage of profiles initially wetted to near saturation throughout their depth. We shall treat this process as if it were unaffected by any water table, which, if it exists, is assumed to lie too far below the zone of interest to be of any direct consequence. Internal drainage beyond the root zone has also been referred to as *deep percolation.*

In the hypothetical case of a very deeply wetted profile, one might justifiably assume that internal drainage, in the absence of any suction gradients, occurs under the influence of gravity alone. If so, the downward flux through any arbitrary plane (say, the bottom of the root zone) should be equal to the hydraulic conductivity and should diminish in time as the conductivity diminishes due to reduction of water content in the infiltration-wetted but since-unreplenished soil above the plane considered. This can be stated formally as follows

Recall the one-dimensional (vertical) form of Darcy's equation:

$$q = -K(\theta) \frac{\partial}{\partial z} (-\psi - z) \tag{13.1}$$

where $q$ is flux, $K(\theta)$ hydraulic conductivity (a function of wetness $\theta$), $z$ depth, and $\psi$ matric suction. With $\partial \psi / \partial z$ assumed to be zero, we have simply

$$q = K(\theta) \tag{13.2}$$

If we happen to know the functional dependence of $K$ on $\theta$, we can formulate $q$ as an explicit function of $\theta$. For instance, if the function is exponential, say,

$$K(\theta) = ae^{b\theta} \tag{13.3}$$

where $a$ and $b$ are constants, we get

$$q = ae^{b\theta} \quad \text{or} \quad \ln q = A + b\theta \tag{13.4}$$

where $A = \ln a$. Thus, even a small decrease in $\theta$ can result in a steep (logarithmic) decrease of the flux $q$.

A simple approach to the internal drainage process is possible if the soil profile can be assumed to drain uniformly, i.e., if the soil is equally wet and its wetness diminishes at an equal rate throughout the draining profile. Observations have shown this to be a reasonable approximation in many cases. In the absence of any flow through the upper soil surface, the flux $q_b$ through any plane at depth $z_b$ must equal the rate of decrease of total water content $W$, where $W = \theta z_b$ (the product of the wetness $\theta$ and depth $z_b$):

$$q_b = K(\theta) = -dW/dt = -z_b \, d\theta/dt \tag{13.5}$$

Note that the downward flux increases in proportion to depth. Note also that it diminishes in time, as does the rate of decrease of soil wetness, in accordance with the functional decrease of conductivity with the remaining soil wetness. Once the characteristic $K(\theta)$ function is established for a given soil, it becomes possible to predict the time dependence of $\theta$ and of drainage rate $q$ for any depth.

## C. Redistribution of Soil Moisture in Partially Wetted Profiles

Perhaps more typical than the condition described in the preceding section (namely, that of a uniformly and deeply wetted profile) are cases in which the end of the infiltration process consists of a wetted zone in the upper part of the profile and an unwetted zone beneath. The postinfiltration movement of water in such a profile can truly be called *redistribution*, as the relatively dry deeper layers (those beyond the infiltration wetting front) draw water from the upper ones. The time-variable rate of redistribution depends not only on the hydraulic properties of the conducting soil (as does the rate of internal drainage, previously described) but also on the initial wetting depth, as well as on the relative dryness of the bottom layers. When the initial wetting depth is small and the underlying soil is relatively dry the suction gradients, augmenting the gravitational gradient, are likely to be strong and hence induce a rapid rate of redistribution. On the other hand, when the initial wetting depth is considerable and the underlying soil itself is wet the suction gradients are small and redistribution occurs primarily under the influence of gravity, as in the case of internal drainage described above.

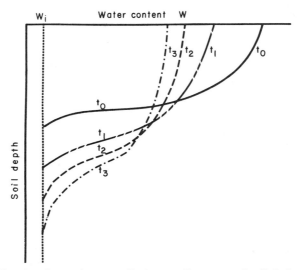

**Fig. 13.1.** The changing moisture profile in a medium-textured soil during redistribution following an irrigation. The moisture profiles shown are for 0, 1, 4, and 14 days after the irrigation. $w_i$ is preirrigation (antecedent) soil wetness.

At first the decrease of soil wetness in the initially wetted zone can be expected to occur more rapidly during the redistribution of moisture in profiles which had been subjected to shallow wetting than in the internal drainage of profiles which had been wetted deeply. Sooner or later, however, the redistribution process "spends itself out," so to speak, and the flux slows down for two reasons: (1) the suction gradients between the wet and dry zones decrease as the former loses, and the latter gains, moisture; (2) as the initially wetted zone quickly desorbs, its hydraulic conductivity decreases correspondingly. With both gradient and conductivity decreasing simultaneously, the flux falls rapidly. The rate of advance of the wetting front decreases accordingly, and this front, which was relatively sharp during infiltration, gradually flattens out and dissipates during redistribution. This is illustrated in Fig. 13.1.

The figure shows that the upper, initially wetted zone drains monotonically, though at a decreasing rate. On the other hand, the sublayer at first wets up but eventually begins also to drain. The time dependence of soil wetness in the upper zone is illustrated in Fig. 13.2 for a sandy soil, in which the unsaturated conductivity falls off rapidly with increasing suction, and for a clayey soil, in which the decrease of conductivity is more gradual and hence redistribution tends to persist longer.

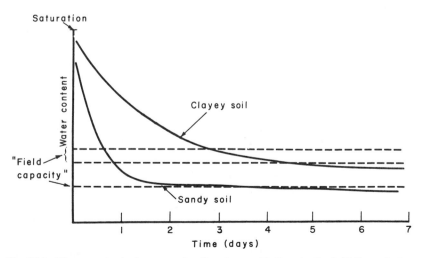

**Fig. 13.2.** The monotonic decrease of soil wetness with time in the initially wetted zone during redistribution.

## D. Analysis of Redistribution Processes

Once again we return to the general equation for flow in a vertical profile, Eq. (12.11):

$$\frac{\partial \theta}{\partial t} = -\frac{\partial}{\partial z}\left( K \frac{\partial \psi}{\partial z} + K \right)$$

In the case of redistribution involving hysteresis, this equation may be written (Miller and Klute, 1967)

$$\left(\frac{d\theta}{d\psi}\right)_h \frac{\partial \psi}{\partial t} = -\frac{\partial}{\partial z}\left[ K_h(\psi) \frac{\partial \psi}{\partial z} \right] - \frac{\partial K_h(\psi)}{\partial z} \tag{13.6}$$

In the above, $\theta$ is volumetric wetness, $t$ time, $K$ conductivity, $z$ depth, $\psi$ suction head, and the subscript $h$ indicates a hysteretic function. After the cessation of infiltration and in the absence of evaporation flux through the soil surface is zero, and hence the hydraulic gradient at the surface must also be zero. Conservation of matter requires that

$$\int_{z=0}^{z=\infty} \theta \, dz = \text{const} \tag{13.7}$$

for all time, provided no sinks are present (e.g., no extraction of water by plant roots).

When redistribution begins, the upper portion of the profile, which was wetted to near saturation during the preceding infiltration process, begins to desorb monotonically. Below a certain depth, however, the soil first wets during redistribution, then begins to drain, and the value of wetness at which this turnabout takes place decreases with depth. Each point in the soil thus follows a different scanning curve, and the conductivity and water-capacity functions vary with position. An approximate analysis of these phenomena was given by Youngs (1958a, b; 1960a,b). He showed that in coarse materials with uniform pore sizes and hence with "abrupt" $\theta(\psi)$ curves redistribution can be slower for shallow than for deep wettings.

The flow equation describing redistribution was analyzed numerically by Staple (1969), Remson *et al.* (1965, 1967), and Rubin (1967). Rubin's analysis was based on the assumptions that, despite hysteresis, there exists at any time and depth in the soil a unique, single-valued relation between wetness and suction, that a soil element that has begun to desorb will not wet up again, and that hysteresis in the relation of conductivity to wetness (but not to suction) can be neglected.

To solve Eq. (12.11), one of the dependent variables (namely, $\psi$) was eliminated:

$$\frac{\partial \theta}{\partial t} = - \frac{\partial}{\partial z}\left(\frac{K}{c}\frac{\partial \theta}{\partial z} + K\right) \tag{13.8}$$

The solution was then sought, subject to the following initial and boundary conditions:

$$
\begin{aligned}
t &= 0, & z &> 0, & \theta &= \theta(z) \\
t &> 0, & z &= 0, & q &= \frac{K}{c}\frac{\partial \theta}{\partial z} + K = 0 \\
t &> 0, & z &= \infty, & \theta &= \theta_i
\end{aligned}
\tag{13.9}
$$

where $q$ is the flux, $\theta(z)$ is a function of soil depth describing the postinfiltration (preredistribution) moisture profile, $\theta_i$ is the soil's preinfiltration wetness, $c$ is the differential water capacity $d\theta/d\psi$, and the other variables are as defined above. The ratio $K/c$ is the diffusivity $D$.

Equation (13.8) was approximated by an implicit difference equation. In order to take hysteresis into account, empirical equations were used to describe the dependence of $\psi$ on $\theta$ in primary wetting in primary drying and in intermediate (transitional) scanning from wetting to drying. Another empirical equation was used to describe the relation of $K$ to $\theta$. Empirical equations cannot be expected to pertain to any but the particular soil considered. The results of the analysis, however, are thought to be generally valid in principle, provided the basic assumptions are met in the real situation, at least approximately.

Rubin's findings appear to indicate that the hysteretic moisture profile is not bounded by the two possible nonhysteretic profiles (one assuming the desorbing branch of the soil-moisture characteristic, the other assuming the sorbing branch). The hysteretic redistribution was shown to be clearly slower than the nonhysteretic one. These results demonstrate the importance of hysteresis in redistribution processes, particularly in coarse-textured soils in which the hysteresis phenomenon is often more pronounced.

A different approach to the analysis of redistribution was taken by Gardner *et al.* (1970a,b). Equation (13.8) can be solved analytically by separation of variables (Gardner, 1962b) in the special case where the empirical relation $D = C\theta^n$ applies (with $C$ and $n$ constants). In this procedure, it is assumed that the solution is of the form $\theta = T(t)Z(z)$, where $T$ is a function of $t$ alone and $Z$ is a function of $z$ alone. It is further assumed that $K = B\theta^m$ (with $B$ and $m$ constants). Their analysis suggests that both where matric suction forces predominate and where, alternatively, the gravity force predominates the time dependence of soil wetness (and of total water content in any depth zone) obeys a relation of the type

$$W, \bar{\theta}, \text{ or } \int \theta \, dz \approx a(b + t)^{-c} \qquad (13.10)$$

where $W$ is the total water content and $\bar{\theta}$ is the mean wetness of the initially wetted zone at time $t$ during redistribution. This theory accords with experimentally measured patterns of redistribution in a sandy loam soil irrigated with different quantities of water (Figs. 13.4 and 13.5). The constants in Eq. (13.10) were shown to be related to the soil's hydraulic conductivity and diffusivity. Of these, the value of $b$ could be neglected after a day or so, so that the simplified equation could be cast into logarithmic form:

$$\log W = \log a - c \log t \qquad (13.11)$$

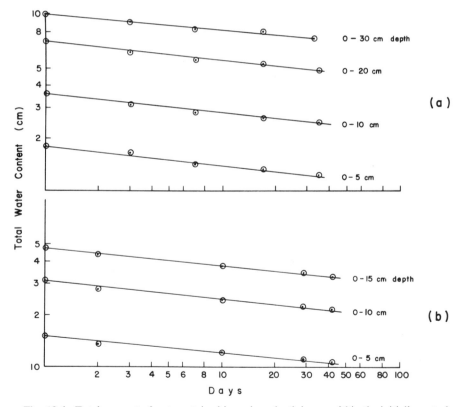

**Fig. 13.4.** Total amount of water retained in various depth layers within the initially wetted zone of a fine sandy loam during redistribution following irrigations of (a) 10 and (b) 5 cm of water. (After Gardner *et al.*, 1970a, b.)

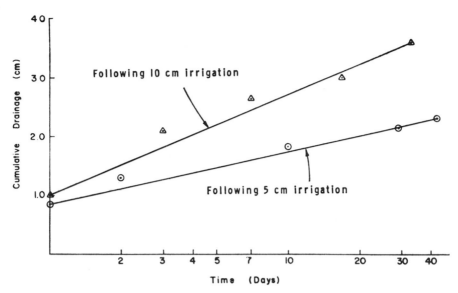

**Fig. 13.5.** Cumulative downward flow through the initial (end of infiltration) wetting front during redistribution following irrigations of 5 and 10 cm of water. (After Gardner *et al.*, 1970a, b.)

The values of $a$ and $c$ can be obtained from a logarithmic graph of $W$ versus $t$ if the data do indeed indicate a straight line, as shown in Fig. 13.5.

Equation (13.10) can also be differentiated with respect to time to give the rate of decrease of water content in the initially wetted zone $(-dW/dt)$, equal to the flux through the initial (end of infiltration) wetting front:

$$-dW/dt = ac/(b + t)^{c+1} \qquad (13.12)$$

This relationship is shown in Figure 13.4.

### E. "Field Capacity"

Early observations that the rate of flow and water-content changes decrease in time (Alway and McDole, 1917; Richards and Moore, 1952; Veihmeyer and Hendrickson, 1931) have been construed to indicate that the flow rate generally becomes negligible within a few days, or even that the flow ceases entirely. The presumed water content at which internal drainage allegedly ceases, termed the *field capacity*, had for a long time been accepted almost universally as an actual physical property, characteristic of and constant for each soil.

Though the field capacity concept originally derived from rather crude measurements of water content in the field (where sampling and measurement errors necessarily limited the accuracy and validity of the results), some workers have sought to explain it somehow in terms of a static equilibrium value or a discontinuity in the capillary water. It was commonly assumed that the application of a certain quantity of water to the soil will fill the deficit to field capacity to a certain and definite depth, beyond which this quantity of water will not penetrate. It became an almost universal practice to calculate the amount of irrigation to be applied at any particular time on the basis of the "deficit to the field capacity" of the soil depth zone to be wetted.

In recent years, with the development of theory and more precise experimental techniques in the study of unsaturated flow processes, the field capacity concept, as originally defined,[1] has been recognized as arbitrary and not an *intrinsic* physical property independent of the way it is measured. When and how can one determine that redistribution has virtually ceased, or that its rate has become negligible or practically zero? Obviously, the criteria for such a determination are subjective, depending as often as not upon the frequency and accuracy with which the soil water content is measured. The common working definition of field capacity (namely, the wetness of the initially wetted zone, say, two days after infiltration) takes no account of such factors as the antecedent (preinfiltration) wetness of the soil, the depth of wetting, the possible presence and depth of a groundwater table, etc.

The redistribution process is in fact continuous and exhibits no abrupt "breaks" or static levels. Although its rate decreases constantly, in the absence of a water table the process continues and equilibrium is approached, if at all, only after very long periods.

The soils for which the field capacity concept is most tenable are the coarse-textured ones, in which internal drainage is initially most rapid but soon slows down owing to the relatively steep decrease of hydraulic conductivity with increasing matric suction. In medium- or fine-textured soils, however, redistribution can persist at an appreciable rate for many days. As an example, we can cite the case of a loessial silt loam in the Negev region of Israel, in which the changes in water content shown in Table 13.1 were observed in the 60–90 cm depth zone following a wetting to a depth exceeding

---

[1] According to Veihmeyer and Hendrickson (1949), the field capacity is "the amount of water held in soil after excess water has drained away and the rate of downward movement has materially decreased, which usually takes place within 2–3 days after a rain or irrigation in pervious soils of uniform structure and texture." By "amount of water" they obviously meant amount per unit volume, or mass, of soil. Apart from that, however, this definition raises more qeustions than it answers. What is "excess water?" How can we be sure just when it has "drained away" or just when downward movement has "materially decreased?" Can we depend universally on the statement that this "usually takes place within 2–3 days after a rain or irrigation," particularly if the amount of water supplied and antecedent soil conditions are left unspecified? And what are "pervious soils?" Finally, we must ask, what about all the world's soils which are not "of uniform structure and texture?"

150 cm (with evaporation effectively prevented by means of a paper mulch). It is seen that the water loss continued incessantly for over five months. The rate of decrease of water content accorded with the function given by Richards *et al.* (1956):

$$-dW/dt = at^{-b} \tag{13.13}$$

where $W$ is the water content, $t$ the time, and $a$ and $b$ constants related to the boundary conditions and conductance properties of the soil. (The exponential constant $b$, which is related to soil diffusivity, is obviously most important, and the greater its value, the steeper the decrease in water content.)

An agriculturist engaged in irrigated farming and accustomed to frequent irrigations is interested mainly in the short-run storage "capacity" of his soil. For him, the field capacity of the loessial soil cited can be taken at about 18%. By way of contrast, an agriculturist engaged in dryland farming is sometimes interested in storing soil water from one season, or even from one year, to the next. For the dry-land farmer, therefore, the field capacity of the same loessial soil cannot be taken at 18% (since the soil does not retain this content beyond a few days) but at 14%, or even less.

A commonly, if tacitly, held assumption is that the redistribution process occurs by itself, independently of other processes of water extraction from the soil. This assumption is seldom, if ever, realistic. When several processes of soil-water extraction (e.g., internal drainage, evaporation, and uptake by plants) occur simultaneously, the reduction in soil water content is obviously more rapid and less apt to cease at any point such as "field capacity."

Various laboratory methods have been proposed for the estimation of field capacity. These include equilibration of presaturated samples with a centrifugal force 1000 times the gravity force (the so-called moisture equivalent) or with a matric suction value of $\frac{1}{10}$ or $\frac{1}{3}$ bar. Although the results of such tests may correlate with measurements of soil moisture storage in the field in certain circumstances, it is a fundamental mistake to expect criteria of this sort to apply universally, since they are solely static in nature while the

**Table 13.1**

|  | Moisture (percent by mass) |
|---|---|
| At the end of infiltration | 29 |
| After   1 day | 20.2 |
| 2 days | 18.7 |
| 7 days | 17.5 |
| 30 days | 15.9 |
| 60 days | 14.7 |
| 156 days | 13.6 |

process they purport to represent is highly dynamic. For example, two texturally and structurally identical profiles will drain quite differently if one of them is uniform throughout all depths whereas the other is underlain by a clay pan. The former profile will tend to drain readily while the latter may remain nearly saturated for long periods.

These and other shortcomings of the field capacity concept were recognized some twenty years ago by L. A. Richards, who proclaimed in a keynote address to the International Congress of Soil Science that "the field capacity concept may have done more harm than good" (Richards, 1960). Alas, the concept has not faded away. Notwithstanding the fundamental impossibility of fixing unequivocally any unique point in time at which internal drainage or redistribution ceases, there remains a universal need for a simple criterion to characterize the ability of soils to retain, or store, moisture (i.e., the upper limit of soil water content which can be depended on, more or less, in the field). However indefinite the field capacity concept, the fact is that the rapid slowing of internal drainage or redistribution is an exceedingly important property of soils, responsible for retaining (albeit temporarily) water available for plant use during periods of no rain or irrigation.

Granted that the field capacity concept is subjective, it is nevertheless considered by many to be necessary.   If so, can it be improved, in principle and in practice? In the first place, since no laboratory system yet devised is capable of duplicating soil-water dynamics in the field, it should be obvious that, to be realistic, field capacity must be measured directly in the field. Too many practitioners still ignore this simple truism, preferring instead to assume that, say, "moisture retention at $\frac{1}{3}$ bar *is* field capacity" when in fact such a value can, at best, be *correlated* with it (a correlation which must be verified in each case and never taken for granted a priori).

Second, the field determination itself must be made reproducible by standardizing a consistent procedure. Such vague specifications as "wet the soil to the depth of interest," and "allow the soil to drain for approximately 2 days" (Peters, 1965) are not good enough. Wetting depth is extremely important, and the preferred depth is the maximal one—considerably beyond the "depth of interest" (presumably, the rooting zone of crops, which itself is too variable to specify). The internal drainage of a profile initally wetted to its *entire* depth is a much more reproducible and reliable criterion than redistribution in a profile wetted to some unspecified partial depth without regard to antecedent conditions.

Third, the measurement of soil-moisture content and depth distribution should be made repeatedly rather than only once at an arbitrary time such as 2 days. Periodically repeated measurements, preferably by a nondestructive method such as neutron gauging, will provide information on the dynamic pattern of internal drainage and allow evaluation of whether any single value of soil moisture at any specifiable characteristic time can be designated as the field capacity. To make this judgment objectively, one must decide at the outset what drainage rate one is willing to consider negligible.[2] For different soils, the characteristic time may vary from a few hours to a few weeks, depending on soil hydraulic properties and on the flux criterion used.

A still better approach is to characterize, for each field to be considered, the internal drainage process (starting from well-defined initial and boundary conditions) as a complete function of time. One way to do this would be to fit the measured function to an empirical equation, such as that proposed by Richards *et al.* (1956) and by Ogata and Richards (1957), as given above in Eq. (13.13):

$$q = -dW/dt = at^{-b}$$

It would then be up to the user to calculate what value of soil moisture gives the best estimate of soil moisture storage to suit his purposes. Assuming, for the sake of argument, that the data obtained in the field accords with the above equation, then, for whatever limiting value of flux $q_n$, one wishes to regard as negligible, one gets

$$t_n = (a/q_n)^{1/b} \tag{13.14}$$

where $t_n$ is the time period required from the cessation of infiltration for the drainage rate to fall to a negligible value. The constants $a$ and $b$ are to be calculated in each case from the complete internal drainage curve as measured in the field.

### F. Summary of Factors Affecting Field Capacity

The field capacity, as commonly measured, may vary between about 4% (by mass) in sands, to about 45% in heavy clay soils, and up to 100% or even more in certain organic soils. Among the factors affecting redistribution and the apparent field capacity are the following:

(1)   Soil texture. Clayey soils retain more water, and retain it longer, than sandy ones. Hence, the finer the texture the higher the apparent field capcity, the slower its attainment, and the less distinct and stable its value. Soil structure may also affect water retention.

(2)   Type of clay present. The higher the content of montmorillonite is, the greater is the content of water adsorbed and retained at any time.

(3)   Organic matter content. Soil organic matter can help retain more water, though the amount of organic matter normally present in mineral soils is too low to retain any significant amounts of water. The effect of organic matter on soil structure can, however, be significant.

---

[2] A likely criterion might be $\frac{1}{10}$ or less of the daily potential evapotranspiration (PET). Thus, if PET is 5 mm/day, one might designate a drainage flux of 0.5 mm/day through the bottom of the root zone as small enough to be disregarded.

(4) Depth of wetting and antecedent moisture. In general (but not always) the wetter the profile is at the outset, and the greater the depth of wetting during infiltration, the slower the rate of redistribution and the greater the apparent field capacity.

(5) The presence of impeding layers in the profile, such as layers of clay, sand, or gravel, can inhibit redistribution and increase the observable field capacity. The rate of outflow from any given layer in the soil thus depends not only on the texture or hydraulic characteristics of that layer, but also on the composition and structure of the entire profile, since the presence at any depth of an impeding layer can retard the movement of water out of the layers above it.

(6) Evapotranspiration. The rate and pattern of the upward extraction of water from the soil can affect the gradients and flow directions in the profile and thus modify the redistribution or internal drainage process.

### Sample Problems

**1.** Assume gravity drainage out the bottom of a uniformly wetted soil profile, 100 cm deep, with an initial wetness of 45%. Further assume the hydraulic conductivity $K$ to be an exponential function of the soil's volume wetness $\theta$, namely, $K = ae^{b\theta}$, where $a = 1.5 \times 10^{-6}$ cm/day, and $b = 35$. Estimate the remaining soil wetness when the drainage flux decreases to $\frac{1}{2}$, then to $\frac{1}{10}$, then to $\frac{1}{100}$ of the evapotranspiration rate (ET), which is 0.5 cm/day.

With gravity alone (no suction gradients), vertical drainage occurs at a rate $q$ equal to the hydraulic conductivity, which is, in turn, a function of the soil's wetness at the depth considered (e.g., at the bottom of the root zone):

$$q = K(\theta) = ae^{b\theta} = 1.5 \times 10^{-6}e^{35\theta}$$

We can now calculate the flux at different assigned wetness values to obtain a plot of log $q$ versus $\theta$, as follows:

When $\theta = 0.45$, $q = 1.5 \times 10^{-6}e^{35 \times 0.45} = 10.38$ cm/day;
when $\theta = 0.40$, $q = 1.5 \times 10^{-6}e^{35 \times 0.40} = 1.80$ cm/day;
when $\theta = 0.35$, $q = 1.5 \times 10^{-6}e^{35 \times 0.35} = 3.15 \times 10^{-1}$ cm/day;
when $\theta = 0.30$, $q = 1.5 \times 10^{-6}e^{35 \times 0.30} = 5.4 \times 10^{-2}$ cm/day;
when $\theta = 0.25$, $q = 1.5 \times 10^{-6}e^{35 \times 0.25} = 9.45 \times 10^{-3}$ cm/day;
when $\theta = 0.20$, $q = 1.5 \times 10^{-6}e^{35 \times 0.20} = 1.65 \times 10^{-3}$ cm/day.

From the $q$–$\theta$ plot we can determine the $\theta$ values at which $q = 2.5 \times 10^{-1}$ cm/day ($\frac{1}{2}$ET), $5 \times 10^{-2}$ cm/day ($\frac{1}{10}$ET), and $5 \times 10^{-3}$ cm/day. The $\theta$ values turn out to be approximately 43, 30, and 23.2%, respectively. Which of the $q$ values is to be considered negligible is a matter of individual decision. Alternatively, we can calculate the $\theta$ remaining at any flux value, as follows:

$q = ae^{b\theta}$. Hence $\ln q = \ln a + b\theta$. Therefore $\theta = (\ln q - \ln a)/b$. Substituting, for example, $q = 5 \times 10^{-2}$ cm/day ($\frac{1}{10}$ET), we get

$$\theta = [-3 - (-13.4)]/35 = 10.4/35 = 0.297$$

which is close to the 30% estimated above. The actual plot of $q$ versus $\theta$ is again left as an instructive exercise to the enterprising student.

    **2.** Use the equation of Richards *et al.* (1956) to estimate the water content of a 1 m deep profile at the end of 1, 2, 3, 4, 4, 5, 6, 7 days of redistribution if the initial water content is 50 cm, factor $a = 4$, and exponent $b = 0.8$. Estimate the time necessary for the downward flux at a depth of 1 m to diminish to 1 cm/day.

    Recalling Eq. (13.13), with $W$ and $t$ as in the preceding problem,

$$dW/dt = -at^{-b}$$

from which we obtain

$$dW = -at^{-b}\, dt$$

Integrating,

$$W = -[a/(-b + 1)]t^{-b+1} + c$$

where $c$, the constant of integration, equals the profile water content at $t = 0$.

    We can now calculate the water content at various times, as follows:

At zero time, $W = 50 - [4/(-0.8 + 1)] \times 0^{0.2} = 50.0$ cm.
After 1 day,  $W = 50 - (4/0.2) \times 1^{0.2} = 30.0$ cm.
After 2 days, $W = 50 - (4/0.2) \times 2^{0.2} = 27.0$ cm.
After 3 days, $W = 50 - (4/0.2) \times 3^{0.2} = 25.1$ cm.
After 4 days, $W = 50 - (4/0.2) \times 4^{0.2} = 23.6$ cm.
After 5 days, $W = 50 - (4/0.2) \times 5^{0.2} = 22.4$ cm.
After 6 days, $W = 50 - (4/0.2) \times 6^{0.2} = 21.4$ cm.
After 7 days, $W = 50 - (4/0.2) \times 7^{0.2} = 20.5$ cm.

To estimate the time $t_n$ needed for the flux to diminish to any specified value $q_n$, we use Eq. (13.14):

$$t_n = (a/q_n)^{1/b}$$

Accordingly,

$$t_n = (4/1)^{1/0.8} = 5.66 \text{ days}$$

# 14    *Groundwater Drainage*

## A. Introduction

It is an interesting fact that some 75% of the earth's fresh water is locked away in polar ice caps and glaciers, while less than 2% is in *surface waters*, such as lakes and streams, and a relatively minute amount is contained in the generally unsaturated topsoil. That leaves nearly 22% of our planet's fresh water, which permeates and saturates porous rocks and subsoils, generally at some depth below the ground surface. It is this amount which we call *groundwater*.

In many regions, groundwater constitutes an important source of fresh water for domestic, agricultural, or industrial use. Since an adequate water supply is a prerequisite for the development of settlements and industries, including the agricultural industry, and since injudicious exploitation of groundwater can deplete the *groundwater reservoir* and diminish its quality, it is important to acquire and disseminate knowledge pertaining to the behavior of groundwater and to methods of managing it. Extraction of groundwater in excess of the *annual recharge* (including the natural percolation of precipitation water, seepage from reservoirs and streams, and artificial injection through wells and by surface ponding) will cause depletion, whereas an excess of recharge over extraction will cause a buildup of groundwater.

In this chapter, we shall be dealing mainly with shallow unconfined groundwater, which occasionally encroaches upon the root zone and must then be drained if optimal growing conditions are to be maintained. For more fundamental expositions of groundwater theory, the reader is referred to publications by Harr (1962), Chow (1964), Todd (1967), Remson *et al.* (1971), Bouwer (1978), and Freeze and Cherry (1979). Problems of groundwater pollution were described by Fried (1976).

## B. Flow of Unconfined Groundwater

While water in unsaturated soil is strongly affected by suction gradients and its movement is subject to very considerable temporal variations in conductivity resulting from changes in soil wetness, groundwater is always under positive hydrostatic pressure and hence saturates the soil. Thus, no suction gradients and no variations in wetness or conductivity normally occur below the water table and the hydraulic conductivity is maximal and fairly constant in time, though it may vary in space and direction. (We are disregarding here possible effects due to overburden pressures and swelling phenomena.)

Despite the differences between the saturated and unsaturated zones, the two are not independent realms but parts of a continuous flow system. Groundwater is sustained (recharged) by percolation through the unsaturated zone, and the position of its surface, the water table, is determined by the relative rates of recharge versus outflow. Reciprocally, the position of the water table affects the moisture profile and flow conditions above it. One problem encountered in attempting to distinguish between the unsaturated and saturated zones is that the boundary between them may not be exactly at the water table but at some elevation above it, corresponding to the upper extent of the *capillary fringe* (at which the suction is equal to the air-entry value for the soil). Frequently, this boundary is diffuse and scarcely definable, particularly when affected by hysteresis.

The various circumstances, or boundary conditions, under which the movement of unconfined groundwater can occur were elucidated by Kirkham (1957), van Schilfgarrde (1957, 1974), and Childs (1969).

Water may seep into or out of the saturated zone through the soil surface or by lateral flow through the sides of a channel or a porous drainage tube. The groundwater table is hardly ever entirely level and may exhibit steep gradients in the drawdown regions near drainage channels, tubes, or wells. Where the land surface elevation varies, as well as where the amount of infiltration water supply varies areally, the water table can change in depth and may in places and at times even intersect the soil surface and emerge as free (ponding) water above it.

If the depth of the water table remains constant, the indication is that the rate of inflow to the groundwater and the rate of outflow are equal. In other words, where there is downward seepage out of the unsaturated zone, this must be offset by downward or horizontal outflow of the groundwater if the water table is to remain stationary. On the other hand, a rise or fall of the water table indicates a net recharge or discharge of groundwater, respectively. Such vertical displacements of the water table can occur periodically, as under a seasonally fluctuating regimen of rainfall or irrigation. The rise and fall of the water table can also be affected by barometric pressure changes, though generally to a minor degree.

Groundwater flow can be geometrically complex where the profile is heterogeneous or anisotropic (Maasland, 1957) or where sources and sinks of water are distributed unevenly. If the profile above the water table consists of a sequence of layers such that a highly conductive one overlies one of low conductivity, then it is possible for the flow rate into the top layer to exceed the transmission rate through the lower layer. In such circumstances, the accumulation of water over the interlayer boundary can result, temporarily at least, in the development of a "perched" (or secondary) water table with positive hydrostatic pressures. If infiltration persists at a relatively high rate, the two bodies of water will eventually merge. If infil-

## C. Analysis of Falling Water Table

Truly steady-state flow processes are rare in unconfined aquifers (as elsewhere). More typical are transient-state processes which involve a change in water-table height. Transient flow under such conditions has been described in terms of the *specific-yield* concept, generally defined as the volume of water extracted from the groundwater per unit area when the water table is lowered a unit distance (or as the ratio of drainage flux to rate of fall of the water table). The assumption that there exists anything like a fixed value of *drainable porosity* and that this fraction of the soil volume drains instantaneously as the water table descends is a gross approximation. In actual fact, the volume of water drained increases gradually with the increasing suction which accompanies the progressive descent of the water table.

According to Luthin (1966), the drainable porosity $f_d$ is not a constant but a function of the capillary pressure $h$ (i.e., the negative pressure head, also called the suction head) and can be written as $f_d(h)$. As the water table drops from $h_1$ to $h_2$, the volume of water $V_w$ drained out of a unit column will be

$$V_w = \int_{h_2}^{h_1} f_d(h)\, dh \qquad (14.1)$$

The function $f_d(h)$ is related to the soil moisture characteristic and is generally a complex function. However, it is sometimes possible to write an approximate expression for the relation of drainable porosity to capillary pressure and still end up with a reasonable prediction of the amount of water that drains out of a profile of soil in which the water table is dropping. The simplest equation to use is that of a straight line $f_d = ah$, where $a$ is the slope of the line. The quantity of water drained is now given by

$$V_w = \int_{h_2}^{h_1} ah\,dh = \frac{a}{2}(h_1^2 - h_2^2) \qquad (14.2)$$

Problems involving falling water tables have been handled by representing the transient process as a succession of steady states (Childs, 1947; Kirkham and Gaskell, 1951; Collis-George and Youngs, 1958; Isherwood, 1959; Bouwer and van Schilfgaarde, 1963). Childs and Poulovassilis (1962) showed that the soil-moisture profile above a descending water table depends upon the time rate of that descent.

The time-dependent vertical drainage of a soil column following a drop of the water table has been studied by Day and Luthin (1956), Youngs (1960), Gardner (1962a), Jensen and Hanks (1967), and Jackson and Whisler (1970).

Gardner assumed a constant mean diffusivity and a linear relation between hydraulic head and soil wetness. His equation is

$$\frac{V_w}{V_{w\infty}} = 1 - \frac{8}{\pi^2}\exp\left(-\frac{\bar{D}\pi^2 t}{4L^2}\right) \qquad (14.3)$$

where $V_w$ is the volume of water per unit area removed during time $t$, $V_{w\infty}$ is the total drainage after infinite time, $\bar{D}$ is the weighted mean diffusivity of the soil, and $L$ is the column length (height).

Young's derivation was based on the capillary-tube model and on the assumption of a downward moving *drainage front*, i.e., the assumption that a distinct demarcation of constant suction exists between the still-saturated and the already drained zones. This implies that an equal quantity of water drains for an equal downward advance of the drainage front. The above assumption is analogous to the Green and Ampt assumption pertaining to infiltration (namely, that there exists a distinct wetting front which remains at constant suction as it advances down the profile). The approximate equation obtained is

$$V_w/V_{w\infty} = 1 - \exp(-q_0 t/V_{w\infty}) \qquad (14.4)$$

where $q_0$ is the initial flux of drainage at the base of the column (the water table) and the other variables are as above.

Youngs' approach was modified by Jackson and Whisler (1970), who assumed that hydraulic conductivity is linearly related to average soil wetness during drainage and that the latter is related to the average length of soil column yet to be drained. They also assumed a constant drainable

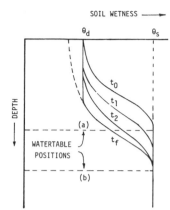

**Fig. 14.1.** Idealized succession of soil moisture profiles following a rapid drop of the water table from position (a) to (b) (schematic). $t_0$ and $t_f$ represent the equilibrium moisture profiles at time zero and at final completion of vertical drainage, respectively. $t_1$ and $t_2$ represent intermediate times during vertical drainage. $\theta_s$ indicates saturation and $\theta_d$ the presumed wetness after drainage; i.e., $\theta_s - \theta_d = $ "drainable" porosity. Note that in reality $\theta_d$ is generally not a constant, but itself depends on proximity to the water table (as shown by the dashed line extension of the $t_f$ profile), on soil profile characteristics, and on the fact that true equilibrium is practically never attained. Moreover, the soil moisture profile above a fluctuating water table also depends on hysteresis and, if the water table is close to the soil surface, on plant activity as well.

porosity. They claimed that their theory accords with experimental data over a much longer time than previously published theories.

In general, the applicability of such approximate theories depends on the validity of the assumptions made in their derivation. Some approximate solutions which provide reasonably accurate predictions of drainage outflow do not give an accurate description of the changing moisture and suction profiles above the water table (Fig. 14.1). More exact solutions of nonsteady drainage problems can be obtained by numerical techniques (e.g., Jensen and Hanks, 1967), but such procedures require the use of a computer for each case considered.

Most of the equations available for the estimation of drainage from soils with falling water tables disregard the possible effect of evapotranspiration. If evapotranspiration occurs while the soil is draining, the amount of drainage will obviously be lessened, and equations such as those cited will give overestimations.

### D. Equations Pertaining to Flow of Unconfined Groundwater

Darcy's law alone is sufficient to describe only steady flow processes. In general, however, Darcy's law must be combined with the mass-conservation law to obtain the general flow equation for homogeneous isotropic media (Hubbert, 1940; de Wiest, 1969).

$$\frac{\partial \theta}{\partial t} = -\left(\frac{\partial q_x}{\partial x} + \frac{\partial q_y}{\partial y} + \frac{\partial q_z}{\partial z}\right)$$

$$\frac{\partial \theta}{\partial t} = K\left(\frac{\partial^2 H}{\partial x^2} + \frac{\partial^2 H}{\partial y^2} + \frac{\partial^2 H}{\partial z^2}\right)$$

(14.5)

where $q_x$, $q_y$, and $q_z$ are the fluxes in the $x$, $y$, and $z$ directions, respectively, $\theta$ is wetness, $t$ time, $K$ conductivity, and $H$ hydraulic head. In a saturated, stable medium, there is no change of wetness (water content) with time, and we obtain

$$K_s\left(\frac{\partial^2 H}{\partial x^2} + \frac{\partial^2 H}{\partial y^2} + \frac{\partial^2 H}{\partial z^2}\right) = 0$$

(14.6)

where $K_s$ is the saturated conductivity. Since $K_s$ is not zero, it follows that

$$\frac{\partial^2 H}{\partial x^2} + \frac{\partial^2 H}{\partial y^2} + \frac{\partial^2 H}{\partial z^2} = 0$$

(14.7)

This equation is known as the *Laplace equation*. The expression $\partial^2/\partial x^2 + \partial^2/\partial y^2 + \partial^2/\partial z^2$, or in vector notation $\nabla^2$, is known as the Laplacian operator. Accordingly, we can write Laplace's equation $\nabla^2 H = 0$.

If, instead of using Cartesian coordinates $(x, y, z)$, we cast Eq. (14.7) into cylindrical coordinates $(r, \alpha, z)$, we obtain

$$\frac{1}{r}\frac{\partial}{\partial r}\left(r\frac{\partial H}{\partial r}\right) + \frac{1}{r^2}\frac{\partial^2 H}{\partial \alpha^2} + \frac{\partial^2 H}{\partial z^2} = 0$$

(14.8)

Laplace's equation also applies to systems other than fluid flow in porous media, namely to the flow of heat in solids and of electricity in electrical conductors. Solutions for boundary values appropriate to the latter systems, some of which are also applicable to soil-water flow, are given by Smythe (1950) and by Carslaw and Jaeger (1959).

The direct analytical solution of Laplace's equation for conditions pertinent to groundwater flow is not generally possible. Therefore, it is often necessary to resort to approximate or indirect methods of analysis. Where flow is restricted to two dimensions, the equation becomes

$$\frac{\partial^2 H}{\partial x^2} + \frac{\partial^2 H}{\partial y^2} = 0$$

(14.9)

which is more readily soluble (Childs, 1969; Domenico, 1972).

Approximate solutions for unconfined groundwater flow problems were reviewed by van Schilfgaarde (1957), who stressed that the simplifications possible in a less than rigorous approach can be of considerable value in practice, but that they require a constant awareness of the limitations which are inherent in the use of such simplifications. His discussion encompasses both steady-state processes (in which the potentials in the system do not change with time) and nonsteady, or transient-state, processes (during which the potentials and fluxes vary).

In the solution of problems relating to unconfined groundwater flow toward a shallow sink (a drainage tube or ditch), it is often convenient to employ the *Dupuit–Forchheimer assumptions* (Forchheimer, 1930) that, in a system of gravity flow toward a shallow sink, all the flow is horizontal and that the velocity at each point is proportional to the slope of the water table but independent of depth. Though these assumptions are obviously not correct in the strict sense and can in some cases lead to anomalous results (Muskat, 1946), they often provide feasible solutions in a form simpler than obtainable by rigorous analysis. They are most suitable where the flow region is of large horizontal extent relative to its depth. Kirkham (1967) showed that the Dupuit–Forchheimer assumptions give exact results if they are applied to a soil having infinite conductivity in the vertical direction. For real soils the results are only approximate.

## E. Groundwater Drainage

The term "drainage" can be used in a general sense to denote outflow of water from soil. More specifically, it can serve to describe the *artificial* removal of excess water, or the set of management practices designed to prevent the occurrence of excess water. The removal of free water tending to accumulate over the soil surface by appropriately shaping the land is termed *surface drainage* and is outside the scope of our present discussion. Finally, *groundwater drainage* refers to the outflow or artificial removal of excess water from within the soil, generally by lowering the water table or by preventing its rise.

Soil saturation *per se* is not necessarily harmful to plants. The roots of very many plants can, in fact, thrive in water, provided it is free of toxic substances and contains sufficient oxygen to allow normal respiration. As is well known, plant roots must respire constantly, since most terrestrial plants are unable to transfer the required oxygen from their canopies to their roots. The problem is that water in a saturated soil seldom can provide sufficient oxygen for root respiration. Excess water in the soil tends to block soil pores and thus retard aeration and in effect strangulate the roots. In *water-logged* soils, gas exchange with the atmosphere is restricted to the surface zone of the soil, while within the profile proper, oxygen may be almost totally absent and carbon dioxide may accumulate. Under anaerobic conditions, various substances are reduced from their normally oxidized states. Toxic concentrations of ferrous, sulfide, and manganous ions can develop. These, in combination with products of the anaerobic decomposition of organic matter (e.g., methane) can greatly inhibit plant growth. At the same time, nitrification is prevented, and various plant and root diseases (especially fungal) are more prevalent.

The occurrence of a high-water-table condition may not always be clearly evident at the very surface, which may be deceptively dry even while the soil is completely water-logged just below the surface zone. Where the effective rooting depth is thus restricted, plants may suffer not only from lack of oxygen in the soil, but also from lack of nutrients. If the water table drops periodically, plants growing in water-logged soils may even, paradoxically, suffer from occasional lack of water, especially when the transpirational demand is very high. High moisture conditions at or near the soil surface cause the soil to be susceptible to compaction by animal and machinery traffic. Necessary operations (e.g., tillage, planting, spraying, and harvesting) are thwarted by poor trafficability (i.e., the ability of the ground to support vehicular traffic). Tractors are bogged down and cultivation tools are clogged by the soft, sticky, wet soil. Furthermore, the surface zone of a wet soil does not warm up readily at springtime, owing to greater thermal inertia and downward conduction and to loss of latent heat by the higher evaporation rate. Consequently germination and early seedling growth are retarded.

Plant sensitivity to restricted drainage is itself affected by temperature. A rise in temperature is associated with a decrease in the solubility of oxygen in water and with an increase in the respiration rate of both plant roots and soil microorganisms. The damage caused by excessive soil moisture is therefore likely to be greater in a warm climate than in a cold one. Moreover, in a warm climate, the evaporation rate and, hence, the hazard of salinity are likely to be greater than in a cool climate. The process of evaporation inevitably results in the deposition of salts at or near the soil surface, and these salts can be removed and prevented from accumulating only if the water table remains deep enough to permit leaching without subsequent resalinization through capillary rise of the groundwater. Irrigated lands, even in arid regions, frequently require drainage. In fact, irrigation without drainage can be disastrous. Once-thriving civilizations based on irrigated agriculture in river valleys (as in Mesopotamia, for instance) have been destroyed through the insidious, and for a time invisible, process of salt accumulation caused by poor drainage.[6]

In many regions, such as coastal plains and river valleys, large tracts of land that are potentially highly productive lie waste, or are of restricted use, because of excessive moisture. Wherever topographic conditions, soil imperviousness, or the presence of shallow groundwater prevent the profile from draining itself adequately, the soil may become an unsuitable medium for plant growth unless drained artificially. Artificial drainage of such lands can result in the reclamation of millions of acres for the production of food and fiber for the world's growing population. In large areas, likewise, drainage is required for the long-term maintenance of soil productivity. Irrigated agriculture cannot long be sustained in many arid regions unless drainage is provided for salinity control as well as for effective soil aeration (Richards, 1954; Fireman, 1957).

On the other hand, the presence of a shallow water table in the soil profile (provided it is not too shallow) can in certain circumstances be beneficial. Where precipitation or irrigation water is scarce, the availability of groundwater within reach of the roots can supplement the water requirements of crops. However, to obtain any lasting benefit from the presence of a water table in the soil, its level and fluctuation must be controlled.

Numerous investigations of groundwater flow and drainage have resulted in a very extensive body of literature on this subject. Reference should be made particularly to the books by Luthin (1957) and by van Schilfgaarde (1974).

In the field, flow conditions can be determined by the measurement of hydraulic head at different points in the soil. This is generally done by means of piezometers, which are borings made to below the water table, to allow repeated measurements of the hydraulic head at various sites (Fig. 14.2). The water levels in piezometers do not indicate the water-table position if there is a vertical component to the flow below the water table, as shown in Fig. 14.3. In this case, the water level in a piezometer depends on the piezometer's depth of penetration. The position of the water table can be estimated from the water level in the shallowest piezometer. The water level difference between any pair of adjacent piezometers inserted to different depths will be proportional directly to the downward (or upward) flux

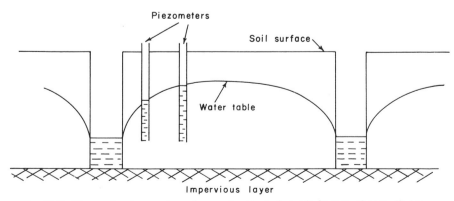

**Fig. 14.2.** Piezometers in a system of groundwater drainage by means of open ditches.

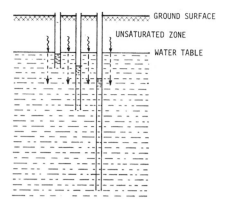

**Fig. 14.3.** Battery of piezometers in an unconfined aquifer with downward flow.

and inversely to the hydraulic conductivity of the intervening soil layer. An example of downward flow below a water table is shown in Fig. 14.4.

### F. Factors Influencing Drainage

The artificial drainage of groundwater is generally carried out by means of *drains*, which may be ditches, pipes, or "mole channels," into which groundwater flows as a result of the hydraulic gradients existing in the soil. The drains themselves are made to flow, by gravity or by pumping, to the *drainage outlet*, which may be a stream, a lake, an evaporation pond, or the sea. In some places, drainage water may be recycled, or reused, for agricultural, industrial, or residential purposes. Because drainage water may contain potentially harmful concentrations of salts, fertilizer nutrients, and pesticide residues, it is not enough to provide means to "get rid" of it; nowadays one must be concerned with the quality of the water to be disposed of and with the long-term consequences of its disposal.

The flow rate from soil to drains depends upon the following factors (among others):

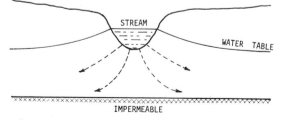

**Fig. 14.4.** Seepage from a stream into a shallow unconfined aquifer overlying an impermeable layer.

(1)   Hydraulic conductivity of the soil, which may differ from one layer or place to another if the soil is heterogeneous and may also vary directionally if the soil is anisotropic. Some soils can be drained easily; others are extremely difficult to drain. Generally, coarse-textured soils drain more readily than fine-textured ones, though of course permeability does not depend on texture alone. The presence of profile layers of low conductivity can greatly retard drainage by affecting the flow pattern in the soil.

(2)   Configuration of the water table and hydraulic pressure of the groundwater. The water table is seldom horizontal or of constant depth. Furthermore, in some cases, the groundwater may be confined and exhibit artesian pressure.

(3)   Depths of the channel or tube drains, relative to the groundwater table and to the soil surface, as well as the slopes of these drains and their outlet elevation.

(4)   Horizontal spacing between drains. According to Kirkham (1949), the discharge per drain becomes constant when the drains are spaced more than about 7 m apart. The total drainage discharged from a field then becomes proportional to the number (i.e., density) of drains installed.

(5)   Character of the drains, whether open ditches or tubes. Open ditches allow a greater seepage surface and are easier to monitor than underground tubes, which are also more expensive to install. However, open ditches break the continuity of the field and are difficult to traverse. They constrain field operations, might scour and collapse, and often allow the proliferation of weeds and pests. Tubes seem therefore to be the preferred mode of drainage.

(6)   Inlet openings in the drain tubes. If segmented tubes are used, small gaps are left between the sections of the tubes to allow inflow from the soil. If continuous tubes are used, they are perforated.

(7)   Envelope materials. Drainage tubes are commonly embedded in gravel to increase the seepage surface and therefore the discharge and to prevent scouring or collapse of the soil at the inlets and possible clogging of the drains by penetration of soil material.

(8)   Diameters of the drains. These must be sufficient to convey the necessary drainage discharge. Installation of wide enough tubes at the outset is no permanent guarantee: drainage tubes tend to clog within a few years, owing to penetration and deposition of sediments and to precipitation of salts (e.g., gypsum) and oxides (e.g., reduced iron and manganese carried in solution may oxidize and precipitate in the drains). Hence the tubes must be flushed out periodically.

(9)   The rate at which water is added to the groundwater (Freeze, 1969) by the excess of infiltration over evapotranspiration or by lateral flow from an external source of water (i.e., outside the field being drained). A steady flow condition, in which the rate of inflow by infiltration is equal to the rate of outflow by drainage, is depicted in Fig. 14.5, in comparison with an unsteady flow condition resulting in a falling water table.

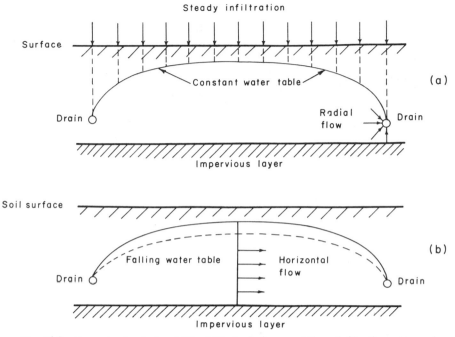

**Fig. 14.5.** Groundwater drainage (a) under steady flow conditions (infiltration rate equals drainage rate and the water table remains at constant depth), and (b) under unsteady flow resulting in a falling water table.

It is a truism which nevertheless bears repeating that water will not spontaneously flow out of the soil into a large cavity or drain unless the pressure of soil water is greater than atmospheric. Drains must be located below the water table to draw water, and the water table cannot be lowered below the drains. Hence the depth and spacing of drains is of crucial importance. Insufficient depth of placement will prevent a set of drains from lowering the water table to the extent necessary. Too great a depth might, on the other hand, deprive the plants of a possibly important source of water during drought periods or whenever evapotranspiration exceeds precipitation.

## G. Drainage Design Equations

Various equations, empirically or theoretically based, have been proposed for the purpose of determining the desirable depths and spacings of drain pipes or ditches in different soil and groundwater conditions. Since field conditions are often complex and highly variable, these equations are generally based upon assumptions which idealize and simplify the flow system. The available equations are therefore approximations which should not be applied blindly. Rather, the assumptions must be examined in the light of all information obtainable concerning the circumstances at hand.

One of the most widely applied equations is that of Hooghoudt (1937), designed to predict the height of the water table which will prevail under a given rainfall or irrigation regime when the conductivity of the soil and the depth and horizontal spacing of the drains are known. This equation, like others of its type, oversimplifies the real field situation, as it disregards additional factors which may have a bearing upon groundwater movement, such as the variable rate of evapotranspiration, soil layering, etc. It is based on the following tacit assumptions:

(1)   the soil is homogeneous and of constant hydraulic conductivity;

(2)   the drains are parallel and equally spaced;

(3)   the hydraulic gradient at each point beneath the water table is equal to the slope of the water table above that point (this gradient is generally directed toward the nearest drain);

(4)   Darcy's law applies;

(5)   an impervious layer exists at a finite depth below the drain;

(6)   the supply of water from above, due to rain or irrigation (presumably, after taking evapotranspiration into account) is at a constant flux $q$.

The shape of the water table between parallel drains is generally described as elliptical.

To derive the Hooghoudt equation, let us examine flow in a profile section of a field having a width of one unit and bounded at its sides by two adjacent drains (tubes or ditches) a distance $S$ apart (Fig. 14.6). Assuming symmetry, we can draw a vertical midplane between the drains which will divide flow toward one drain from flow to the other. Now let us consider flow toward one of the drains through any arbitrary vertical plane located a distance $x$ from that drain. The quantity of water passing through this plane per unit time must be equal to the percolation flux $q$ multiplied by the width from the arbitrary plane to the midplane between the drains. This width is $\frac{1}{2}S - x$. Accordingly, the horizontal flow per unit time through the arbitrary plane is

$$Q = -q(\tfrac{1}{2}S - x) \tag{14.10}$$

At the same time, $Q$ can be obtained from Darcy's law. If we assume the effective gradient to be equal to the slope of the water table $(dh/dx)$ at the arbitrary vertical plane, we get

$$Q = -Kh\,dh/dx \tag{14.11}$$

where $K$ is the hydraulic conductivity and $h$ is the height of the water table above an impervious layer which is assumed to form the "floor" of the

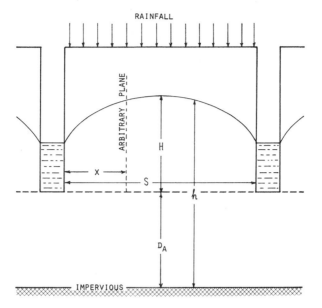

**Fig. 14.6.** Model used in derivation of Hooghoudt's equation.

flow system. Now we can equate the two equations:

$$q(\tfrac{1}{2}S - x) = Kh\, dh/dx \qquad (14.12)$$

or

$$\tfrac{1}{2}qS\, dx - qx\, dx = Kh\, dh$$

which can be integrated to yield

$$qSx - qx^2 = Kh^2 \qquad (14.13)$$

Assuming that at $x = 0$ (i.e., at the drain) $h = d_a$ (the height of the drain above the impervious floor), while at $x = \tfrac{1}{2}S$ (the midplane) $h = H + d_a$ (where $H$ is the maximal height of the water table above the drains), we obtain Hooghoudt's equation for the elliptical shape of the water table between drains:

$$S^2 = (4KH/q)(2d_a + H) \qquad (14.14)$$

This equation has been used widely for determining the desirable spacing and depth of drains needed to maintain the water table below a certain level. That level, as well as the average infiltration flux and hydraulic conductivity, must be known a priori. A depth must also be known, or assumed, for the impervious layer.

In the event that $H$ is negligible compared to $d_a$, we can write

$$Q = -Kd_a \, dH/dx \tag{14.15}$$

Equating this with (14.10), we obtain

$$q(\tfrac{1}{2}S - x) = Kd_a \, dH/dx, \qquad \tfrac{1}{2}qS \, dx - qx \, dx = Kd_a \, dH$$

Integration yields

$$qSx - qx^2 = 2KHd_a + \text{const} \tag{14.16}$$

Since at $x = 0$, $H = 0$, the constant of integration is also equal to 0. Therefore

$$H = qx(S - x)/2Kd_a \tag{14.17}$$

At the midpoint between drains, where $x = \tfrac{1}{2}S$, we get for the maximum height $H_{max}$ of the water-table mound

$$H_{max} = qS^2/8Kd_a \tag{14.18}$$

We thus see that the height of rise of the water table between drains is related directly to the recharging flux $q$ and to the square of the distance $S$ between drains and inversely related to the soil's hydraulic conductivity $K$.

An equation to describe the time rate of water-table drop at the midpoint between drains following an abrupt drop of the water table at the drains, known as the *Glover equation*, is as follows:

$$S^2 = (\pi^2 K \bar{h}_i t/f_d) \ln(4H_i/\pi H) \tag{14.19}$$

wherein $\bar{h}_i$ is the average initial depth of the water-bearing stratum, $f_d$ the assumed drainable porosity, $H_i$ the initial height of the midpoint water table above the drains, and $H$ the height at any time $t$.

Among the most serious weaknesses of the approach described are the assumptions of an impervious layer at some definable shallow depth and

**Table 14.1**

PREVALENT DEPTHS AND SPACINGS OF DRAINAGE TUBES
IN VARIOUS SOIL TYPES

| Soil type | Hydraulic conductivity (cm/day) | Spacing of drains (m) | Depth of drains (m) |
|---|---|---|---|
| Clay | 0.15 | 10–20 | 1–1.5 |
| Clay loam | 0.15–0.5 | 15–25 | 1–1.5 |
| Loam | 0.5–2.0 | 20–35 | 1–1.5 |
| Fine, sandy loam | 2.0–6.5 | 30–40 | 1–1.5 |
| Sandy loam | 6.5–12.5 | 30–70 | 1–2 |
| Peat | 12.5–25 | 30–100 | 1–2 |

the disregard of that portion of the total flow which occurs above the water table (Donnan, 1947; Bouwer, 1959). Corrections to account for that flow were described by van Schilfgaarde (1974).

Other equations, derived by alternative and in some cases more rigorous procedures, have been offered by, among others, Kirkham (1958), Ernst (1962), and the U.S. Bureau of Reclamation (Luthin, 1966). The ranges of depth and spacing generally used for the placement of drains in field practice are shown in Table 14.1.

It is of interest to note that in Holland, the country with the most experience in drainage, a common criterion for drainage is to provide for the removal of about 7 mm/day and to prevent a water-table rise above 50 cm from the soil surface. In more arid regions, because of the greater evaporation rate and groundwater salinity, the water table must generally be kept very much deeper. In the Imperial Valley of California, for instance, the drain depth ranges from about 150 to 300 cm and the water-table depth midway between drains is about 120 cm. For medium- and fine-textured soils the depth should be greater still where the salinity risk is high. Since there is a practical limit to the depth of drain placement, it is the density of drain spacing which must be increased under such circumstances. By setting adjacent lines closer together, we can ensure that their drawdown curves will intersect at a lower midpoint level.

## Sample Problems

**1.** An initially saturated vertical soil column is drained by dropping the water table abruptly to a level 1 m below its original height, as depicted in Fig. 14.7. Plot the fractional amount of drainable water removed as a function of time if the weighted mean diffusivity of the draining soil is $10^{-2}$ cm²/sec. Use Eq. (14.3).

According to Gardner (1962a), the volume of water $V_w$ drained per unit area, as a fraction of the total drainable water $V_{w\infty}$ (i.e., the cumulative volume per unit area which drains after infinite time), is a function of time $t$, weighted mean diffusivity $\bar{D}$ and column length $L$, according to

$$V_w/V_{w\infty} = 1 - (8/\pi^2) \exp(-\bar{D}\pi^2 t/4L^2)$$

We can now substitute the appropriate values of $\bar{D}$ and $L$ and assign successive values of $t$ to obtain actual solutions, as follows:

At $t = 1$ hr $= 3600$ sec,

$$V_w/V_{w\infty} = 1 - 8/(9.87 \times e^\varepsilon)$$

where $\varepsilon = 10^{-2} \times 9.87 \times 3600/(4 \times 100^2) = 8.88 \times 10^{-3}$. Hence

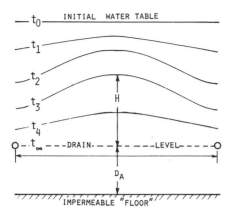

**Fig. 14.7.** Idealized succession of water table positions during drainage of an initially saturated profile.

$$V_w/V_{w\infty} = 0.197 = 19.7\%$$

At $t = 4$ hr $= 14400$ sec,

$$\varepsilon = 10^{-2} \times 9.87 \times 14400/(4 \times 100^2) = 3.55 \times 10^{-2}$$

$$V_w/V_{w\infty} = 1 - 8/(9.87 \times e^{3.55 > 10^{-2}}) = 0.218 = 21.8\%$$

At $t = 1$ day $= 86400$ sec,

$$\varepsilon = 10^{-2} \times 9.87 \times 86400/(4 \times 100^2) = 0.213$$

$$V_w/V_{w\infty} = 1 - 8/(9.87 \times e^{0.213}) = 0.345 = 34.5\%$$

At $t = 2$ days $= 172800$ sec,

$$\varepsilon = 10^{-2} \times 9.87 \times 172800/(4 \times 100^2) = 0.426$$

$$V_w/V_{w\infty} = 1 - 8/(9.87 \times e^{0.426}) = 0.471 = 47.1\%$$

At $t = 4$ days $= 345600$ sec,

$$\varepsilon = 0.852 \qquad V_w/V_{w\infty} = 1 - 8/(9.87 \times e^{0.852}) = 0.654 = 65.4\%$$

At $t = 8$ days $= 691200$ sec,

$$\varepsilon = 1.704 \qquad V_w/V_{w\infty} = 1 - 8/(9.87 \times e^{1.704}) = 0.853 = 85.3\%$$

*Note:* The actual plotting of these data is left as an exercise for students, who may also wish to compare Gardner's theory with one or more of the other theories extant. Note that the equation used is not suitable for very small values of $t$, for which the fraction of water drained is overestimated (since $V_w/V_{w\infty}$ does not tend to zero as $t$ approaches zero).

**2.** Use the Hooghoudt equation to compare the necessary drain spacings $S$ for two soils with hydraulic conductivity $K$ values of $10^{-4}$ and $10^{-5}$ cm/sec. Assume the allowable maximum water-table mound $H_{max}$ to be 1 m above the drains, which are 2 m ($d_a$) above an impervious stratum. Total rainfall is 1200 mm and total evapotranspiration 1000 mm during a 6 month growing season.

Recall Eq. (14.18):

$$H_{max} = qS^2/8Kd_a$$

which we can solve for $S$:

$$S = (8Kd_a H_{max}/q)^{1/2}$$

where $q$ is the average percolation flux. Substituting the given values, we have

for Soil A,

$$S = \left[ \frac{8 \times 10^{-4} \text{ cm/sec} \times 200 \text{ cm} \times 100 \text{ cm}}{(120 \text{ cm} - 100 \text{ cm})/15{,}760{,}000 \text{ sec}} \right]^{1/2}$$

in which the number of seconds by which we divide the denominator represents a 6 month period. Hence,

$$S = 3550.8 \text{ cm} = 35.5 \text{ m}$$

for Soil B,

$$S = \left[ \frac{8 \times 10^{-5} \text{ cm/sec} \times 200 \text{ cm} \times 100 \text{ cm}}{(120 - 100) \text{ cm}/15{,}760{,}000 \text{ sec}} \right]^{1/2} = 1122.6 \text{ cm} = 11.23 \text{ m}$$

# 15   *Evaporation from Bare-Surface Soils*

## A. Introduction

Evaporation in the field can take place from plant canopies, from the soil surface, or, more rarely, from a free-water surface. Evaporation from plants, called *transpiration*, is the principal mechanism of soil-water transfer to the atmosphere when the soil surface is covered with vegetation. Soil-water uptake and transpiration by plants is, however, the subject of our next chapter. When the surface is at least partly bare, evaporation can take place from the soil as well as from plants. Since it is generally difficult to separate these two interdependent processes, they are commonly lumped together and treated as if they were a single process, called *evapotranspiration*.

In the absence of vegetation, and when the soil surface is subject to radiation and wind effects, evaporation occurs directly and entirely from the soil. This process is the subject of our present chapter. It is a process which, if uncontrolled, can involve very considerable losses of water in both irrigated and unirrigated agriculture. Under annual field crops, the soil surface may remain largely bare throughout the periods of tilllage, planting, germination, and early seedling growth, periods in which evaporation can deplete the moisture of the surface soil and thus hamper the growth of young plants during their most vulnerable stage. Rapid drying of a seedbed can thwart germination and thus doom an entire crop from the start. The problem can also be acute in young orchards, where the soil surface is often kept bare continuously for several years, and in dryland farming in arid zones, where the land is regularly fallowed for several months to collect and conserve rainwater from one season to the next.

Evaporation of soil water involves not only loss of water but also the danger of soil salinization. This danger is felt most in regions where irrigation water is scarce and possibly brackish and where annual rainfall is low, as well as in regions with a high groundwater table.

## B. Physical Conditions

Three conditions are necessary if the evaporation process from a given body is to persist. First, there must be a continual supply of heat to meet the latent heat requirement (which is about 590 cal/gm of water evaporated at 15°C). This heat can come from the body itself, thus causing it to cool, or as is more commonly the case, it can come from the outside in the form of radiated or advected energy. Second, the vapor pressure in the atmosphere over the evaporating body must remain lower than the vapor pressure at the surface of that body[1] (i.e., there must be a vapor-pressure gradient between the body and the atmosphere), and the vapor must be transported away, by diffusion or convection, or both. These two conditions—namely, supply of energy and removal of vapor—are generally external to the evaporating body and are influenced by meteorological factors such as air temperature, humidity, wind velocity, and radiation, which together determine the *atmospheric evaporativity* (the maximal flux at which the atmosphere can vaporize water from a free-water surface).[2]

The third condition is that there be a continual supply of water from or through the interior of the body to the site of evaporation. This condition depends upon the content and potential of water in the body as well as upon its conductive properties, which together determine the maximal rate at which the body can transmit water to the evaporation site. Accordingly, the actual evaporation rate is determined either by external evaporativity or by the soil's own ability to deliver water, whichever is the lesser (and hence the limiting factor).

If the top layer of soil is initially quite wet, as it typically is at the end of an infiltration episode, the process of evaporation into the atmosphere will generally reduce soil wetness and thus increase matric suction at the surface. This, in turn, will generally cause soil water to be drawn upward from the layers below, provided they are sufficiently moist.

---

[1] Even in a humid region, the atmosphere on a clear day is likely to be at a relative humidity equivalent to a negative water potential of many score or even hundreds of bars.

[2] Atmospheric evaporativity, also called "the evaporative demand of the atmosphere," is not entirely independent of the properties of the evaporating surface. For instance, the net supply of energy for evaporation is affected by the reflectivity, emissivity, and thermal conductivity of the surface soil. Hence, the evaporative demand acting on a soil will not be exactly equal to evaporation from a free-water surface. The latter itself depends on the size and depth of the water body considered.

Among the various sets of conditions under which evaporation may occur are the following:

(1)   A shallow groundwater table may be present at a constant or variable depth—or it may be absent (or too deep to affect evaporation). Where a groundwater table occurs close to the surface, continual flow may take place from the saturated zone beneath through the unsaturated soil to the surface. If this flow is more or less steady, continued evaporation can occur without materially changing the soil-moisture content (though cumulative salinization may take place at the surface). In the absence of shallow groundwater, on the other hand, the loss of water at the surface and the resulting upward flow of water in the profile will necessarily be a transient-state process causing the soil to dry.

(2)   The soil profile may be uniform (homogeneous and isotropic). Alternatively, soil properties may change gradually in various directions, or the profile may consist of distinct layers differing in texture or structure.

(3)   The profile may be shallow (of *finite depth*), resting on bedrock or some other impervious floor, or it may be deep (*semi-infinite*). In intermediate cases, the soil profile may be effectively semi-infinite for a time, then become finite as the downward-propagating effect of the evaporation process reaches the bottom boundary.

(4)   The flow pattern may be one dimensional (vertical), or it may be two or three dimensional, as in the presence of vertical cracks which form secondary evaporation planes inside the profile.

(5)   Conditions may be nearly isothermal or strongly nonisothermal. In the latter case, temperature gradients and conduction of heat and vapor through the system may interact with liquid water flow.

(6)   External environmental conditions may remain constant or fluctuate. Such fluctuation, furthermore, can be predictably periodic and regular (e.g., diurnal or seasonal) or it can be highly irregular (e.g., spells of cool or warm weather).

(7)   Soil moisture flow may be governed by evaporation alone or by both evaporation (at the top of the profile) and internal drainage, or redistribution, down below.

(8)   The soil may be stable or unstable. For instance, the surface zone may become denser under traffic or under raindrop impact and subsequent shrinkage. Additionally, the soil surface may become encrusted or infused with salt, which then precipitates as the soil solution evaporates. This is quite apart from the fact that, as a soil dries, its thermal properties inevitably change, including its thermal conductivity and reflectivity.

(9)   The surface may or may not be covered by a layer of mulch (e.g., plant residues) differing from the soil in hydraulic, thermal, and diffusive properties.

(10) Finally, the evaporation process may be continuous over a prolonged period of time or it may be interrupted by regularly recurrent or sporadic episodes of rewetting (e.g., intermittent rainfall or scheduled irrigation).

To be studied systematically, each of the listed circumstances, as well as others not listed but perhaps equally relevant, must be formulated in terms of a specific set of initial and boundary conditions. A proper formulation of an evaporation process should account for spatial and temporal variability, as well as for interactions with the above-ground and below-ground environment. We shall now proceed to describe a few of the circumstances under which evaporation of soil moisture may occur.

### C. Evaporation in the Presence of a Water Table

The steady-state upward flow of water from a water table through the soil profile to an evaporation zone at the soil surface was first studied by Moore (1939). Theoretical solutions of the flow equation for this process were given by several workers, including Philip (1957d), Gardner (1958), Anat *et al.* (1965), and Ripple *et al.* (1972).

The equation describing steady upward flow is

$$q = K(\psi)(d\psi/dz - 1) \tag{15.1}$$

or

$$q = D(\theta)\,d\theta/dz - K(\psi) \tag{15.2}$$

where $q$ is flux (equal to the evaporation rate under steady-state conditions), $\psi$ suction head, $K$ hydraulic conductivity, $D$ hydraulic diffusivity, $\theta$ volumetric wetness, and $z$ height above the water table. The equation shows that flow stops ($q = 0$) when $d\psi/dz = 1$. Another form of Eq. (15.1) is

$$q/K(\psi) + 1 = d\psi/dz \tag{15.3}$$

Integration should give the relation between depth and suction or wetness:

$$z = \int \frac{d\psi}{1 + q/K(\psi)} = \int \frac{K(\psi)}{K(\psi) + q}\,d\psi \tag{15.4}$$

or

$$z = \int \frac{D(\theta)}{K(\theta) + q}\,d\theta \tag{15.5}$$

In order to perform the integration in Eq. (15.4) we must know the functional relation between $K$ and $\psi$, i.e., $K(\psi)$. Similarly, the functions $D(\theta)$ and

$K(\theta)$ must be known if Eq. (15.5) is to be integrated. An empirical equation for $K(\psi)$, given by Gardner (1958), is

$$K(\psi) = a(\psi^n + b)^{-1} \qquad (15.6)$$

where the parameters $a,b$, and $n$ are constants which must be determined for each soil. Accordingly, Eq. (15.1) becomes

$$e = q = \frac{a}{\psi^n + b}\left(\frac{d\psi}{dz} - 1\right) \qquad (15.7)$$

where $e$ is the evaporation rate.

With Eq. (15.6), Eq. (15.4) can be used to obtain suction distributions with height for different fluxes, as well as fluxes for different surface-suction values. The theoretical solution is shown graphically in Fig. 15.1 for a fine sandy loam soil with an $n$ value of 3.

The curves show that the steady rate of capillary rise and evaporation depend on the depth of the water table and on the suction at the soil surface.

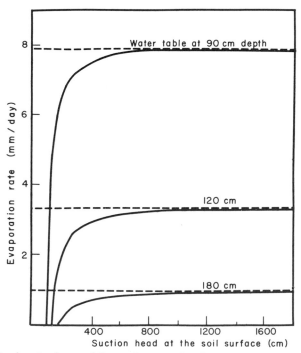

**Fig. 15.1.** Steady rate of upward flow and evaporation from a water table as function of the suction prevailing at the soil surface. The soil is a fine sandy loam, with $n = 3$. (After Gardner, 1958.)

This suction is dictated largely by the external conditions, since the greater the atmospheric evaporativity, the greater the suction at the soil surface upon which the atmosphere is acting. However, increasing the suction at the soil surface, even to the extent of making it infinite, can increase the flux through the soil only up to an asymptotic maximal rate which depends on the depth of the water table. Even the driest and most evaporative atmosphere cannot steadily extract water from the surface any faster than the soil profile can transmit from the water table to that surface. The fact that the soil profile can limit the rate of evaporation is a remarkable and useful feature of the unsaturated flow system. The maximal transmitting ability of the profile depends on the hydraulic conductivity of the soil in relation to the suction.

Disregarding the constant $b$ of Eq. (15.6), Gardner (1958) obtained the function

$$q_{max} = Aa/d^n \tag{15.8}$$

where $d$ is the depth of the water table below the soil surface, $a$ and $n$ are constants from Eq. (15.6), $A$ is a constant which depends on $n$, and $q_{max}$ is the limiting (maximal) rate at which the soil can transmit water from the water table to the evaporation zone at the surface.

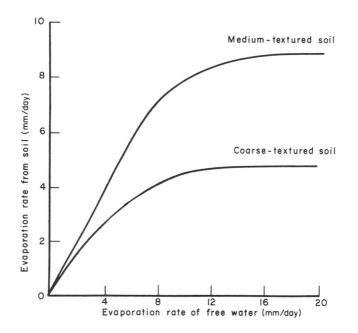

**Fig. 15.2.** Theoretical relation between the rate of evaporation from coarse- and medium-textured soils (water table depth, 60 cm) and the rate of evaporation from a free-water surface. (After Gardner, 1958.)

We can now see how the actual steady evaporation rate is determined either by the external evaporativity or by the water-transmitting properties of the soil, depending on which of the two is lower, and therefore limiting. Where the water table is near the surface, the suction at the soil surface is low and the evaporation rate is determined by external conditions. However, as the water table becomes deeper and the suction at the soil surface increases, the evaporation rate approaches a limiting value regardless of how high external evaporativity may be.

Equation (15.8) suggests that the maximal evaporation rate decreases with water-table depth more steeply in coarse-textured soils (in which $n$ is greater) than in fine-textured soils. Nevertheless, a sandy loam soil can still evaporate water at an appreciable rate (as shown in Fig. 15.1) even when the water table is as deep as 180 cm. Figure 15.2 illustrates the effect of texture on the limiting evaporation rate.

### D. Hazard of Salinization Due to High Water Table

The rise of water from a shallow water table can in some cases serve the useful purpose of supplying water to the root zone of crops. On the other hand, this process also entails the hazard of salinization, especially where the groundwaters are brackish and potential evaporativity is high. In fine-textured soils, the danger of salinization can be appreciable even where the water table is several meters deep. The tendency for water to be drawn from the water table toward the soil surface will persist as long as the suction head prevailing at the surface is greater than the depth of the water table. The gradual and irreversible salinization of the soil may have been the process responsible for the destruction of once-thriving agricultural civilizations based on the irrigation of river valleys with high-water-table conditions. Excessive irrigation tends to raise the water table and thus aggravate the salinization problem.

Lowering the water table by drainage can decisively reduce the rate of capillary rise and evaporation. Drainage is a costly operation, however, and it is therefore necessary, ahead of time, to determine the optimal depth to which the water table should be lowered. Among the important considerations in this regard is the necessity to limit the rate of capillary rise to the surface (Gardner, 1958). In the soil described by Fig. 15.1, for example, the maximal rate of profile transmission to the surface is 8 mm/day when the water table is at a depth of 90 cm. Since potential evaporativity is seldom greater than this, it follows that a water-table rise above the depth of 90 cm would not be likely to increase the evaporation rate. On the other hand, a lowering of the water table to a depth of 180 cm can decrease the evaporation rate to 1 mm/day.[3] An additional lowering of the water table to a depth of 360 cm will reduce the maximal evaporation rate to 0.12 mm/day, while any further lowering of the water table can cause only a negligible reduction of evaporation and might in any case be prohibitively expensive to carry out.

Uniform profiles occur only rarely. As mentioned in our preceding section, the soil often consists of several more or less distinct layers. Layered conditions may in fact exist even when the soil is texturally uniform, owing to structural differentiation. Holmes *et al.* (1960) pointed out the possible effects of surface layers with different tilths upon evaporation in the presence of a water table. Talsma (1963) found that tillage reduced the evaporation rate by half and decreased suction below the tilled zone. He attributed these effects to the formation of layered conditions and to the resulting shift in the actual evaporation zone. Gardner (1958) and Gardner and Fireman (1958) showed that a dry layer at the surface reduces the steady evaporation rate in hyperbolic relation to its thickness. Water movement through such a layer occurs mainly by vapor diffusion.

The problem of salt movement and accumulation in soils is treated by Hillel (1980, Chapter 9). A specific elucidation of soil salinity as related to high groundwater conditions and to drainage can be found in the book edited by van Schilfgaarde (1974).

## E. Evaporation in the Absence of a Water Table (Drying)

Steady evaporation from soils is not a widespread occurrence, since, even where high-water-table conditions exist, water-table depths (as well as external conditions) seldom remain constant for very long. More commonly, soil-moisture evaporation occurs under unsteady conditions and results in a net loss of water from the soil, i.e., it results in drying. The process of drying involves considerable losses of water, especially in arid regions, where these losses can amount to 50% or more of total precipitation (Hide, 1954, 1958). In this section, we shall consider the drying of initially wetted soil profiles which do not contain a water table anywhere near enough to the soil surface to have any bearing on the evaporation process.

We begin once again by assuming that external conditions, and hence atmospheric evaporativity, are constant. Under such conditions (which, incidentally, happen to be the easiest to set up in laboratory experiments), the soil drying process has been observed to occur in three recognizable stages (Fisher, 1923; Pearce *et al.*, 1949):

    (1)  An initial *constant-rate stage*, which occurs early in the process,

---

[3] This is a consequence of the equation $q_{max} = Aad^{-n}$. When $n = 3$, the doubling of $d$ (from 90 to 180 cm) will reduce $q_{max}$ by $2^3$, i.e., by a factor of 8.

while the soil is wet and conductive enough to supply water to the site of evaporation at a rate commensurate with the evaporative demand. During this stage, the evaporation rate is limited by, and hence also controlled by, external meteorological conditions (i.e., radiation, wind, air humidity, etc.) rather than by the properties of the soil profile. As such, this stage, being *weather controlled*, is analogous to the *flux-controlled stage* of infiltration (in contrast with the *profile-controlled stage;* see Sections B and H, on rain infiltration, in Chapter 12). The evaporation rate during this stage might also be influenced by soil surface conditions, including surface reflectivity and the possible presence of a mulch, insofar as these can modify the effect of the meterorological factors acting on the soil. In a dry climate, this stage of evaporation is generally brief and may last only a few hours to a few days.

(2)  An intermediate *falling-rate stage*, during which the evaporation rate falls progressively below the potential rate (the evaporativity). At this stage, the evaporation rate is limited or dictated by the rate at which the gradually drying soil profile can deliver moisture toward the evaporation zone. Hence it can also be called the *soil profile-controlled stage*. This stage may persist for a much longer period than the first stage.

(3)  A residual *slow-rate stage*, which is established eventually and which may persist at a nearly steady rate for many days, weeks, or even months. This stage apparently comes about after the surface-zone has become so desiccated that further liquid-water conduction through it effectively ceases. Water transmission through the desiccated layer thereafter occurs primarily by the slow process of vapor diffusion, and it is affected by the vapor diffusivity of the dried surface zone and by the adsorptive forces acting over molecular distances at the particle surfaces. This stage is often called the *vapor diffusion stage* and can be important where the surface layer is such that it becomes quickly desiccated (e.g., a loose assemblage of clods).

Whereas the transition from the first to the second stage is generally a sharp one, the second stage generally blends into the third stage so gradually that the last two cannot be separated so easily. A qualitative explanation for the occurrence of these stages follows.

During the initial stage, the soil surface gradually dries out and soil moisture is drawn upward in response to steepening evaporation-induced gradients. The rate of evaporation can remain nearly constant as long as the moisture gradients toward the surface compensate for the decreasing hydraulic conductivity (resulting from the decrease in water content). In terms of Darcy's law, $q = K(d\psi/dz)$, we can restate the above principle by noting that the flux $q$ remains constant because the gradient $d\psi/dz$ increases sufficiently to offset the decrease of $K$. Sooner or later, however, the soil surface approaches equilibrium with the overlying atmosphere (i.e., becomes approximately air dry). From this moment on, the moisture gradients toward the surface cannot increase any more, and in fact, must tend to decrease as the soil in depth loses more and more moisture. Since, as the evaporation process continues, both the gradients and the conductivities at each depth near the surface are decreasing at the same time, it follows that the flux toward the surface and the evaporation rate inevitably decrease as well. As shown in Fig. 15.3, the end of the first, i.e., the beginning of the second, stage of drying can occur rather abruptly. The pattern of evaporation under different evaporative conditions is also shown in terms of cumulative evaporation in Fig. 15.4.

The tendency of the moisture gradients toward the soil surface to become steeper during the first stage of the process, as the surface becomes progressively drier, and their tendency to become less and less steep during the second stage, after the surface has dried to its final "air-dry" value, is depicted in Fig. 15.5. Continuation of the evaporation process for a prolonged period is sometimes accompanied by the downward movement into the profile of a "drying front" and the development of a distinct desiccated zone, through which water can move from the still-moist underlying layers only by vapor diffusion.

The length of time the initial stage of drying lasts depends upon the intensity of the meteorological factors that determine atmospheric evaporativity, as well as upon the conductive properties of the soil itself. Under similar external conditions, the first stage of drying will be sustained longer in a clayey than in a sandy soil, since clayey soils retain higher wetness and conductivity values as suction develops in the upper zone of the profile.

When external evaporativity is low, the initial, constant-rate stage of drying can persist longer. This fact has led to the hypothesis that an initially high evaporation rate may in the long run reduce cumulative moisture loss to the atmosphere. This hypothesis was raised in a number of Russian papers and cited by Lemon (1956). Gardner and Hillel (1962), on the other hand, concluded that the higher initial drying rate will in fact result in a higher cumulative loss at any time. The total water loss resulting from any finite initial evaporation rate will gradually approach (but never surpass) the total

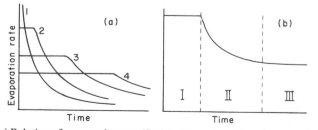

**Fig. 15.3.** (a) Relation of evaporation rate (flux) to time under different evaporativities (curves 1–4 are in order of decreasing initial evaporation rate). (b) Relation of relative evaporation rate (actual rate as a fraction of the potential rate) to time, indicating the three stages of the drying process.

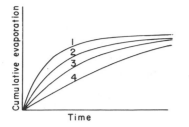

**Fig. 15.4.** Relation of cumulative evaporation to time (curves 1–4 are in order of decreasing initial evaporation rate).

loss in the extreme case where the initial evaporation rate is infinitely high. It is at present debatable whether this pattern holds, strictly speaking, for all profile and environmental conditions or whether the presence of nonuniformities (profile layers, mulches, hysteresis effects, etc.) may modify it.

## F. Analysis of the First and Second Stages of Drying

Since the gravitational effect is in general relatively negligible for evaporation,[4] it is possible to base an analysis on the hydraulic diffusivity and water-

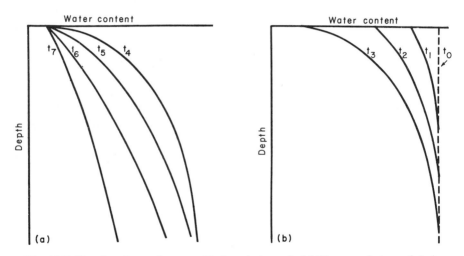

**Fig. 15.5.** The changing moisture profile in a drying soil. (a) The second stage of drying, in which the moisture gradients decrease as the deeper layers lose moisture by continued upward movement. (b) The first stage, during which the gradients toward the surface become steeper.

[4] As we have already pointed out, the equivalent suction exercised by the atmosphere can amount to many (often hundreds of) bars. This suction, distributed over a few centimeters of soil depth, generally constitutes a force many times greater than gravity. For instance, a suction difference of 1 bar over 1 cm of soil is a head gradient of 1000, three orders of magnitude greater than the gravitational head gradient.

content gradient relationships. Both stages of drying depend upon the diffusivity: the first stage for its duration, the second stage for its rate.

A mathematical study of the constant-rate stage of drying of both finite-length and semi-infinite soil columns was carried out by Covey (1963), who neglected gravity and used an exponential dependence of diffusivity upon wetness. For homogeneous soil columns, initially uniformly wet, Covey devised a criterion for determining when the column behaves as though it were infinitely long and when finiteness of length becomes important. When the drying rate is slow, a condition is soon reached in which all depths lose water at about the same rate per unit volume, and the profile dries nearly uniformly with depth. If $e$ is the rate of evaporation (volume of water lost per unit area per unit time) from a soil initially uniformly wet to a depth $L$, then after a short time, which depends on the diffusivity, the rate of water loss per unit volume will be $e/L$, and we can then write for the flow equation (Gardner and Hillel, 1962)

$$-\frac{e}{L} = \frac{\partial \theta}{\partial t} = \frac{\partial}{\partial z}\left(D\frac{\partial \theta}{\partial z}\right) \qquad (15.9)$$

where $z$ is height above the bottom of the profile, $D$ diffusivity, $\theta$ volume wetness, and $t$ time. The first integral of this equation can be written directly:

$$-ez/L = D\, d\theta/dz + const \qquad (15.10)$$

The constant of integration is zero if there is no flow through the bottom boundary of the profile ($z = 0$), since at that location $d\theta/dz$ is zero. To perform the second integration, $D$ must be a known function of $\theta$. Using an exponential function ($D = D_f \exp aC$, where $a$ is a constant, and $C$ is the soil wetness minus the final air-dry wetness at which $D = D_f$) Gardner and Hillel evaluated the time and remaining water content at the end of the first stage of drying, when the actual evaporation rate ceases being equal to the potential rate. This occurs when $C$ falls to zero at $z = L$. The total water content $W$ of the profile is then approximated by

$$W = (L/a)\ln(1 + eaL/2D_f) \qquad (15.11)$$

At the end of the first stage of drying, the soil surface is more or less at its constant, "final" moisture and the profile in depth has a moisture distribution which depends upon the preceding evaporation rate. In early studies of the second (falling-rate) stage of drying, the initial evaporation rate was assumed to be infinitely high. Subject to an infinite evaporativity, the soil surface is brought instantly to its final state of dryness, so that the first stage of drying ends essentially as soon as it begins. Under such conditions, the profile moisture distribution is obviously not the same as it would be at the onset of the second stage after a prolonged period of constant evaporation at a moderate rate, but the effect of this initial distribution tends to become small after a time, beyond which the evaporation rate depends only on the total remaining water content of the soil. Hence an analysis of the second stage of drying based on the assumption of an initially infinite evaporation rate can be a valid approximation in many cases (Gardner and Hillel, 1962).

For semi-infinite soil columns subjected to infinite evaporativity at the surface, a solution of the flow equation, neglecting gravity, indicates that the cumulative evaporation $E$ is related linearly to the square root of time according to the equation (Gardner, 1959)

$$E = 2(\theta_i - \theta_f)\sqrt{\bar{D}t/\pi} \qquad (15.12)$$

The evaporative flux $e$, being the time derivative of $E$, is thus inversely proportional to the square root of time:

$$e = dE/dt = (\theta_i - \theta_f)\sqrt{\bar{D}/\pi t} \qquad (15.13)$$

In these equations, $\theta_i$ is the initial profile wetness, $\theta_f$ is the final (surface) wetness, and $\bar{D}$ is the weighted mean diffusivity which can be used to characterize the drying process.[5]

Using the same exponential diffusivity function as mentioned above (namely, $D = D_0 \exp[\beta(\theta - \theta_f)/(\theta_i - \theta_f)]$), Gardner obtained an approximate solution for the flow equation by means of an iterative procedure (based on Crank, 1956, p. 152). He also reported experimental results indicating that cumulative evaporation from semi-infinite columns is indeed linear with the square root of time, as predicted by Eq. (15.12). Accordingly, the time required for soil moisture at a specified depth to fall to a given level of dryness is proportional to that depth squared.

Gardner (1959), as well as Klute et al. (1965), studied falling-rate evaporation from finite-length columns. In this case, the flux, though initially proportional to $t^{-1/2}$, decreases more steeply with time as the wetness at the bottom of the column is reduced. The evaporation rate thereafter becomes roughly proportional to the product of the average diffusivity and the water content. A field study of evaporation from a sandy soil was reported by

---

[5] Equations (15.11) and (15.12) are similar to those given for horizontal infiltration. The major difference is, however, that $\bar{D}$ must be weighted differently for drying than for wetting. According to Crank (1956), the weighted mean diffusivity for desorption processes is

$$\bar{D} = \frac{1.85}{(\theta_i - \theta_0)^{1.85}} \int_{\theta_0}^{\theta_i} D(\theta)(\theta_i - \theta)^{0.85} \, d\theta$$

This weighting function yields lower values of $\bar{D}$ for corresponding $\theta$ values than the approximate weighting function for sorption processes. The need to weight the diffusivity differently can be ascribed to the fact that in infiltration the maximal flux occurs at the wet end of the column, where diffusivity is highest, whereas in drying the greatest flux is through the dry end, where diffusivity is lowest. This fact makes sorption processes inherently faster than desorption processes and contributes to soil-moisture conservation.

Black *et al.* (1969), who also found that the evaporation rate declined as the square root of time.

Gardner and Hillel (1962) used the following equation to predict the rate of evaporation from finite-length columns during the decreasing-rate stage of drying:

$$e = -dW/dt = D(\bar{\theta})W\pi^2/4L^2 \qquad (15.14)$$

where $\bar{\theta}$ is the average volumetric wetness obtained by dividing the total water content of the soil $W$ by the depth of wetting $L$ and $D(\theta)$ is the known diffusivity function. The cumulative evaporation can be obtained by integrating (5.17) with respect to time.

Rose (1966), using the sorptivity concept of Philip (1957d), described evaporation in terms of the following equation:

$$E = st^{1/2} + bt \qquad (15.15)$$

where $s$, the sorptivity (or desorptivity, since we are dealing with a drying process) is positive and $b$ is negative. This approach also applies only to the falling-rate stage of drying.

The numerical procedure of Hanks and Bowers (1962), first developed for infiltration, has been applied to drying by Hanks and Gardner (1965), who studied the effects of different $D-\theta$ relations as well as of layering. The use of computer-based simulation techniques in the analysis of drying processes under various initial and boundary contitions was demonstrated by Hillel (1977).

## G. Reduction of Evaporation from Bare Soils

In principle, the evaporation flux from the soil surface can be modified in three basic ways: by controlling energy supply to the site of evaporation (e.g., modifying the albedo through color or structure of the soil surface, shading the surface) by reducing the potential gradient, or the force driving water upward through the profile (e.g., lowering the water table, if present, or warming the surface so as to set up a downward-acting thermal gradient), or by decreasing the conductivity or diffusivity of the profile, particularly of the surface zone (e.g., tillage and mulching practices).

The actual choice of means for reduction of evaporation depends on the stage of the process one wishes to regulate: whether it be the first stage, in which the effect of meteorological conditions on the soil surface dominates the process, or the second stage, in which the rate of water supply to the surface, determined by the transmitting properties of the profile, becomes the rate-limiting factor. Methods designed to affect the first stage do not necessarily serve during the second stage, and vice versa.

Covering or *mulching* the surface with vapor barriers or with reflective materials can reduce the intensity with which external factors, such as radiation and wind, act upon the surface (Hanks *et al.*, 1961; Bond and Willis, 1969). Thus, such surface treatments can retard evaporation during the initial stage of drying. A similar effect can result from application of materials which lower the vapor pressure of water (Law, 1964). Retardation of evaporation during the first stage can provide the plants with a greater opportunity to utilize the moisture of the uppermost soil layers, an effect which can be vital during the germination and establishment phases of plant growth. The retardation of initial evaporation can also enhance the process of internal drainage, and thus allow more water to migrate downward into the deeper parts of the profile, where it is conserved longer and is less likely to be lost by evaporation (Hillel, 1968). We shall discuss the practice of mulching in greater detail later in this section.

An opposite approach to evaporation control during the first stage is to induce a temporarily higher evaporation rate so as to rapidly desiccate the surface, thus hastening the end of the first stage and using the hysteresis effect to help arrest or retard subsequent outflow. Heating the surface by darkening its color or by means of flaming, microwave, or plasma-jetting devices has been suggested but not proven in practice. Shallow cultivation designed to pulverize the soil at the surface often has the immediate effect of causing the loosened layer to dry faster and more completely but may over a period of time help conserve the moisture of the soil underneath. The problem here is how to obtain a favorable balance between the short-run loss and the long-run gain.

During the second stage of drying, the effect of surface treatments is likely to be only slight and reduction of the evaporation rate and of eventual water loss will depend on decreasing the diffusivity or conductivity of the soil profile in depth. Deep tillage, for instance, by possibly increasing the range of variation of diffusivity with changing water content, may reduce the rate at which the soil can transmit water toward the surface during the second stage of the drying process. However, the evaporation rate is usually much lower during the second stage than during the first, and it is questionable whether it might be worthwhile to invest in the control of evaporation in this stage.

An irrigation regime having an excessively high irrigation frequency can cause the soil surface to remain wet and the first stage of evaporation to persist most of the time, resulting in a maximum rate of water loss. Water loss by evaporation from a single deep irrigation is generally smaller than from several shallow ones with the same total amount of water. However, water losses due to percolation are likely to be greater from deep irrigations than from shallow ones. New water application methods such as *drip* (or *trickle*) *irrigation*, which concentrate the water in a small fraction of the area while maintaining the greater part of the soil surface in a dry state, are likely to reduce the direct evaporation of soil moisture very significantly.

**Sample Problems**

1. The average daily rate of evaporation from a saturated uniform soil with a high water-table condition is 0.9 cm. Prior studies have shown that the soil's behavior conforms to Eq. (15.8), with the composite coefficient $Aa = 4.88$ cm$^2$/sec and the exponential constant $n = 3$. Estimate the threshold depth beyond which the water table must be lowered if evaporation is to be reduced and the water-table depth at which the evaporation rate will fall to 10% of the potential rate. Assume steady-state conditions. Plot the expectable daily evaporation rate as a function of water-table depth.

We begin with Eq. (15.8), which, according to Gardner (1958) estimates the maximal evaporation rate possible from a soil with a high water-table condition:

$$q_{max} = Aa/d^n$$

where $d$ is depth of the water table below the soil surface. Accordingly, the maximal depth of the water table still capable of supplying the surface with the steady flux needed to sustain the potential evaporation rate is

$$d = \left(\frac{Aa}{q_{max}}\right)^{1/n} = \left[\frac{4.88 \text{ cm}^2/\text{sec}}{(0.9 \text{ cm/day})/(86400 \text{ sec/day})}\right]^{1/n}$$

$$\cong 77.7 \text{ cm below the soil surface}$$

Note that any water-table position higher than $-77.7$ cm could theoretically allow a flux greater than the climatically determined potential rate; hence for a shallow water table the actual evaporation rate is flux controlled at the surface. For deeper water-table conditions, however, it becomes profile controlled.

Now, to calculate the water-table depth $d_{0.1}$ for which $q$ will be 10% of the potential rate, we can write

$$d_{0.1} = \left[\frac{4.88 \text{ cm}^2/\text{sec}}{(0.1 \times 0.9 \text{ cm/day})/(86400 \text{ sec/day})}\right]^{1/3} \cong 167.3 \text{ cm}$$

To plot the $q_{max}$–$d$ curve, we can obtain the following data:

From $d = 0$ to $d = 77.7$ cm, $q_{max} = 0.9$ cm/day.
At $d = 80$ cm, $q_{max} = 4.88/80^3 = 0.82$ cm/day.
At $d = 100$ cm, $q_{max} = 4.88/100^3 = 0.42$ cm/day.
At $d = 150$ cm, $q_{max} = 4.88/150^3 = 0.12$ cm/day.
At $d = 200$ cm, $q_{max} = 4.88/200^3 = 0.05$ cm/day.
At $d = 250$ cm, $q_{max} = 4.88/250^3 = 0.027$ cm/day.

Again, the actual plotting of these data is left to the enterprising student.

**2.**   Calculate the 10 day time course of evaporation and evaporation rate during the drying process of an infinitely deep, initially saturated, uniform column of soil subjected to an infinitely high evaporativity. Assume the initial volume wetness $\theta_i$ to be 48% and the air-dry value $\theta_f$ to be 3%, and assume the weighted mean diffusivity to be 100 cm$^2$/day.

The problem posed can be described approximately by the theory of Gardner (1959), Eq. (15.12):

$$E = 2(\theta_i - \theta_f)\sqrt{\bar{D}t/\pi}$$

where $E$ is cumulative evaporation. We now solve successively for days 1, 2, 3, etc., to obtain the time course of cumulative evaporation. Thus

after day 1, $E = 2(0.48 - 0.03)\sqrt{100 \times 1/3.1416} = 5.08$ cm;

after day 2, $E = 2 \times 0.45\sqrt{100 \times 2/3.1416} = 7.18$ cm;

after day 3, $E = 2 \times 0.45\sqrt{100 \times 3/3.1416} = 8.80$ cm;

after day 4, $E = 2 \times 0.45\sqrt{100 \times 4/3.1416} = 10.16$ cm;

after day 5, $E = 2 \times 0.45\sqrt{100 \times 5/3.1416} = 11.36$ cm;

after day 6, $E = 2 \times 0.45\sqrt{100 \times 6/3.1416} = 12.44$ cm;

after day 7, $E = 2 \times 0.45\sqrt{100 \times 7/3.1416} = 13.44$ cm;

after day 8, $E = 2 \times 0.45\sqrt{100 \times 8/3.1416} = 14.37$ cm;

after day 9, $E = 2 \times 0.45\sqrt{100 \times 9/3.1416} = 15.24$ cm;

after day 10, $E = 2 \times 0.45\sqrt{100 \times 10/3.1416} = 16.06$ cm.

To estimate the mean evaporation rate during each day, we can use Eq. (15.13) to calculate the midday rate:

$$e = dE/dt = (\theta_i - \theta_f)\sqrt{\bar{D}/\pi t}$$

Midday 1: $e = 0.45\sqrt{100/\pi \times 0.5} = 3.59$ cm/day.

Midday 2: $e = 0.45\sqrt{100/\pi \times 1.5} = 2.07$ cm/day.

Midday 3: $e = 0.45\sqrt{100/\pi \times 2.5} = 1.60$ cm/day.

Midday 4: $e = 0.45\sqrt{100/\pi \times 3.5} = 1.36$ cm/day.

Midday 5: $e = 0.45\sqrt{100/\pi \times 4.5} = 1.20$ cm/day.

Midday 6: $e = 0.45 \sqrt{100/\pi \times 5.5} = 1.08$ cm/day.

Midday 7: $e = 0.45 \sqrt{100/\pi \times 6.5} = 1.00$ cm/day.

Midday 8: $e = 0.45 \sqrt{100/\pi \times 7.5} = 0.93$ cm/day.

Midday 9: $e = 0.45 \sqrt{100/\pi \times 8.5} = 0.87$ cm/day.

Midday 10: $e = 0.45 \sqrt{100/\pi \times 9.5} = 0.82$ cm/day.

Note that the sum of the daily rates thus estimated does not equal the cumulative evaporation for the 10 day period, since the midday rates calculated for the first three days (when the rate is descending steeply as a curvilinear function of time) underestimate the effective mean rates. Note also that the representation of a continuously drying system in terms of a single, constant "weighted mean diffusivity" is itself an oversimplification. Any more realistic treatment, however, necessarily involves much more complex mathematics.

**3.** Consider the drying of a uniform soil wetted to a depth of 100 cm at which an impervious horizon exists, allowing no drainage. The initial volume wetness is 24%, and the initial diffusivity is 400 cm²/day. Assume a diffusivity function of the type $D = a \exp b\theta$, where $a = 1.2$ and $b = 20$. Estimate the evaporation rate as it varies during the first 10 days of the drying process, under an evaporativity of 1.2 cm/day.

The case described can be represented by the finite-column model of Gardner and Hillel, Eq. (15.14):

$$e = -dW/dt = D(\bar{\theta})W\pi^2/4L^2$$

where $e$ is evaporation rate (cm/day), $W$ is total profile water content (cm), $t$ is time (days), $D$ is diffusivity (cm²/day), which is a function of mean wetness $\bar{\theta}$, and $L$ is length of the wetted profile (cm). The mean wetness $\bar{\theta}$ is related to total profile water content $W$ by

$$\bar{\theta} = W/L \qquad \text{or} \qquad W = \bar{\theta}L$$

Substituting the appropriate values into Eq. (15.14), we calculate the first day's evaporation rate:

$$e_1 = [(400 \text{ cm}^2/\text{day}) \times (0.24 \times 100 \text{ cm}) \times 9.87]/4 \times 100^2 \text{ cm}^2$$
$$= 2.37 \text{ cm/day}$$

This is much greater than the evaporativity, which is "only" 1.2 cm/day. Hence we assume that the evaporation rate is at first controlled by, and equal to, the evaporativity. So we deduct 1.2 cm from the water content of the profile and proceed to calculate the second day's evaporation rate. Our $W$ is now  24  cm − 1.2 cm = 22.8 cm,  and  our  new  $\theta$ = 22.8/100 = 0.228. Hence,  our  new  $D = 1.2\ e^{20 \times 0.228}$ = 238 cm²/day. The second day's evaporation rate is therefore

$$e_2 = [(238\ cm^2/day) \times 22.8\ cm \times 9.87]/40{,}000\ cm^2 = 1.34\ cm$$

This is still higher than the potential evaporation rate (the evaporativity), hence we once again deduct 1.2 cm from the water content and proceed to calculate the third day's evaporation rate, updating our variables as follows: $W$ = 22.8 cm − 1.2 cm = 21.6 cm,   $\theta$ = 21.6/100 = 0.216,   and   $D$ = $1.2e^{20 \times 0.216}$ = 90 cm²/day.

$$e_3 = [(90\ cm^2/day) \times 21.6\ cm \times 9.87]/40{,}000\ cm^2 = 0.48\ cm$$

This evaporation rate is less than the evaporativity, so we can conclude that the constant-rate, flux-controlled phase of the process has ended and we have entered the falling-rate, profile-controlled phase. Toward the fourth day's events we calculate that $W$ = 21.6 − 0.48 = 21.12 cm, $\theta$ = 21.12/100 = 0.2112, and $D = 1.2\ e^{20 \times 0.2112}$ = 82 cm²/day:

$$e_4 = [(82\ cm^2/day) \times 21.12\ cm \times 9.87]/40{,}000\ cm^2 = 0.43\ cm$$

Again, we update:  $W$ = 21.12 − 0.43 = 20.69,  $\theta$ = 0.207,  and  $D$ = $1.2\ e^{20 \times 0.207}$ = 75 cm²/day.

$$e_5 = [(75\ cm^2/day) \times 20.69\ cm \times 9.87]/40{,}000\ cm^2 = 0.38\ cm$$

Now  $W$ = 20.69 − 0.38 = 20.31, $\theta$ = 0.2031, and $D = 1.2\ e^{20 \times 2031}$ = 70 cm²/day.

$$e_6 = [(70\ cm^2/day) \times 20.31\ cm \times 9.87]/40{,}000\ cm^2 = 0.35\ cm$$

Now  $W$ = 20.31 − 0.35 = 19.96, $\theta$ = 0.1996, and $D = 1.2e^{20 \times 0.1996}$ = 65 cm²/day.

$$e_7 = [(65\ cm^2/day) \times 19.96\ cm \times 9.87]/40{,}000\ cm^2 = 0.32\ cm$$

Now $W$ = 19.96 − 0.32 = 19.64 cm, $\theta$ = 0.1964, and $D = 1.2e^{20 \times 0.1964}$ = 61 cm²/day.

$$e_8 = [(61\ cm^2/day) \times 19.64\ cm \times 9.87]/40{,}000\ cm^2 = 0.30\ cm$$

Now $W$ = 19.64 − 0.30 = 19.34 cm, $\theta$ = 0.1934, and $D = 1.2e^{20 \times 0.1934}$ = 57 cm²/day.

$$e_9 = [(57\ cm^2/day) \times 19.34\ cm \times 9.87]/40{,}000\ cm^2 = 0.27\ cm$$

Now $W$ = 19.34 − 0.27 = 19.07 cm, $\theta$ = 0.1907, and $D = 1.2e^{20 \times 0.1907}$ = 54 cm²/day.

$$e_{10} = [(54\ cm^2/day) \times 19.07\ cm \times 9.87]/40{,}000\ cm^2 = 0.25\ cm$$

The total 10 day cumulative evaporation is thus estimated to be

1.2 + 1.2 + 0.48 + 0.43 + 0.38 + 0.35 + 0.32 + 0.30 + 0.27 + 0.25
  = 5.18 cm.

This amount of evaporation is, interestingly, only one-third of the amount we calculated in Problem 2 for the same period from an infinitely deep, initially saturated column of soil subject to infinite evaporativity.

# 16    *Uptake of Soil Moisture by Plants*

## A. Introduction

Nature, despite its celebrated laws of conservation, can in some ways be exceedingly wasteful, or so it appears at least from our own partisan viewpoint. One of the most glaring examples is the way it requires plants to draw quantities of water from the soil far in excess of their essential metabolic needs. In dry climates, plants growing in the field may consume hundreds of tons of water for each ton of vegetative growth. That is to say, the plants must inevitably transmit to an unquenchably thirsty atmosphere most (often well over 90%) of the water they extract from the soil. The loss of water vapor by plants, a process called *transpiration*, is not in itself an essential physiological function, nor a direct result of the living processes of the plants. In fact, plants can thrive in an atmosphere saturated or nearly saturated with vapor and hence requiring very little transpiration. Rather than by plant growth per se, transpiration is caused by the vapor pressure gradient between the normally water-saturated leaves and the often quite dry atmosphere. In other words, it is exacted of the plants by the *evaporative demand* of the climate in which they live.

In a sense, the plant in the field can be compared to the wick in an old-fashioned lamp. Such a wick, its bottom dipped into a reservoir of fuel while its top is subject to the burning fire which consumes the fuel, must constantly transmit the liquid from bottom to top under the influence of physical forces imposed upon the passive wick by the conditions prevailing at its two ends. Similarly, the plant has its roots in the soil-water reservoir while its leaves are subject to the radiation of the sun and the sweeping action of the wind which require it to transpire unceasingly.

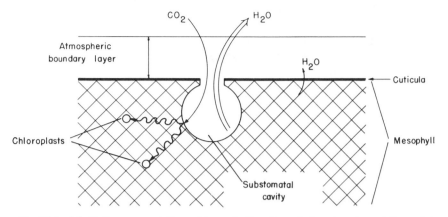

**Fig. 16.1.** Schematic representation of transpiration through the stomate and the cuticle, and of the diffusion of $CO_2$ into the stomate and through the mesophyll to the chloroplasts. (After Rose, 1966.)

This analogy, to be sure, is a gross oversimplification, since plants are not all that passive and in fact are able at times to limit the rate of transpiration by shutting the stomates of their leaves (Fig. 16.1). However, for this limitation of transpiration most plants pay, sooner or later, in reduced growth potential, since the same stomates which transpire water also serve as foci for the uptake of the carbon dioxide needed in photosynthesis. Furthermore, reduced transpiration often results in warming of the plants and consequently in increased respiration and hence in further reduction of net photosynthesis.

To grow successfully, a plant must achieve a water economy such that the demand made upon it is balanced by the supply available to it. The problem is that the evaporative demand of the atmosphere is almost continuous, whereas rainfall occurs only occasionally and irregularly. To survive during dry spells between rains, the plant must rely upon the diminishing reserves of water contained in the pores of the soil, which itself loses water by direct evaporation and internal drainage.[1]

How efficient is the soil as a water reservoir for plants? How do plants draw water from the soil, and to what limit can soil water continue to sustain plant growth? How is the actual rate of transpiration determined by the interaction of plant, soil, and meteorological factors? These and related questions are the topics of this chapter.

---

[1] Perhaps we can identify with the plight of arid zone plants more fully if we imagine ourselves to be living under a government which taxes away 99% of our income while requiring us to keep our reserves in a bank which is embezzled daily.

## B.  The Soil–Plant–Atmosphere Continuum

Current approaches to the problem of soil-water extraction and utilization by plants are based on recognition that the field with all its parts—soil, plant, and atmosphere taken together—forms a physically integrated, dynamic system in which various flow processes occur interdependently like links in a chain. This unified system has been called the SPAC (soil-plant–atmosphere continuum) by J. R. Philip (1966b). The universal principle which operates consistently throughout the system is that water flow always takes place spontaneously from regions of higher to regions of lower potential energy. Former generations of soil physicists, plant physiologists, and meteorologists, each group working separately in what it considered to be its separate and exclusive domain, tended to obscure this principle by expressing the energy state of water in different terms (e.g., the "tension" of soil water, the "diffusion pressure deficit" of plant water, and the "vapor pressure" or "relative humidity" of atmospheric water) and hence failed to communicate readily across their self-imposed interdisciplinary boundaries.

As we have now come to understand, the various terms used to characterize the state of water in different parts of the soil–plant–atmosphere system are merely alternative expressions of the *energy level*, or *potential* of water. Moreover, the very occurrence of differences, or gradients, of this potential between locations in the system constitutes the force inducing flow within and between the soil, the plant, and the atmosphere. This principle applies even though different components of the overall potential gradient are effective in varying degrees in different parts of the soil–plant–atmosphere system.[2]

In order to describe the interlinked processes of water transport throughout the SPAC, we must evaluate the pertinent components of the energy potential of water and their effective gradients as they vary in space and time. As an approximation, the flow rate through each segment of the system can be assumed to be proportional directly to the operating potential gradient, and inversely to the segment's resistance. The flow path includes liquid water movement in the soil toward the roots, liquid and perhaps vapor movement across the root-to-soil contact zone, absorption into the roots and across their membranes to the vascular tubes of the xylem, transfer through the xylem up the stem to the leaves, evaporation in the intercellular spaces within the leaves, vapor diffusion through the substomatal cavities

---

[2] For example, osmotic potential differences have little effect on liquid water movement in the soil but strongly affect flow from soil to plant. In the gaseous domain, vapor diffusion is proportional to the vapor pressure gradient rather than to the potential gradient as such. Vapor pressure and potential are exponentially, rather than linearly, related.

and out the stomatal perforations to the quiescent boundary air layer in contact with the leaf surface, and through it, finally, to the turbulent atmosphere which carries away the water thus extracted from the soil.

Since we have dealt extensively in this text with the principles governing flow in the soil domain but scarcely at all with the plant domain, it is time for us to digress from our primary concern with the soil per se and attempt to describe, albeit briefly, some pertinent aspects of plant–water relations.

## C. Basic Aspects of Plant–Water Relations

Close observation of how higher plants are built reveals much about how they function in their terrestrial environment (Epstein, 1973). To begin with, it is intriguing to compare the shape of such plants to the shape of familiar higher animals. The characteristic feature of animal bodies, in contrast with plants, is their minimal area of external surface exposure. Apart from a few protruding organs needed for mobility and sensory perception, animals are rather compact and bulky in appearance. Not so is the structure of plants, whose vital functions require them to maximize rather than minimize surface exposure both above and below ground. The aerial canopies of plants frequently exceed the area of covered ground by several fold. Such a large surface helps the plants to intercept and collect sunlight and carbon dioxide, two resources which are diffuse rather than concentrated.

Even more striking is the shape of roots, which proliferate and ramify throughout a large volume of soil while exposing an enormous surface area: a single annual plant can develop a root system with a total length of several hundred kilometers and with a total surface area of several hundred square meters. (Estimates of total length and surface area of roots are ten times larger if root hairs are taken into account.) The need for such exposure becomes apparent if we consider the primary function of roots, which is to continuously gather water and nutrients from a medium that often provides only a meager supply of water per unit volume and that generally contains soluble nutrients only in very dilute concentrations. And while the atmosphere is a well-stirred and thoroughly mixed fluid, the soil solution is a sluggish and unstirred fluid which moves toward the roots at a grudgingly slow pace, so that the roots have no choice but to move toward it. Indeed, roots forage constantly through as large a soil volume as they can, in a constant quest for more water and nutrients. Their movement and growth, involving proliferation in the soil region where they are present and extension into ever new regions, are affected by a host of factors additional to moisture and nutrients, e.g., temperature, aeration, mechanical resistance, the possible presence of toxic substances, and the primary roots' own *geotropism* (i.e., their preference for a vertically downward direction of growth).

Green plants are the earth's only true *autotrophs*, able to create new living matter from purely inorganic raw materials. We refer specifically to the synthesis of basic sugars (subsequently elaborated into more complex compounds) by the combination of atmospheric carbon dioxide and soil-derived water, accompanied by the conversion of solar radiation into chemical energy in the process of *photosynthesis*, usually described by the following deceptively simple formulas:

$$6CO_2 + 6H_2O + (\text{sunlight energy}) \rightarrow C_6H_{12}O_6 + 6O_2$$
$$\underset{\text{glucose}}{nC_6H_{12}O_6} \rightleftharpoons \underset{\text{starch}}{(C_6H_{10}O_5)_n} + nH_2O \tag{16.1}$$

We, along with the entire animal kingdom, owe our lives to this process, which not only produces our food but also releases into the atmosphere the elemental oxygen which we need for our respiration. Plants also respire, and the process of respiration represents a reversal of photosynthesis in the sense that some photosynthetic products are reoxidized to yield the original constituents (water, carbon dioxide, and energy) as separate entities. Thus

$$C_6H_{12}O_6 + 6O_2 \rightarrow 6CO_2 + 6H_2O + (\text{thermal energy}) \tag{16.2}$$

In examining these formulas, we note immediately the central role of water as a major metabolic agent in the life of the plant, i.e., as a source of hydrogen atoms for the reduction of carbon dioxide in photosynthesis and as a product of respiration. Water is also the solvent and hence conveyor of transportable ions and compounds into, within, and out of the plant. It is, in fact, a major structural component of plants, often constituting 90% or more of their total "fresh" mass. Much of this water occurs in cell vacuoles under positive pressure, which keeps the cells turgid and gives rigidity to the plant as a whole.

Although all plants are absolutely dependent on water, different types of plants differ in adaptation to environments with varying degrees of water availability or abundance. *Hydrophytes*, or *aquatic plants*, inhabit water-saturated domains. Plants adapted to drawing water from shallow water tables are called *phreatophytes*. Plants which grow in aerated soils, generally in semihumid to semiarid climates, are called *mesophytes*. Most crop plants belong in this category. Mesophytes control their water economy by developing extensive root systems and optimizing the ratio of roots to shoots (the former supply water and nutrients, the latter photosynthesize and transpire) and by regulating the aperture of their stomates. On the dry end of the scale are *xerophytes*, which are adapted to growing in desert environments. Such plants generally exhibit special features (called *xeromorphic*) designed to minimize water loss (e.g., thickened epidermis and a waxy cuticle, recessed stomates and reduced leaf area, and specialized water storage or *succulent* tissues).

Only a very small fraction (generally less than 1%) of the water absorbed by plants is used in photosynthesis, while most (often over 98%) is lost as vapor, in a process known as *transpiration*. This process is made inevitable by the exposure to the atmosphere of a large area of moist cell surfaces, necessary to facilitate absorption of carbon dioxide and oxygen, and hence transpiration has been described as a "necessary evil" (Sutcliffe, 1968). Mesophytes are extremely sensitive, and vulnerable, to lack of sufficient water to replace the amount lost in transpiration. Water deficits impair plant growth and, if extended in duration, can be fatal. It was once believed that transpiration is beneficial in that it induces a greater uptake of nutrients from the soil. However, it now appears that the two processes of water uptake and nutrient uptake are largely independent. The absorption of nutrients into roots and their transmission up the shoots are not merely a passive consequence of the transpiration-induced water flow but are effected by metabolically active processes (Epstein, 1977). A more likely beneficial effect of transpiration is prevention of excessive heating of leaves by solar radiation. Indeed, when water deficits occur and transpiration is reduced by stomatal closure, plant-canopy temperature rises measurably.

## D. Root Uptake, Soil-Water Movement, and Transpiration

The rate of water uptake from a given volume of soil depends on rooting density (the effective length of roots per unit volume of soil). soil conductivity, and the difference between average soil-water suction and root suction. If the initial soil-water suction is uniform throughout all depths of the rooting zone, but the active roots are not uniformly distributed, the rate of water uptake should be highest where the density of roots is greatest. However, more rapid uptake will result in more rapid depletion of soil moisture, and the rate will not remain constant very long.

Nonuniformity of water uptake from different soil depths has been found in the field, as shown in Fig. 16.2 (after Ogata *et al.*, 1960). In a nonuniform root system, suction gradients can form which may induce water movement from one layer to another in the soil profile itself. In general, the magnitude of this movement is likely to be small relative to the water uptake rate by the plants, but in some cases it can be considerable. As a rough approximation, it is sometimes possible to divide the rooting system into two layers: an *upper layer*, in which root density is greatest and nearly uniform and in which water depletion is similarly uniform, and a *lower layer*, in which the roots are relatively sparse and in which the rate of water depletion is slow as long as the water content of the upper layer is fairly high. The water content of the lower layer is depleted by two sometimes simultaneous processes: uptake by the roots of that layer, and direct upward flow in the soil itself, caused by suction gradients, from the still-moist lower layer to the more-rapidly depleted upper layer.

A mature root system occupies a more or less constant soil volume of fixed depth so that uptake should depend mainly upon the size of this volume, its water content and hydraulic properties, and the density of the roots. On the other hand, in young plants, root extension and advance into deeper and moister layers can play an important part in supplying plant water requirements (Kramer and Coile, 1940; Wolf, 1968).

One possible reason for the differences observed between the responses of pot-grown and of field-grown plants to the soil-water regime is the differ-

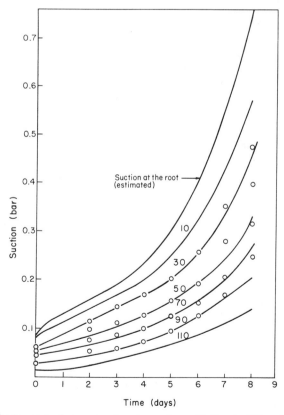

**Fig. 16.2.** The increase of average soil-water suction with time in different depths in a field of alfalfa. The numbers by the curves represent different depths (cm) within the rooting zone. (After Ogata *et al.*, 1960; Gardner, 1964.)

ence in root distribution with depth. In a pot, root density can be farily uniform, while in the field, it generally varies with depth. Furthermore, the roots present in different layers may exhibit different water uptake and transmission properties. For instance, the roots of the deeper layers may offer greater resistance to water movement within the plant than the roots of the upper layers (e.g., Wind, 1955). The possible contribution of moist sublayers underlying the rooting zone can be especially significant where a high water table is present.

Rose and Stern (1967) presented an analysis of the time rate of water withdrawal from different soil depth zones in relation to soil wetness and hydraulic properties, and to the rate of plant-root uptake. The water conservation equation for a given depth of soil (assuming flow to be vertical only) for a given period of time (from $t_1$ to $t_2$) can be written

$$\int_{t_1}^{t_2} (i - v_z - q)\, dt - \int_0^z \int_{t_1}^{t_2} \frac{\partial \theta}{\partial t}\, dz\, dt = \int_0^z \int_{t_1}^{t_2} r_z\, dz\, dt \qquad (16.3)$$

where $i$ is rate of water supply (precipitation or irrigation), $q$ evaporation rate from the soil surface, $v_z$ the vertical flux of water at depth $z$, $\theta$ the volumetric soil wetness, and $r_z$ the rate of decrease of soil wetness due to water uptake by roots.

The average rate of uptake by roots at the depth $z$ is

$$r_z = \int_{t_1}^{t_2} r_z\, dt / (t_2 - t_1) \qquad (16.4)$$

The pattern of soil-water extraction by a root system can be determined by repeated calculations based on the preceding equations for successive small intervals of time and depth. The total (cumulative) water uptake by the roots $R_z$ is given by

$$R_z = \int_0^z r_z\, dz \qquad (16.5)$$

These relationships were used by Rose and Stern to describe the pattern of soil-water extraction by a cotton crop in the field. The results indicated that nearly all water extraction by the crop took place from the top 30 cm during the early stages of growth and that, although the zone of uptake extended downward during later stages, essentially all of the seasonal uptake took place from the top 100 cm.

A similar and detailed field study of water extraction by a root system was carried out by van Bavel *et al.* (1968a, b), using the *instantaneous profile method* of obtaining the hydraulic properties of a complete profile in situ (Watson, 1966; Hillel *et al.*, 1972). (This method requires frequent, independent, and simultaneous measurements of hydraulic head and soil wet-

ness in an internally draining profile, coupled with measurements of evaporation.) The calculated root extraction rates agreed reasonably well with
separate measurements of transpiration obtained with lysimeters. Soil-
water movement within the root zone of a sorghum crop indicated initially
a net downward outflow from the root zone, but this movement later reversed itself to indicate a net upward inflow from the wet subsoil to the root
zone above.

Denmead and Shaw (1962) presented experimental confirmation of the
effect of dynamic conditions on water uptake and transpiration. They
measured the transpiration rates of corn plants grown in containers and
placed in the field under different conditions of irrigation and atmospheric
evaporativity. Under an evaporativity of 3–4 mm/day, the actual transpiration rate began to fall below the potential rate at an average soil-water
suction of about 2 bar. Under more extreme meteorological conditions,
with an evaporativity of 6–7 mm/day, this drop already began at a soil
water suction value of 0.3 bar. On the other hand, when the potential
evaporativity was very low (1.4 mm/day), no drop in transpiration rate was
noticed until average soil-water suction exceeded 12 bar. The volumetric
water contents at which the transpiration rates fell varied between 23%,
under the lowest evaporativity, and 34%, under the highest evaporativity
measured. This is illustrated in Figs. 16.3 and 16.4.

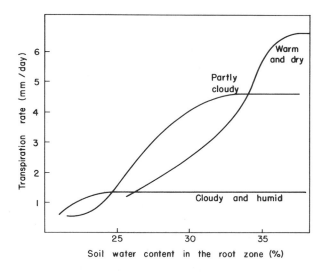

**Fig. 16.3.** Relation of actual transpiration rate to soil-water content, under different
meteorological conditions. (After Denmead and Shaw, 1962.)

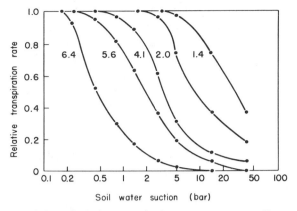

**Fig. 16.4.** The relation of relative transpiration rate to average soil-water suction, under different meteorological conditions. The numbers represent different rates of potential evapotranspiration. (After Denmead and Shaw, 1962.)

### E. Classical Concepts of Soil-Water Availability to Plants

The concept of *soil-water availability*, while never clearly defined in physical terms, has for many years excited controversy among adherents of different schools of thought. Veihmeyer and Hendrickson (1927, 1949, 1950, 1955) claimed that soil water is equally available throughout a definable range of soil wetness, from an upper limit (*field capacity*) to a lower limit (the *permanent wilting point*), both of which are characteristic and constant for any given soil. They postulated that plant functions remain unaffected by any decrease in soil wetness until the permanent wilting point is reached, at which plant activity is curtailed abruptly. This schematized model, though based upon arbitrary limits,[3] enjoyed widespread acceptance for many years, particularly among workers in the field of irrigation management.

---

[3] We have already pointed out (Chapter 2) that the field capacity concept (Israelsen and West, 1922; Veihmeyer and Hendrickson, 1927), though useful in some cases, lacks a universal physical basis (Richards, 1960). The wilting point, if defined simply as the value of soil wetness of the root zone at the time plants wilt, is not easy to recognize, since wilting is often a temporary phenomenon, which may occur in midday even when the soil is quite wet. The *permanent wilting percentage* (Hendrickson and Veihmeyer, 1945) is based upon the *wilting coefficient* concept of Briggs and Shantz (1912) and has been defined as the root-zone soil wetness at which the wilted plant can no longer recover turgidity even when it is placed in a saturated atmosphere for 12 hr. This is still an arbitrary criterion, since plant-water potential may not reach equilibrium with the average soil moisture potential in such a short time. In any case, plant response depends as much on the evaporative demand (its variation and peak intensity) as on soil wetness (which itself is a highly variable function of space and time).

Other investigators, however (notably Richards and Wadleigh, 1952), produced evidence indicating that soil-water availability to plants actually decreases with decreasing soil wetness, and that a plant may suffer water stress and reduction of growth considerably before the wilting point is reached. Still others, seeking to compromise between the opposing views, attempted to divide the so-called "available range" of soil wetness into "readily available" and "decreasingly available" ranges, and searched for a "critical point" somewhere between field capacity and wilting point as an additional criterion of soil-water availability.

These different hypotheses, in vogue until quite recently, are represented graphically in Fig. 16.5.

None of these schools was able to base its hypotheses upon a comprehensive framework that could take into account the array of factors likely to influence the water regime of the soil–plant–atmosphere system as a whole. Rather, they tended to draw generalized conclusions from a limited set of experiments conducted under specific and sometimes poorly defined conditions. Over the years, a great mass of empirical data has been collected, which for a long time no one knew how to explain, correlate, or resolve into a systematic theory based on universal physical principles.

The picture was further confused by failure to distinguish between different types of plant response to soil moisture. While the transpiration rate may be, for a time, relatively independent of soil-water content changes in the root zone, other forms of plant activity may not be. Photosynthesis, vegetative growth, flowering, fruiting, and seed or fiber production may be related quite differently to the content or state of soil water.

It has long been recognized that soil wetness per se is not a satisfactory criterion for availability. Hence, attempts were made to correlate the water

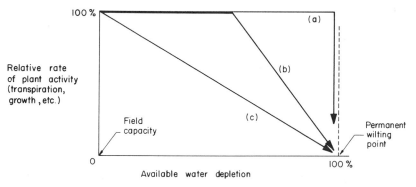

**Fig. 16.5.** Three classical hypotheses regarding the availability of soil water to plants: (a) equal availability from field capacity to wilting point, (b) equal availability from field capacity to a "critical moisture" beyond which availability decreases, and (c) availability decreases gradually as soil moisture content decreases.

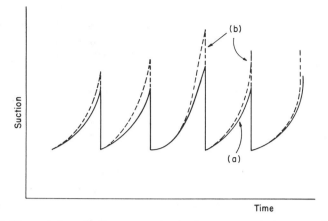

**Fig. 16.6.** The variation of soil-water suction in the root zone during successive irrigation cycles: (a) the average suction (tensiometric measurement) and (b) the suction of the soil in contact with the root.

status of plants with the energy state of soil water, i.e., with the soil-water potential (variously termed tension, suction, soil moisture stress, etc., indeed a bewildering variety of alternative terms!). The soil-water "constants" were therefore defined in terms of potential values (e.g., $-\frac{1}{10}$ or $-\frac{1}{3}$ bar for field capacity, $-15$ bar for permanent wilting), which could be applied universally, rather than in terms of soil wetness (Richards and Weaver, 1944; Slater and Williams, 1965). However, even though the use of energy concepts represented a considerable advance over the earlier notions, it still fell short of taking into account the dynamic nature of soil–plant–water relations.

A fundamental experimental difficulty encountered in any attempt at an exact physical description of soil-water uptake by plants is the inherently complicated nature of the space–time relationships involved in this process. Roots grow in different directions and spacings, and as yet we have no experimental method to measure the microscopic gradients and fluxes of water in their immediate vicinity. The conventional methods for measurement of the content or potential of soil water are based on the sampling or sensing of a relatively large volume and are therefore oblivious to the microgradients toward the roots. Water suction of the soil in contact with the roots can be much greater than the average suction, as illustrated in Fig. 16.6.

An additional difficulty in describing the system physically arises from the fact that, up to the present, no satisfactory way has been found to grow plants in a soil of constant water potential. Rather, it is necessary to irrigate the soil periodically, thus refilling its effective reservoir. In a variable soil-moisture regime, plants may be influenced more by the extreme values of the water potential they experience than by the average value (see Fig. 16.6.) Furthermore, root distribution is not generally uniform or constant within the root zone. Nor does the water-extraction pattern necessarily correspond to the root-distribution pattern. Hence, the correlation of plant response to soil-water conditions requires integrating each of the two over both space and time. In the process, the actual relationship can become quite obscure.

## F. Newer Concepts of Soil-Water Availability to Plants

In the last two decades, a fundamental change has taken place in our conception of soil–plant–water relationships. With the development of our theoretical understanding of the state and movement of water in the soil, plant, and atmosphere, and with the concurrent development of experimental techniques allowing more exact measurement of the interrelationships of potential, conductivity, water content, and flux, both in the soil and in the plant, the way has been opened for a more holistic approach to the problem. It has become increasingly clear that in a dynamic system such static concepts as "the soil-water constants" (i.e., field capacity, permanent wilting point, critical moisture, capillary water, gravitational water, etc.) are often physically meaningless, as they are based on the supposition that processes in the field bring about static levels of soil-water content or potential. In fact, as we are now very much aware, flow takes place almost incessantly, though in varying fluxes and directions, and static situations are exceedingly rare.

These developments have led to abandonment of the classical concept of available water in its original sense. Clearly, there is no fundamental qualitative difference between the water at one value of soil wetness or potential and another (e.g., across such arbitrary dividing points as $\frac{1}{10}$, $\frac{1}{3}$, or 15 bar of suction), nor is the amount and rate of water uptake by plants an exclusive function of the content or potential of soil water. The amount and rate of water uptake depend on the ability of the roots to absorb water from the soil with which they are in contact, as well as on the ability of the soil to supply and transmit water toward the roots at a rate sufficient to meet transpiration requirements. These, in turn, depend on *properties of the plant* (rooting density, rooting depth, and rate of root extension, as well as the physiological ability of the plant to continue drawing water from the soil at the rate needed to avoid wilting while maintaining its vital functions even while its own water potential decreases); *properties of the soil* (hydraulic conductivity–diffusivity–matric suction–wetness relationships); and also to a considerable extent the *meteorological conditions* (which dictate the rate at which the plant is required to transpire and hence the rate at which it must extract water from the soil in order to maintain its own hydration).

From a physical point of view, evapotranspiration can be viewed as a continuous stream flowing from a periodically replenished *source* of limited capacity and variable potential, namely, the reservoir of soil moisture, to a *sink* of virtually unlimited capacity (though of variable strength or evaporative potential)—the atmosphere. As long as the rate of root uptake of soil moisture balances the rate of canopy loss by transpiration, the stream continues unabated while the plant remains fully hydrated. The moment the uptake rate falls below transpiration, the plant itself must begin to lose moisture. This imbalance cannot continue for any length of time without resulting in loss of turgidity and hence in wilting of the plant.

The concept of *potential transpiration* (Penman, 1949) is an attempt to characterize the evaporative flux extracted from a stand of plants fully covering the ground surface when the supply of soil water is not limiting.[2] Accordingly, it is the meteorological conditions rather than soil or plant conditions which exercise the greatest influence on the transpiration rate as long as the soil is wet enough. However, as soil wetness is diminished, even though not completely depleted, actual transpiration begins to fall below the potential rate either because the soil cannot supply water fast enough and/or because the roots can no longer extract it fast enough to meet the meteorological demand. The point at which this condition is reached depends in a combined way upon the weather, the plants, and the soil. Any attempt to extricate one from the others and attribute phenomena to soil conditions alone is at best futile, and at worst misleading.

## G. Irrigation, Water-Use Efficiency, and Water Conservation

Any concept of efficiency is a measure of the output obtainable from a given input. Irrigation or water-use efficiency can be defined in different ways, however, depending on the nature of the inputs and outputs considered. For example, one can define as an economic criterion of efficiency the financial return in relation to the money invested in the installation and operation of water supply and delivery system. The problem is that costs and prices fluctuate from year to year and vary widely from place to place, and may not be universally comparable. Perhaps a more objective criterion for the relative merits of alternative irrigation systems is an agronomic one, namely, a comparison of the marketable yield per unit of land area or, better yet, per unit amount of water applied. This is merely an illustration of how the measure of efficiency depends on the point of view.

A widely applicable expression of efficiency is the *crop water-use efficiency*, which has been defined (Viets, 1962) as the amount of dry matter produced per unit volume of water taken up by the crop from the soil. As most of the water taken up by plants in the field is transpired (in arid regions, 99% or so!), while generally only a small amount is retained, the plant water-use efficiency is in effect the reciprocal of what has long been known as the transpiration ratio (Briggs and Shantz, 1912), originally defined as the mass of water transpired per unit mass of dry matter produced.

What we shall refer to as *technical efficiency* is what irrigation engineers call irrigation efficiency. It is generally defined as the net amount of water added to the root zone divided by the amount of water taken from some source (Hillel and Rawitz, 1972). As such, this criterion of efficiency can be applied to complex regional projects, or to individual farms, or to specific fields. In each case, the difference between the net amount of water added to the root zone and the amount withdrawn from the source represents the loss incurred in conveyance and distribution.

In practice, many (perhaps most) irrigation projects operate in an inherently inefficient way. In many of the surface irrigation schemes, one or a few farms may be allocated large flows representing the entire discharge of a lateral canal for a specified period of time. Since water is delivered to the consumer only at fixed times and charges may be assessed per delivery regardless of the actual amount used, customers tend to take as much water as they can while they can. This often results in overirrigation, which not only wastes water but also causes project-wide and perhaps even region-wide problems connected with the disposal of return flow, water-logging of soils, leaching of nutrients, and excessive elevation of the water table requiring expensive drainage to rectify. Although it is difficult to arrive at reliable statistics, it has been estimated that the average irrigation efficiency in such schemes is less than 50%. Since it is a proven fact that, with proper management irrigation efficiencies of 80–90% can be achieved in actual practice, there is obviously room, and need, for much improvement.

Particularly difficult to change are management practices which lead to deliberate waste not necessarily because of insurmountable technical problems or lack of knowledge but simply because it appears more convenient or even more economical in the short run to waste water rather than to apply proper management practices of water conservation. Such situations typically occur when the price of irrigation water is lower than the cost of labor or of the automated equipment needed to avoid overirrigation. Very often the price of water does not reflect its true cost but is kept deliberately low by government subsidy, which can be self-defeating (Hillel and Rawitz, 1972).

Where open and unlined distribution ditches are used, uncontrolled seepage and evaporation, as well as transpiration by riparian phreatophytes, are also a major cause of water loss. Even pipeline distribution systems do not always prevent loss. Leaky joints resulting from poor workmanship, corrosion, poorly maintained valves, or mechanical damage by farm machinery may cause large losses. At times the damage is not immediately obvious, as during the failure of a buried pipe.

Surface runoff resulting from the excessive application of water ideally should not occur. Sprinkler irrigation systems should be designed to apply water at rates which never exceed soil infiltrability. In the case of gravity irrigation systems, however, it is often virtually impossible to achieve uniform water distribution over the field without incurring some runoff ("tail water"). Only when provision is made to collect irrigation and rainwater surpluses at the lower end of the field and guide them as controlled return flow can this water be considered anything but a loss.

Evaporative losses associated with water application include any evaporation from open water surfaces of border checks or furrows, evaporation of water droplets during their flight from sprinkler to ground surface, wind drift of droplets away from the target area, and evaporation from wetted crop canopies or from the wet soil immediately after the irrigation, not all of which can be avoided.

In the open field, little can be done to reduce transpiration if the conditions required for high yields are to be maintained. Attempts to use chemical sprays known as "antitranspirants" have generally failed, and attempts to control wind movement above a crop by windbreaks may or may not produce the desired effect economically. It appears at present that the greatest promise for increasing water-use efficiency lies in allowing the crop to transpire freely at the climatic limit by alleviating any water shortages while at the same time avoiding waste and obviating all other environmental constraints to attainment of the fullest possible production potential of the crop. This is particularly important in the case of the new and superior varieties which have been developed in recent years and which can provide high yields only if water stress is prevented and such other factors as soil fertility (availability of nutrients), aeration, salinity, and soil tilth are also optimized. Plant diseases and pests, as well as insufficient fertility of the soil, may depress yields without a proportionate decrease in transpiration and water use. All management practices can thus influence water-use efficiency, and none can be considered in isolation from the others.

# 17  *Water Balance and Energy Balance in the Field*

## A. Introduction

Any attempt to control the quantity and availability of soil moisture to plants must be based on a thorough understanding and a quantitative knowledge of the dynamic balance of water in the soil. The *field-water balance*, like a financial statement of income and expenditures, is an account of all quantities of water added to, subtracted from, and stored within a given volume of soil during a given period of time. The various soil-water flow processes that we have attempted to describe in earlier chapters of this book as separate phenomena (e.g., infiltration, redistribution, drainage, evaporation, water uptake by plants) are in fact strongly interdependent, as they occur sequentially or simultaneously.

The water balance is merely a detailed statement of the *law of conservation of matter*, which states simply that matter can neither be created nor destroyed but can only change from one state or location to another. Since no significant amounts of water are normally decomposed, or composed, in the soil, the water content of a soil profile of finite volume cannot increase without addition from the outside (as by infiltration or capillary rise), nor can it diminish unless transported to the atmosphere by evapotranspiration or to deeper zones by drainage.

The field water balance is intimately connected with the *energy balance*, since it involves processes that require energy. The energy balance is an expression of the classical *law of conservation of energy*, which states that, in a given system, energy can be absorbed from, or released to, the outside, and that along the way it can change form, but it cannot be created or destroyed.

The content of water in the soil affects the way the energy flux reaching the field is partitioned and utilized. Likewise, the energy flux affects the state and movement of water. The water balance and energy balance are inextricably linked, since they are involved in the same processes within the same environment. A physical description of the soil–plant–atmosphere system must be based on an understanding of both balances together. In particular, the evaporation process, which is often the principal consumer of both water and energy in the field, depends, in a combined way, on the simultaneous supply of water and energy.

## B. Water Balance of the Root Zone

In its simplest form, the water balance merely states that, in a given volume of soil, the difference between the amount of water added $W_{in}$ and the amount of water withdrawn $W_{out}$ during a certain period is equal to the change in water content $\Delta W$ during the same period:

$$\Delta W = W_{in} - W_{out} \tag{17.1}$$

When gains exceed losses, the water-content change is positive; and conversely, when losses exceed gains, $\Delta W$ is negative.

To itemize the accretions and depletions from the soil storage reservoir, one must consider the disposition of rain or irrigation reaching a unit area of soil surface during a given period of time. Rain or irrigation water applied to the land may in some cases infiltrate into the soil as fast as it arrives. In other cases, some of the water may pond over the surface. Depending on the slope and microrelief, a portion of this water may exit from the area as surface run-off ("overland flow") while the remainder will be stored temporarily as puddles in surface depressions. Some of the latter evaporates and the rest eventually infiltrates into the soil after cessation of the rain. Of the water infiltrated, some evaporates directly from the soil surface, some is taken up by plants for growth or transpiration, some may drain downward beyond the root zone, whereas the remainder accumulates within the root zone and adds to soil moisture storage. Additional water may reach the defined soil volume by runoff from a higher area, or by upward flow from a water table or from wet layers present at some depth. The pertinent volume or depth of soil for which the water balance is computed is determined arbitrarily. Thus, in principle, a water balance can be computed for a small sample of soil or for an entire watershed. From an agricultural or plant ecological point of view, it is generally most appro-

priate to consider the water balance of the root zone per unit area of field.
The root zone water balance is expressed in integral form thusly:

$$\text{(change in storage)} = \text{(gains)} - \text{(losses)}$$

$$(\Delta S + \Delta V) = (P + I + U) - (R + D + E + T)$$

(17.2)

wherein $\Delta S$ is change in root zone soil moisture storage, $\Delta V$ increment of
water incorporated in the plants, $P$ precipitation, $I$ irrigation, $U$ upward
capillary flow into the root zone, $R$ runoff, $D$ downward drainage out of
the root zone, $E$ direct evaporation from the soil surface, and $T$ transpiration by plants. All quantities are expressed in terms of volume of water
per unit area (equivalent depth units) during the period considered.

The time rate of change in soil moisture storage can be written as follows
(assuming the rate of change of plant-water content to be relatively unimportant):

$$dS/dt = (p + i + u) - (r + d + e + t_r)$$

(17.3)

Here each of the lowercase letters represents the instantaneous time rate
of change of the corresponding integral quantity in the first equation. The
change in root zone soil moisture storage can be obtained by integrating
the change in soil wetness over depth and time as follows:

$$S = \int_0^z \int_{t_1}^{t_2} \left( \frac{\partial \theta}{\partial t} \right) dz \, dt$$

(17.4)

where $\theta$ is the volumetric soil wetness, measurable by sampling or by means
of a neutron meter. Note that $t$ is time whereas $t_r$ is transpiration rate in
our notation.

The largest composite term in the "losses" part of Eq. (17.2) is generally
the evapotranspiration $E + T$. It is convenient at this point to refer to the
concept of "potential evapotranspiration" (designated $E_{to}$), representing
the climatic "demand" for water. Potential evapotranspiration from a
well-watered field depends primarily on the energy supplied to the surface
by solar radiation, which is a climatic characteristic of each location (depending on latitude, season, slope, aspect, cloudiness, etc.) and varies little
from year to year. $E_{to}$ depends secondarily on atmospheric advection, which
is related to the size and orientation of the field and the nature of its upwind
"fetch" or surrounding area. Potential evapotranspiration also depends
upon surface roughness and soil thermal properties, characteristics which
vary in time (van Bavel and Hillel, 1976). As a first approximation and
working hypothesis, however, it is often assumed that $E_{to}$ depends entirely
on the external climatic inputs and is independent of the transient properties
of the field itself.

Actual evapotranspiration, $E_{ta}$ is generally a fraction of $E_{to}$ depending on the degree and density of plant canopy coverage of the surface, as well as on soil moisture and root distribution. $E_{ta}$ from a well-watered stand of a close growing crop will generally approach $E_{to}$ during the active growing stage, but may fall below it during the early growth stage, prior to full capoy coverage, and again toward the end of the growing season, as the matured plants begin to dry out (Hillel and Guron, 1973). For the entire season, $E_{ta}$ may total 60–80% of $E_{to}$ depending on water supply: the drier the soil moisture regime, the lower the actual evapotranspiration. The relation of yield to *ET* is still a matter of some controversy.

Another important, indeed essential, item of the field-water balance is the drainage out of the root zone *D*. A certain amount of drainage is required for aeration and for leaching out excess salts so as to prevent their accumulation in the root zone, a particular hazard of arid zone farming. Where natural drainage is lacking or insufficient, artificial drainage becomes a prerequisite for sustainable agriculture.

The various items entering into the water balance of a hypothetical rooting zone are illustrated in Fig. 17.1. In this representation, only vertical flows are considered within the soil. In a larger sense, any soil layer of interest forms a part of an overall hydrologic cycle, illustrated in Fig. 17.2, in which the flows are multidirectional.

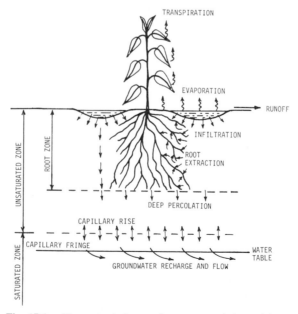

**Fig. 17.1.** The water balance of a root zone (schematic).

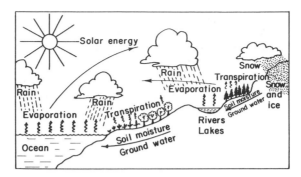

**Fig. 17.2.** The hydrologic cycle (schematic). (After Bertrand, 1967.)

## C. Evaluation of the Water Balance

Simple and readily understandable though the field water balance may seem in principle, it is still rather difficult to measure in practice. A single equation can be solved if it has only one unknown. Often the largest component of the field water balance, and the one most difficult to measure directly, is the evapotranspiration $E + T$, also designated $E_t$. To obtain $E_t$ from the water balance (Deacon *et al.*, 1958) we must have accurate measurements of all other terms of the equation. It is relatively easy to measure the amount of water added to the field by rain and irrigation $(P + I)$, though it is necessary to consider possible nonuniformities in areal distribution. The amount of run-off generally is (or at least should be) small in agricultural fields, and particularly in irrigated fields, so that it can sometimes be regarded as negligible in comparison with the major components of the water balance.

For a long period, e.g., an entire season, the change in water content of the root zone is likely to be small in relation to the total water balance. In this case, the sum of rain and irrigation is approximately equal to the sum of evapotranspiration $E_t$ and deep percolation $D$. For shorter periods, the change in soil-water storage $\Delta S$ can be relatively large and must be measured. This measurement can be made by sampling periodically, or by use of specialized instruments.[1]

During dry spells, without rain or irrigation, $W_{in} = 0$, so that the sum of $D$ and $E_t$ now equals the reduction in root-zone water storage $\Delta S$:

$$-\Delta S = D + E_t \qquad (17.5)$$

[1] Of the various methods for measuring the content of water in soil, the neutron meter is the most satisfactory at present since it measures wetness on the volume or depth fraction basis directly and since it samples a large volume and minimizes sampling errors (with repeated measurements made at the same site and depth).

Common practice in irrigation is to measure the total water content of the root zone just prior to an irrigation, and to supply the amount of water necessary to replenish the soil reservoir to some maximal water content, generally taken to be the "field capacity." Some ecologists and irrigationists have tended to assume that the deficit of soil moisture which develops between rains or irrigations is due to evapotranspiration only, thus disregarding the amount of water which may flow through the bottom of the root zone, either downward or upward. This flow is not always negligible and often constitutes a tenth or more of the total water balance (Robins *et al.*, 1954; Nixon and Lawless, 1960; Rose and Stern, 1967a, b).

It should be obvious that measurement of root-zone or subsoil water content by itself cannot tell us the rate and direction of soil-water movement (van Bavel *et al.*, 1968a, b). Even if the water content at a given depth remains constant, we cannot conclude that the water there is immobile, since it might be moving steadily through that depth. Tensiometric measurements can, however, indicate the directions and magnitudes of the hydraulic gradients through the profile (Richards, 1965) and allow us to compute the fluxes from knowledge of the hydraulic conductivity versus suction or wetness for the particular soil. More direct measurements of the deep percolation component of the field water balance may eventually become possible with the development of water flux meters (Cary, 1968). Such devices have not yet proven to be practical, however.

The most direct method for measurement of the field water balance is by use of lysimeters (van Bavel and Myers, 1962; Pruitt and Angus, 1960; King *et al.*, 1956; Pelton, 1961; McIlroy and Angus, 1963; Forsgate *et al.*, 1965; Rose *et al.*, 1966; Harrold, 1966; Black *et al.*, 1968; Hillel *et al.*, 1969). These are generally large containers of soil, set in the field to represent the prevailing soil and climatic conditions and allowing more accurate measurement of physical processes than can be carried out in the open field. From the standpoint of the field water balance, the most efficient lysimeters are those equipped with a weighing device and a drainage system, which together allow continuous measurement of both evapotranspiration and percolation. Lysimeters may not provide a reliable measurement of the field water balance, however, when the soil or above-ground conditions of the lysimeter differ markedly from those of the field itself.

## D. Radiation Exchange in the Field

By *radiation* we refer to the emission of energy in the form of electromagnetic waves from all bodies above 0°K. *Solar* (sun) *radiation* received on the earth's surface is the major component of its energy balance. Green plants are able to convert a part of the solar radiation into chemical energy. They do this in the process of photosynthesis, upon which all life on earth ultimately depends. For these reasons, it is appropriate to introduce a discussion of the energy balance with an account of the radiation balance.

Solar radiation reaches the outer surface of the atmosphere at a nearly constant flux of about 2 cal/min cm$^2$ perpendicular to the incident radiation.[2] Nearly all of this radiation is of the wavelength range of 0.3–3 μm (3000–30,000 Å), and about half of this radiation consists of visible light (i.e., 0.4–0.7 μm in wavelength). The solar radiation corresponds approximately to the emission spectrum of a blackbody[3] at a temperature of 6000°K. The earth, too, emits radiation, but since its surface temperature is about 300°K, this *terrestrial radiation* is of much lower intensity and greater wavelength than solar radiation[4] (i.e., in the wavelength range of 3–50 μm). Between these two radiation spectra, the sun's and the earth's, there is very little overlap, and it is customary to refer to the first as *short-wave* and to the second as *long-wave radiation* (Sellers, 1965).

In passage through the atmosphere, solar radiation changes both its flux and spectral composition. About one-third of it, on the average, is reflected back to space (this reflection can become as high as 80% when the sky is completely overcast with clouds). In addition, the atmosphere absorbs and scatters a part of the radiation, so that only about half of the original flux density of solar radiation finally reaches the ground.[5] A part of the reflected and scattered radiation also reaches the ground and is called *sky radiation*. The total of direct solar and sky radiations is termed *global radiation*.

*Albedo* is the reflectivity coefficient of the surface toward short-wave radiation. This coefficient varies according to the color, roughness, and inclination of the surface, and is of the order of 5–10% for water, 10–30% for a vegetated area, 15–40% for a bare soil, and up to 90% for fresh snow.

---

[2] 1 cal/cm$^2$ = 1 langley (Ly). 58 Ly ≈ 1 mm evaporation equivalent (latent heat = 580 cal/gm).

[3] A blackbody is one which absorbs all radiation reaching it without reflection, and emits at maximal efficiency. According to the *Stefan–Boltzmann law*, the total energy emitted by a body $J_t$, integrated over all wavelengths, is proportional to the fourth power of the absolute temperature $T$. This law is usually formulated as $J_t = \varepsilon\sigma T^4$ (where $\sigma$ is a constant, and $\varepsilon$ the emissivity coefficient). For a perfect blackbody, $\varepsilon = 1$.

[4] According to *Wien's law*, the wavelength of maximal radiation intensity is inversely proportional to the absolute temperature: $\lambda_m T = 2900$ (where $\lambda_m$ is the wavelength in microns and $T$ is the temperature on the Kelvin scale). *Planck's law* gives the intensity distribution of energy emitted by a blackbody as a function of wavelength and temperature: $E_\lambda = C_1/\lambda^5[\exp(C_2/\lambda T) - 1]$, where $E_\lambda$ is the energy flux emitted in a particular wavelength range, and $C_1$, $C_2$ are constants.

[5] In arid regions, where the cloud cover is sparse, the actual radiation received at the soil surface can exceed 70% of the "external" radiation. In humid regions, this fraction can be 40% or lower.

In addition to these incoming and reflected short-wave radiation fluxes, there is also a long-wave radiation (heat) exchange. The earth's surface emits radiation, and at the same time the atmosphere absorbs and emits long-wave radiation, part of which reaches the surface. The difference between the outgoing and incoming fluxes is called the *net long-wave radiation*. During the day, the net long-wave radiation may be a small fraction of the total radiation balance, but during the night, in the absence of direct solar radiation, the heat exchange between the land surface and the atmosphere dominates the radiation balance.

The overall difference between total incoming and total outgoing radiation (including both the short-wave and long-wave components) is termed *net radiation*, and it expresses the rate of radiant energy absorption by the field.

$$J_n = J_s^{\downarrow} - J_s^{\uparrow} + J_1^{\downarrow} - J_1^{\uparrow} \qquad (17.6)$$

where $J_n$ is the net radiation, $J_s^{\downarrow}$ the incoming flux of short-wave radiation from sun and sky, $J_s^{\uparrow}$ the short-wave radiation reflected by the surface, $J_1^{\downarrow}$ the long-wave radiation from the sky, and $J_1^{\uparrow}$ the long-wave radiation reflected and emitted by the surface. At night, the short-wave fluxes are negligible, and since the long-wave radiation emitted by the surface generally exceeds that received from the sky, the nighttime net radiation flux is negative.

The reflected short-wave radiation is equal to the product of the incoming short-wave flux and the reflectivity coefficient (the albedo $\alpha$):

$$J_s^{\uparrow} = \alpha J_s^{\downarrow}$$

Therefore,

$$J_n = J_s^{\downarrow}(1 - \alpha) - J_1 \qquad (17.7)$$

where $J_1$ is the net flux of long-wave radiation, which is given a negative sign. (Since the surface of the earth is usually warmer than the atmosphere, there is generally a net loss of thermal radiation from the surface.) As a rough average, $J_n$ is typically of the order of 55–70% of $J_s^{\downarrow}$ (Tanner and Lemon, 1962).

## E. Total Energy Balance

Having balanced the gains and losses of radiation at the surface to obtain the net radiation, we next consider the transformation of this energy.

Part of the net radiation received by the field is transformed into heat, which warms the soil, plants, and atmosphere. Another part is taken up by the plants in their metabolic processes (e.g., photosynthesis). Finally, a major part is generally absorbed as latent heat in the twin processes of evaporation and transpiration. Thus,

$$J_n = LE + A + S + M \tag{17.8}$$

where $LE$ is the rate of energy utilization in evapotranspiration (a product of the rate of water evaporation $E$ and the latent heat of vaporization $L$), $A$ is the energy flux that goes into heating the air (called *sensible heat*), $S$ is the rate at which heat is stored in the soil, water, and vegetation, and $M$ represents other miscellaneous energy terms such as photosynthesis and respiration.

The energy balance is illustrated in Fig. 17.3.

Where the vegetation is short (e.g., grass or field crops), the storage of heat in the vegetation is negligible compared with storage in the soil (Tanner, 1960). (The situation might be different, of course, in the case of the voluminous, and massive, vegetation of a forest.) The heat stored in the soil under short and sparse vegetation may be a fairly large portion of the net radiation at any one time during the day, but the net storage over a 24 hr period is usually small (since the nighttime loss of soil heat negates the daytime gain). For this reason, mean soil temperature generally does not change appreciably from day to day. The daily soil-storage term has been variously reported to be of the order of 5–15% of $J_n$ (Decker, 1959; Tanner and Pelton, 1960). This obviously depends on season. In spring and summer, this term is positive, but it becomes negative in autumn.

In the past, the miscellaneous energy terms [$M$ in Eq. (17.8)] were believed to be a negligible portion of the energy balance. Measurements of carbon

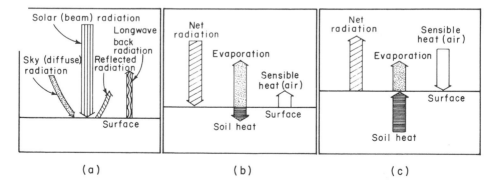

**Fig. 17.3.** Schematic representation of (a) the radiation balance and (b) the daytime and (c) the nighttime energy balance. [Net radiation = (solar radiation + sky radiation) − (reflected radiation + back radiation).] It is to be remembered that the daytime net radiation during the growing season is much greater than at night. (After Tanner, 1968.)

dioxide exchange over active crops in the natural environment, however, have revealed that photosynthesis may in some cases account for as much as 5% of the daily net radiation where there is a large mass of active vegetation particularly under low-light conditions. In general, though, $M$ is much less than that (Lemon, 1960).

Overall, the amount of energy stored in soil and vegetation and that fixed photochemically account for a rather small portion of the total daily net radiation, with the major portion going into latent and sensible heat. The proportionate allocation between these terms depends on the availability of water for evaporation, but in most agriculturally productive fields the latent heat predominates over the sensible heat term.

## F. Transport of Heat and Vapor to the Atmosphere

The transport of sensible heat and water vapor (which carries latent heat) from the field to the atmosphere is affected by the turbulent movement of the air in the atmospheric boundary layer.[6] The sensible heat flux $A$ is proportional to the product of the temperature gradient $dT/dz$ and the turbulent transfer coefficient for heat $k_a$ ($cm^2/sec$):

$$A = -c_p \rho_a k_a \, dT/dz \qquad (17.9)$$

where $c_p$ is the specific heat capacity of air at constant pressure (cal/cm °C), $\rho_a$ the density of air, $T$ temperature (°C), and $z$ height (cm).

The rate of latent heat transfer by water vapor from the field to atmosphere, $LE$, is similarly proportional to the product of the vapor pressure gradient and the appropriate turbulent transfer coefficient for vapor.

If we assume that the transfer coefficients for heat and water vapor are equal, then the ratio of the sensible heat transport to the latent heat transport becomes

$$\beta = A/LE \approx \xi_c \, \Delta T/\Delta e \qquad (17.10)$$

where $\Delta T/\Delta e$ is the ratio of the temperature gradient to the vapor pressure gradient in the atmosphere above the field, and $\xi_c$ is the psychometric constant $\approx 0.66$ mbar/°C.

The ratio $\beta$ is called the *Bowen ratio*, and it depends mainly on the temperature and moisture regimes of the field. When the field is wet, the relative humidity gradients between its surface and the atmosphere tend to be

---

[6] "A laminar boundary layer," generally less than 1 mm thick, is recognized in immediate contact with the surface of an evaporating body. Through this layer, transport occurs by diffusion. Beyond this, turbulent transport becomes predominant in the "turbulent boundary layer."

large, whereas the temperature gradients tend to be small. Thus, $\beta$ is rather small when the energy is consumed mainly in evaporation. When the field is dry, on the other hand, the relative humidity gradients toward the atmosphere are generally small, and the temperature gradients tend to be steep, so that the Bowen ratio becomes large. In a recently irrigated field, $\beta$ may be smaller than 0.2, while in a dry field in which the plants are under a water stress (with stomatal resistance coming into play), the surface may warm up and a much greater share of the incoming energy will be lost to the atmosphere directly as sensible heat. Under extremely arid conditions, in fact, $LE$ may tend to zero and $\beta$ to infinity. With advection (Section G), sensible heat may be transferred from the air to the field, and the Bowen ratio can become negative.

Whether or not water-vapor transport to the atmosphere from a vegetated field becomes restricted obviously depends not only upon the soil-water content per se, but on a complex interplay of factors in which the characteristics of the plant cover (i.e., density of the canopy, root distribution, and physiological responses to water stress) play an important role.

The assumption that the transfer coefficients for heat and vapor are equal (or at least proportional) is known as the principle of *similarity* (Tanner, 1968). Transfer through the turbulent atmospheric boundary layer takes place primarily by means of *eddies*, which are ephemeral, swirling microcurrents of air, whipped up by the wind. Eddies of varying size, duration, and velocity fluctuate up and down at varying frequency, carrying both heat and vapor. While the instantaneous gradients and vertical fluxes of heat and vapor will generally fluctuate, when a sufficiently long averaging period is allowed (say, 15–60 min), the fluxes exhibit a stable statistical relationship over a uniform field.[7] This is not the case at a low level over an inhomogeneous surface of spotty vegetation and partially exposed soil. Under such conditions, cool, moist packets of air may rise from the vegetated spots while warm, dry air may rise from the dry soil surface, with the latter rising more rapidly owing to buoyancy.[8]

Using the Bowen ratio, the latent and sensible heat fluxes can be written (recalling that $J_n = S + A + LE$, and that $\beta = A/LE$):

$$LE = (J_n - S)/(1 + \beta) \qquad 17.11)$$

$$A = \beta(J_n - S)/(1 + \beta) \qquad 17.12)$$

---

[7] It is reasonable to assume that momentum and carbon dioxide, as well as vapor and heat, are carried by the same eddies.

[8] An index of the relative importance of buoyancy (thermal) versus frictional forces in producing turbulence is the Richardson number $R_i = g(dT/dz)/[T(du/dz)^2]$, where $dT/dz$ is the temperature, and $g$ the acceleration of gravity. The air profile tends to be stable when $R_i$ is positive and unstable (buoyant) when $R_i$ is negative (Sellers, 1965).

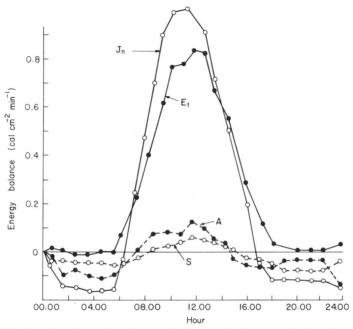

**Fig. 17.4.** The diurnal variation of net radiation $J_n$ and of energy utilization by evapotranspiration $E_t$, sensible heating of the atmosphere $A$, and heating of the soil $S$. Alfalfa-brome hay on Plainfield sand, 4 September 1957. (After Tanner, 1960.)

Thus, $LE$ can be obtained from micrometeorological measurements in the field (i.e., $J_n$, $S$, and $\beta$) without necessitating measurements of soil-water fluxes or plant activity.

The diurnal variation of the components of the energy balance is illustrated in Fig. 17.4. The diurnal as well as the annual patterns of the components of the energy balance differ for different conditions of soil, vegetation, and climate (Sellers, 1965).

## G. Advection

The equations given for the energy balance apply to *extensive uniform areas* in which all fluxes are vertical or nearly so. On the other hand, any *small field* differing from its surrounding area is subject to lateral effects and can exchange energy in one form or another with neighboring areas. Specifically, winds sweeping over a small field can transport heat into or out of it. This phenomenon, called *advection*, can be especially important in arid regions, where small irrigated fields are often surrounded by an expanse of dry land. Under such conditions, the warm and dry incoming air can transfer sensible heat (which is transformed into latent heat of vaporization) down to the crop (Graham and King, 1961; Halstead and Covey, 1957; Rosenberg, 1974).

The extraction of sensible heat from a warm mass of air flowing *over* the top of a field, and the conversion of this heat to latent heat of evaporation, is called the *oasis effect*. The passage of warm air *through* the vegetative cover has been called the *clothesline effect* (Tanner, 1957). A common sight in arid regions is the poor growth of the plants near the windward edge of a field, where penetration of warm, dry wind contributes energy for evapotranspiration. Where advective heat inflow is large, evapotranspiration from rough and "open" vegetation (e.g., widely spaced row crops or trees) can greatly exceed that from smooth and close vegetation (e.g., mowed grass).

The effects of advection are likely to be small in very large and uniform fields but very considerable in small plots which differ markedly from their surroundings. With advection, latent heat "consumption" can be larger than net radiation. Hence, values of evapotranspiration and of irrigation requirements obtained from small experimental plots are not typically representative of large fields, unless these plots are "guarded" in the upwind direction by an expanse, or *fetch*, of vegetation of similar roughness characteristics and subject to a similar water regime. It should be obvious from the preceding that a small patch of vegetation, particularly if it consists of a spaced stand of shrubs or trees, can at times evaporate water in excess of the evaporation from a free water surface such as a lake, a pond, or a pan.

Advection is not confined to small fields. Large-scale or "macrometeorological" advective effects also occur and were described by Slatyer and McIlroy (1961), who pointed out that even in relatively humid regions, advection caused by the movement of weather systems may temporarily cause latent heat consumption to exceed average net radiation. A case in point is the periodic invasion of semihumid regions along the Mediterranean littoral by searing desert winds, variously called sharkiyeh, sirocco, or khamsin.

## H.  Potential Evapotranspiration (Combination Formulas)

The concept of *potential evapotranspiration* is an attempt to characterize the micrometeorological environment of a field in terms of an evaporative power, or demand; i.e., in terms of the maximal evaporation rate which the atmosphere is capable of exacting from a field of given surface properties. The concept probably derives from the common observation that when a wet object is exposed and dried gradually in the open air, progressively longer increments of time are generally required to remove equal increments of water. The evaporation rate obviously depends both on the environment and on the state of wetness of the object itself. Intuitively, therefore, one might suppose that there ought to be a definable evaporation rate for the special case in which the object is maintained perpetually in as wet a state as possible, and that this evaporation rate should depend only on the meteorological environment. More specifically, Penman (1956) defined potential evapotranspiration as "the amount of water transpired in unit time by a short green crop, completely shading the ground, of uniform height and never short of water." As such, it is a useful standard of reference for the comparison of different regions and of different measured evapotranspiration values within a given region.

To obtain the highest possible yields of many agricultural crops, irrigation should be provided in an amount sufficient to prevent water from becoming a limiting factor. Knowledge of the potential evapotranspiration can therefore serve as a basis for planning the irrigation regime. In general, the actual evapotranspiration $E_{ta}$ from various crops will not equal the potential value $E_{to}$, but in the case of a close-growing crop the maintenance of optimal soil moisture conditions for maximal yields will generally result in $E_{ta}$ being nearly equal to, or a nearly constant fraction of, $E_{to}$, at least during the active-growth phase of the crop season.

Various empirical approaches have been proposed for the estimation of potential evapotranspiration (e.g., Thornthwaite, 1948; Blaney and Criddle, 1950). The method proposed by Penman (1948) is physically based and hence inherently more meaningful. His equation, based on a combination of the energy balance and aerodynamic transport considerations, is a major contribution in the field of agricultural and environmental physics.

$$LE = \frac{(\Delta/\xi)J_n + LE_a}{\Delta/\xi + 1} \qquad (17.13)$$

where $LE_a = 0.35(e_a - e)(0.5 + U_2/100)$ (mm/day); $e_a$ = saturated vapor pressure at mean air temperature (mm Hg); $e$ = mean vapor pressure in air; $U_2$ = mean wind speed in miles per day at 2m above ground. This equation permits a calculation of the potential evapotranspiration rate from measurements of the net radiation, and of the temperature, vapor pressure, and wind velocity taken at one level above the field.

Actual evapotranspiration from an actively growing crop in the field generally constitutes a fraction, often in the range between 60% and 90%, of the potential evapotranspiration as determined by the Penman equation or by evaporation pans. The Penman formulation avoids the necessity of determining the value of $T_s$, the surface temperature, just as it disregards the possible fluctuations in the direction and magnitude of the soil heat flux term. Moreover, it makes no provision for surface roughness or air instability (buoyancy) effects. Finally, the Penman theory takes no explicit account of advection. To correct for the differences between potential evapotranspiration from rough surfaces and potential evaporation from smooth water $E_0$, Penman used the following empirical factors determined in Southern England:

$E_0$(bare soil)/$E_0$(water) = 0.9

$E_0$(turf)/$E_0$(water) = 0.6 in winter,   ranging to 0.8 in summer

It should be emphasized that the representation of potential evapotranspiration purely as an externally imposed "forcing function" is a rather gross approximation. In actual fact, the field participates, as it were, in determining its evapotranspiration rate even when it is well endowed with water, through the effect of its radiant reflectivity, aerodynamic roughness, thermal capacity and conductivity, etc. The often stated principle that all well-watered fields, regardless of their specific characteristics, are subject to, or exhibit, the same potential evapotranspiration is only more or less correct.

The Penman formulation was modified by van Bavel (1966) to allow for short-term variations in soil heat flux and for differences among various surfaces. His method for predicting potential evapotranspiration requires the additional measurements of net radiation and soil heat flux. A roughness height parameter is used to characterize the aerodynamic properties of the surface, i.e., to take account of the fact that, all other things being equal, potential evapotranspiration from a corn field should exceed that from a lawn, which, in turn, should be greater than that from a smooth, bare soil.

Potential evapotranspiration $LE_0$ is given by

$$LE_0 = \frac{(\Delta/\xi)(J_n - S) + k_v d_a}{(\Delta/\xi) + 1} \tag{17.14}$$

wherein $\Delta$ is the slope of the saturation vapor pressure versus temperature curve at mean air temperature, $\xi$ the psychrometric constant, $J_n$ net radiation, $S$ soil heat flux, $d_a$ the vapor pressure deficit at elevation $Z_a$ (namely, $(e_s - e_a)$), and $k_v$ the transfer coefficient for water vapor (a function of wind speed and surface roughness). For the dependence of $k_v$ on mean wind speed $U_2$, Penman (1948) suggested empirically $k_v = 20(1 + U_2/100) = 20 + U_2/5$, where $e_s$ and $e_a$ are given in millimeters of mercury.

Further improvements of the physically based prediction of evapotranspiration can result from inclusion of air stability or buoyancy effects, in recognition of the fact that vapor transfer is enhanced whenever the thermal structure of the air becomes unstable (Szeicz et al., 1973). The advent of remote-sensing infrared thermometry has made possible continuous monitoring of surface temperature, and hence also allows a better estimation of the vapor pressure at the surface.

The ratio $\Delta/\xi$ and the saturation vapor pressure $e_s$ at various temperatures are available in standard tables in many texts on physical meteorology and environmental physics (e.g., Sellers, 1965; Slatyer and McIlroy, 1961; Monteith, 1973). A summary is given in the accompanying table.

| $T$ (°C) | 10 | 15 | 20 | 25 | 30 | 35 |
|---|---|---|---|---|---|---|
| $\Delta/\xi$ | 1.23 | 1.64 | 2.14 | 2.78 | 3.57 | 4.53 |
| $e_s$ (mm Hg) | 9.20 | 12.78 | 17.53 | 23.75 | 31.82 | 42.18 |
| $e_s$ (mbar) | 12.27 | 17.04 | 23.37 | 31.67 | 42.43 | 56.24 |

**Sample Problems**

**1.** The total incoming global radiation $J_s$ (sun and sky) received by a particular field on a given day is 500 cal/cm², or langleys. The albedo $a$ is 15%. The net outgoing long-wave radiation balance $J_l$ amounts to 10 cal/cm². The sensible heat transfer to the air $A$ is 12 cal/cm², the net heat flow into the soil $S$ is 6 cal/cm², and the metabolic uptake of energy $M$ is 8 cal/cm². Calculate the net radiation, the amount of energy available for latent heat transfer (evapotranspiration), and the day's evapotranspiration in millimeters of water. On the following day, the sensible heat transfer is reversed and evapotranspiration totals 7.5 mm. If everything else remains the same, calculate the amount of advected energy taken up by the field.

To calculate the net radiation $J_n$ we write the radiation balance, Eq. (17.7):

$$J_n = J_s^{\downarrow}(1 - a) - J_l = 500 \text{ cal/cm}^2(1 - 0.15) - 10 \text{ cal/cm}^2 = 415 \text{ cal/cm}^2$$

The latent heat term $LE$ can be calculated from the overall energy balance, Eq. (17.8), when all other terms are known:

$$J_n = LE + A + S + M \qquad \text{or} \qquad LE = J_n - A - S - M$$

Using the values given, we have

$$LE = 415 - 12 - 6 - 8 = 389 \text{ cal/cm}^2$$

Since roughly 580 cal are required at prevailing temperatures to vaporize 1 gm or 1 cm³ of water, i.e., $L = 580$ cal, the amount of evaporation is

$$LE/L = 389 \text{ cal/cm}^2/580 \text{ cal/cm}^3 = 0.67 \text{ cm} = 6.7 \text{ mm}$$

On the following day, with a positive influx of sensible heat by advection, evapotranspiration amounts to 7.5 mm, and hence the latent heat term is

$$LE = 0.75 \text{ cm} \times 580 \text{ cal/cm}^3 = 435 \text{ cal/cm}^2$$

The energy balance is therefore

$$\text{income} = \text{disposal}: \quad J_n + A = LE + S + M$$

and the advected energy is

$$A = LE + S + M - J_n = 435 + 6 + 8 - 415 = 34 \text{ cal/cm}^2$$

**2.** On a given day in early spring the daily net radiation $J_n$ is 350 cal/cm$^2$, the mean air temperature $T_a$ at standard height (2 m) is 15°C, the mean vapor pressure $e_a$ at that height is 8 mm Hg, and the mean wind speed $u_2$ is 15 mile/day. On a given day in the late spring the net radiation is 420 cal/cm$^2$, mean air temperature 20°C, mean vapor pressure 9 mm Hg, and mean wind speed 20 mile/day. Finally, on a given day in summer $J_n$ is 500 cal/cm$^2$, $T_a$ is 25°C, $e_a$ is 10 mm Hg, and $u_2$ is 25 mile/day. Estimate the potential evapotranspiration using Eq. (17.14). Assume the net soil heat flux $S$ to be zero in all cases.

Potential evapotranspiration $LE_0$ is given by Eq. (17.14):

$$LE_0 = \frac{(\Delta/\xi)(J_n - S) + k_v d_a}{(\Delta/\xi) + 1}$$

where $d_a$ is the mean vapor pressure deficit $(e_s - e_a)$ at standard height. Recall that $\Delta/\xi$ and $e_s$ at several temperatures are tabulated in Section H, and assume that $k_v = 20 + U_2/5$ (Penman, 1948). Accordingly, for the early spring day we get

$$LE_0 = \frac{1.64 \times (350 - 0) + (20 + 15/5)(12.78 - 8)}{1.64 + 1} = 259 \text{ cal/cm}^2 \text{ day}$$

For the late spring day,

$$LE_0 = \frac{2.14 \times (420 - 0) + (20 + 20/5)(17.53 - 9)}{2.14 + 1} = 351 \text{ cal/cm}^2 \text{ day}$$

and for the summer day,

$$LE_0 = \frac{2.78 \times (500 - 0) + (20 + 25/5)(23.75 - 10)}{2.78 + 1} = 459 \text{ cal/cm}^2 \text{ day}$$

Remembering that approximately 580 cal are required to vaporize 1 gm of water, and assuming a water density of 1 gm/cm$^3$, we can use 58 cal/cm$^2$ day as the latent heat flux equivalent to the evaporation of 1 mm of water per day. Hence the values of potential evapotranspiration are estimated to be

$$259/58 = 4.5 \text{ mm for the early spring day}$$

$$351/58 = 6.1 \text{ mm for the late spring day}$$

$$459/58 = 7.9 \text{ mm for the summer day}$$

# Bibliography

Aase, J. K., and Kemper, W. D. (1968). Effect of ground color and microwatersheds on corn growth. *J. Soil Water Conserv.* **23,** 60–62.

Abdalla, A. M., Hetteriaratchi, D. R. P., and Reece, A. R. (1969). The mechanics of root growth in granular media. *J. Agr. Eng. Res.* **14,** 236–248.

Abramowitz, M., and Stegun, I. A. (1964). Handbook of mathematical functions. Nat. Bur. Stand. Appl. Math. Ser., Vol. 55, U.S. Government Printing Office, Washington, D.C.

Acevedo, E., Hsiao, T. C., and Henderson, D. W. (1971). Immediate and subsequent growth response of maize leaves to changes in water status. *Plant Physiol.* **48,** 631–636.

Adam, K. M., Bloomsburg, G. L., and Corey, A. T., Diffusion of trapped gas from porous media, *Water Resour.* **5,** 840–849.

Adams, J. E. (1970). Effect of mulches and bed configuration. *Agron. J.* **62,** 785–790.

Adams, J. E., and Hanks, R. J. (1964). Evaporation from soil shrinkage cracks. *Soil Sci. Soc. Am. Proc.* **28,** 281–284.

Adrian, D. D., and Franzini, J. B. (1966). Impedance to infiltration by pressure build-up ahead of the wetting front. *J. Geophys. Res.* **71,** 5857–5862.

Alexander, M. (1961, 1977). "Introduction to Soil Microbiology," Wiley, New York.

Allison, L. E. (1956). Soil and plant responses to VAMA and HPAN soil conditioners in the presence of high exchangeable sodium. *Soil Sci. Soc. Am. Proc.* **20,** 147–151.

Alway, F. J., and McDole, G. R. (1917). Relation of the water-retaining capacity of a soil to its hygroscopic coefficient. *J. Agr. Res.* **9,** 27–71.

American Society of Agronomy Monograph No. 9 (1965). "Methods of Soil Analysis, Physical and Mineralogical Properties," Part I.

Amerman, C. R., Hillel, D. I., and Peterson, A. E. (1970). A variable-intensity sprinkling infiltrometer. *Soil Sci. Soc. Am. Proc.* **34,** 830–832.

Anat, A., Duke, H. R., and Corey, A. T. (1965). Steady upward flow from water tables. Colorado State Univ. Hydrol. Paper No. 7, June.

Anderson, D. M., and Tice, A. R. (1971). Low-temperature phases of interfacial water in clay-water systems. *Soil Sci. Soc. Am Proc.* **35,** 47–54.

Anderson, M. S. (1926). Properties of soil colloidal material. U.S. Dept. Agr. Bull. 1452.

Arbhabhirama, A., and Kridakorn, C. (1968). Steady downward flow to a water table. *Water Resour. Res.* **4,** 116–121.

Aslyng, H. C. (1963). Soil physics terminology. *Int. Soc. Soil Sci. Bull.* **23,** 7.

ASTM (American Society for Testing Materials) (1956). *Symp. Vane Shear Testing Soils* Spec. Tech. Publ. 193. Am. Soc. Testing Mater., Philadelphia, Pennsylvania.

ASTM (American Society for Testing Materials) (1958a). Book of Standards, Part II. pp. 217–224. Am. Soc. Testing Mater., Philadelphia, Pennsylvania.

ASTM (American Society for Testing Materials) (1958b). "Procedures for Soil Testing." Philadelphia, Pennsylvania.

Atterberg, A. (1911). Die Plastizitat der Tone. *Int. Mitt. Bodenk.* **1**, 10–43.

Atterberg, A. (1912). Die Konsistenz und die Bindigheit der Boden. *Int. Mitt. Bodenk.* **2**, 148–189.

Bachmat, Y., and Elrick, D. E. (1970). Hydrodynamic instability of miscible fluids in a vertical porous column. *Water Resources Res.* **6**, 156–171.

Bailey, A. C., and Vanden Berg, G. E. (1968). Yielding by compaction and shear in unsaturated soils. *Trans. Am. Soc. Agr. Eng.* **11**, 307–311, 317.

Barber, E. S. (1965). Stress distribution. *In* "Methods of Soil Analysis," Part I. Monograph 9, Am. Soc. Agron., Madison, Wisconsin.

Barley, K. P. (1962). The effect of mechanical stress on the growth of roots. *J. Exp. Bot.* **13**, 95–110.

Barrs, H. D. (1968). Determination of water deficits in plant tissues. *In* "Water Deficits and Plant Growth" (T. T. Kozlowski, ed.), pp. 235–368. Academic Press, New York.

Bear, J. (1969). "Dynamics of Fluids in Porous Media," 453–457. Elsevier, Amsterdam.

Bear, J. (1972). "Dynamics of Fluids in Porous Media." Elsevier, New York.

Bear, J., Zaslavsky, D., and Irmay, S. (1968). "Physical Principles of Water Percolation and Seepage." UNESCO, Paris.

Bekker, M. G. (1956). "Theory of Land Locomotion." Univ. Michigan Press, Ann Arbor, Michigan.

Bekker, M. G. (1960). "Off-the-Road Locomotion: Research and Development in Terramechanics." Univ. Michigan Press, Ann Arbor, Michigan.

Bekker, M. G. (1961). Mechanical properties of soil and problems of compaction. *Trans Am. Soc. Agr. Eng.* **4**, 231–234.

Belmans, C., Feyen, J., and Hillel, D. (1979). An attempt at experimental validation of macroscopic-scale models of soil moisture extraction by roots. *Soil Sci.* **127**, 174–186.

Bernacki, H., Haman, J., and Kanafojski, Cz. (1967). "Agricultural Machines: Theory and Construction." PWRIL, Warsaw, Poland. (Translated from Polish by U.S. Department of Agriculture and National Science Foundation, TT-69-50019).

Beskow, G. (1935). Soil freezing and frost heaving with special applications to roads and railroads. *Swed. Geol. Soc.* **26**, C, 375. Translation (with special supplement of progress 1935–1946) by J. O. Osterberg. Tech. Inst., Northwestern Univ., Evanston, Illinois.

Betrand, A. R. (1965). Rate of water intake in the field. *In* "Methods of Soil Analysis" (C. A. Black, ed.). Monograph No. 9, Am. Soc. Agron., Madison, Wisconsin.

Betrand, A. R. (1967). Water conservation through improved practices. *In* "Plant Environment and Efficient Water Use." Am. Soc. Agron., Madison, Wisconsin.

Biggar, J. W., and Nielsen, D. R. (1976). Spatial variability of the leaching characteristics of a field soil. *Water Resour. Res.* **12**, 78–84.

Birkeland, P. W. (1974). "Pedology, Weathering, and Geomorphological Research." Oxford Univ. Press, London and New York.

Birkle, D. E., Letey, J., Stolzy, L. H., and Szuszkiewicz, T. E. (1964). Measurement of oxygen diffusion rates with the platinum microelectrode. *Hilgardia* **35**, 555–556.

Bishop, A. W., and Blight, G. E. (1963). Some aspects of effective stress in saturated and partly saturated soils. *Geotechnique* **13**, 177–197.

Bishop, A. W., and Henkel, D. J. (1964). "The Measurement of Soil Properties in the Triaxial Test." Arnold, London.

Black, C. A. (ed.) (1965). "Methods of Soil Analysis," Part I. Am. Soc. Agron., Madison, Wisconsin.

Black, T. A., Gardner, W. R., and Thurtell, G. W. (1969). The prediction of evaporation, drainage and soil water storage for a bare soil. *Soil Sci. Soc. Am. Proc.* **33**, 655–660.

Black, T. A., Tanner, C. B., and Gardner, W. R. (1970). Evaporation from a snap bean crop. *Agron. J.* **62**, 66–69.

Black, T. A., Thurtell, G. W., and Tanner, C. B. (1968). Hydraulic load-cell lysimeter, construction, calibration, and tests. *Soil Sci. Soc. Am. Proc.* **32**, 632–639.

Blake, G. R. (1965). Bulk density. *In* "Methods of Soil Analysis," pp. 374–390. Am. Soc. of Agron., Madison, Wisconsin.

Blake, G. R., and Page, J. B. (1948). Direct measurement of gaseous diffusion in soils. *Soil Sci. Soc. Am. Proc.* **13**, 37–42.

Blaney, H. F., and Criddle, W. D. (1950). Determining water requirements in irrigated areas from climatological and irrigation data. U.S. Soil Conservat. Serv. Tech. Publ. 96

Bodman, G. B., and Colman, E. A. (1944). Moisture and energy conditions during downward entry of water into soils. *Soil Sci. Soc. Am. Proc.* **8**, 116–122.

Bodman, G. B., and Constantin, G. K. (1965). Influence of particle size distribution in soil compaction. *Hilgardia* **36**, 567–591.

Bodman, G. B., Johnson, D. E., and Kruskal, W. H. (1958). Influence of VAMA and of depth of rotary hoeing upon infiltration of irrigation water. *Soil Sci. Soc. Am. Proc.* **22**, 463–468.

Boersma, L. (1965a). Field measurement of hydraulic conductivity below a water table. *In* "Methods of Soil Analysis," pp. 222–223. *Monograph 9,* Am. Soc. Agron., Madison, Wisconsin.

Boersma, L. (1965b). Field measurement of hydraulic conductivity above a water table. *In* "Methods of Soil Analysis," pp. 234–252. Monograph 9, Am. Soc. Agron., Madison, Wisconsin.

Bolt, G. H. (1956). Physico-chemical analysis of the compressibility of pure clays. *Geotechnique* **8**, 86–90.

Bolt, G. H. (1976). Soil physics terminology. *Bull. Int. Soc. Soil Sci.* **49**, 26–36.

Bolt, G. H., and Bruggenwert, M. G. M. (ed.) (1976). "Soil Chemistry." Elsevier, Amsterdam.

Bolt, G. H., and Frissel, M. J. (1960). Thermodynamics of soil moisture. *Neth. J. Agr. Sci.* **8**, 57–78.

Bolt, G. H., and Peech, M. (1953). The application of the Gouy theory to soil-water systems. *Soil Sci. Soc. Am. Proc.* **17**, 210–213.

Boltzmann, L. (1894). Zur integration des diffusiongleichung bei variabeln diffusions coefficienten. *Ann. Phys.* **53**, 959–964.

Bomba, S. J. (1968). Hysteresis and time-scale invariance in a glass-bead medium. PhD Thesis, Univ. Wisconsin, Madison, Wisconsin.

Bond, J. J., and Willis, W. O. (1969). Soil water evaporation: Surface residue rate and placement effects. *Soil Sci. Soc. Am. Proc.* **33**, 445–448.

Bouma, J., Hillel, D. I., Hole, F. D. and Amerman. C. R. (1971). Field measurement of unsaturated hydraulic conductivity by infiltration through artificial crust. *Soil Sci. Soc. Am. Proc.* **32**, 362–364.

Bouwer, H. (1959). Theoretical aspects of flow above the water table in tile drainage of shallow homogeneous soil. *Soil Sci. Soc. Am. Proc.* **23**, 200–203.

Bouwer, H. (1961). A double tube method for measuring hydraulic conductivity of soil in sites above a water table. *Soil Sci. Am. Proc.* **25**, 334–342.

Bouwer, H. (1962a). Field determination of hydraulic conductivity above a water table with the double tube method. *Soil Sci. Am. Proc.* **26**, 330–335.

Bouwer, H. (1962b). Analyzing groundwater mounds by resistance network analog. *J. Irrig. Drain. Div. Proc. Am. Soc. Civ. Eng.* **88,** IR3, 15–36.

Bouwer, H. (1964). Resistance network analogs for solving ground-water problems. *Ground-Water* **2,** (3), 1–7.

Bouwer, H. (1978). "Groundwater Hydrology." McGraw-Hill, New York.

Bouyoucos, G. J. (1937). Evaporating the water with burning alcohol as a rapid means of determining moisture content of soils. *Soil Sci.* **44,** 377–383.

Bouyoucos, G. J., and Mick, A. H. (1940). An electrical resistance method for the continuous measurement of soil moisture under field conditions. Michigan Agr. Exp. Sta. Tech. Bull. 172.

Bower, C. A., and Goertzen, J. O. (1959). Surface area of soils and clays by an equilibrium ethylene glycol method. *Soil Sci.* **87,** 289–292.

Bower, H., and van Schilfgaarde, J. (1963). Simplified method of predicting fall of water table in drained land. *Trans. Am. Soc. Agr. Eng.* **6,** 196, 288–291.

Brady, N. C. (1974). "The Nature and Properties of Soils." MacMillan, New York.

Braester, C. (1973). Linearized solution of infiltration at constant rate. *In* "Physical Aspects of Soil, Water and Salts in Ecosystems" (A. Hadas *et al.,* eds.). Springer-Verlag, Berlin and New York.

Bresler, E. (1972a). Interacting diffuse layers in mixed mono-divalent ionic systems. *Soil Sci. Soc. Am. Proc.* **36,** 891–896.

Bresler, E. (1972b). Control of soil salinity. *In* "Optimizing the Soil Physical Environment Toward Greater Crop Yields" (D. Hillel, ed.) pp. 102–132. Academic Press, New York.

Bresler, E. (1973). Simultaneous transport of solutes and water under transient unsaturated flow conditions. *Water Resour. Res.* **9,** 975–986.

Bresler, E. (1978). Theoretical modeling of mixed-electrolyte solution flows for unsaturated soils. *Soil Sci.* **125,** 196–203.

Bresler, E., and Hanks, R. J. (1969). Numerical Method for estimating simultaneous flow of water and salt in unsaturated soils. *Soil Sci. Soc. Am. Proc.* **33,** 827–832.

Bresler, E., and Miller, R. D. (1975). Estimation of pore blockage induced by freezing of unsaturated soil. *Proc. Conf. on Soil-Water Prob. In Cold Reg. Calgary.* pp. 162–175.

Bresler, E., Kemper, W. D., and Hanks, R. J. (1969). Infiltration, redistribution, and subsequent evaporation of water from soil as affected by wetting rate and hysteresis. *Soil Sci. Soc. Am. Proc.* **33,** 832–840.

Briggs, L. J., and Shantz, H. L. (1921). The relative wilting coefficient for different plants. *Bot. Gaz.* (*Chicago*) **53,** 229–235.

Bronowski, J. (1977). "A Sense of the Future." MIT Press, Cambridge, Massachusetts.

Brooks, R. H., and Corey, A. T. (1966). Properties of porous media affecting fluid flow. *Proc. Am. Soc. Civ. Eng., J. Irrigation Drainage Div.* **IR2,** 61–68.

Brooks. R. H., and Corey, A. T. (1964). Hydraulic properties of porous media. Hydrology Paper No. 3. Colorado State Univ., Ft. Collins, Colorado.

Brown, P. A. (1970). Measurement of water potential with thermocouple psychrometers: construction and application. USDA Forest Service Research Rep., INT-80.

Browning, G. M. (1950). Principles of solid physics in relation to tillage. *Agr. Eng.* **31,** 341–344.

Bruce, R R., and Klute, A. (1956). The measurement of soil-water diffusivity. *Soil Sci. Soc. Am. Proc.* **20,** 458–562.

Bruce, R. R., and Whisler, F. D. (1973). Infiltration of water into layered field soils. *In* "Physical Aspects of Soil, Water and Salts in Ecosystems" (A. Hadas *et al.,* eds.). Springer-Verlag, Berlin and New York.

Brunauer, S., Emmett, P. H., and Teller, E. (1938). Adsorption of gases in multimolecular layers. *J. Am. Chem. Soc.* **60**, 309–319.

Brustkern, R. L., and Morel-Seytoux, H. J. (1970). Analytical treatment of two-phase infiltration. *J. Hydraul. Div. ASCE* **96**, 2535–2548.

Buckingham, E. (1904). Contributions to our knowledge of the aeration of soils. U. S. Bur. Soils Bull. 25.

Buckingham, E. (1907). Studies on the movement of soil moisture. U.S. Dept. of Agr. Bur. of Soils, Bull. 38.

Buras, N. (1974). Water management systems. *In* "Drainage for Agriculture" (J. van Schilfgaarde, ed.). Monograph 17, Am. Soc. Agron., Madison, Wisconsin.

Burdine, N. T. (1952). Relative permeability calculations from pore-size distribution data. *Trans. AIME* **198**, 35–42.

Burger, H. C. (1915). *Phys. Zs.* **20**, 73–76.

Burrows, W. C. (1963). Characterization of soil temperature distribution from various tillage-induced microreliefs. *Soil Sci. Soc. Am. Proc.* **27**, 350–353.

Burwell, R. E., and Larson, W. E. (1969). Infiltration as influenced by tillage-induced random roughness and pore space. *Soil Sci. Soc. Am. Proc.* **33**, 449–452.

Busscher, W. J. (1979). Simulation of infiltration from a continuous and intermittent subsurface source. *Soil Sci.* **128**, (in press).

Buswell, A. M., and Rodebush, W. H. (1956). *Water. Sci. Am.* **202**, 1–10.

Byers, G. L., and Webber, L. R. (1957). Tillage practices in relation to crop yields, power requirements, and soil properties. *Can. J. Soil Sci.* **37**, 71–75.

Cameron, D. R. (1978). Variability of soil water retention curves and predicted hydraulics conductivities on a small plot. *Soil Sci.* **126**, 364–371.

Campbell, G. S. (1977). "An Introduction to Environmental Biophysics." Springer-Verlag, New York.

Cannell, R. Q. (1977). Soil aeration and compaction in relation to root growth and soil management. *Adv. Appl. Biol.* **2**, 1–86.

Carman, P. C. (1939). *J. Agr. Sci.* **29**, 262.

Carmen, P. C. (1956). "Flow of Gases Through Porous Media." Academic Press, New York.

Carslaw, J. S., and Jaeger, J. C. (1959). "Conduction of Heat in Solids." Oxford Univ. Press (Clarendon), London and New York.

Carson, J. E. (1961). Soil Temperature and Weather Conditions. Rep. No. 6470, Argonne National Laboratories, Argon.

Carvallo, H. O., Cassell, D. K., Hammond, J., and Bauer, A. (1976). Spatial variability of *in situ* unsaturated hydraulic conductivity of Maddock sandy loam. *Soil Sci.* **121**, 1–7.

Cary, J. W. (1963). Onsager's relations and the non-isothermal diffusion of water vapor. *J. Phys. Chem.* **67**, 126–129.

Cary, J. W. (1964). An evaporation experiment and its irreversible thermodynamics. *Int. J. Heat Mass Transfer* **7**, 531–538.

Cary, J. W. (1966). Soil moisture transport due to thermal gradients: Practical aspects. *Soil Sci. Soc. Am. Proc.* **30**, 428–433.

Cary, J. W. (1968). An instrument for *in situ* measurement of soil moisture flow and suction. *Soil Sci. Soc. Am. Proc.* **32**, 3–5.

Cary, J. W., and Taylor, S. A. (1962a). The interaction of the simultaneous diffusions of heat and water vapor. *Soil Sci. Soc. Am. Proc.* **26**, 413–416.

Cary, J. W., and Taylor, S. A. (1962b). Thermally driven liquid and vapor phase transfer of water and energy in soil. *Soi. Sci. Soc. Am. Proc.* **26**, 417–420.

Casagrande, A. (1937). Seepage through dams. *J. New England Water Works Assoc.* **51**, 131–172.

Casagrande, A. (1948). Classification and identification of soils. *Trans. Am. Soc. Civil Eng.* **113**, 901–903.

Casagrande, A., and Fadum, R. E. (1940). Notes on soil testing for engineering purposes. Harvard Univ. Grad. School of Eng. Soil Mech. Ser. No. 8, pp. 37–49.

Cassel, D. K., and Bauer, A. (1975). Spatial variability in soils below depth of tillage: Bulk density and fifteen atmosphere percentage. *Soil Sci. Soc. Am. Proc.* **39**, 247–250.

Cedergren, H. R. (1967). "Seepage Drainage and Flow Nets." Wiley, New York.

Chahal, R. S. and Miller, R. D. (1965). Supercooling of water in glass capillaries. *Brit. J. Appl. Phys.* **16**, 231–239.

Chancellor, W. J. (1976). Compaction of soil by agricultural equipment. Bull. 1881, Div. Agr. Sci., Univ. California, Richmond, California.

Chancellor, W. J., and Schmidt, R. H. (1962). Soil deformation beneath surface loads. *Trans. Am. Soc. Agr. Eng.* **5**, 204–246, 249.

Chapman, H. D. (1965). Cation exchange capacity. *In* "Methods of Soil Analysis" (C.A. Black, ed.). Monograph 9, Am. Soc. Agron., Madison, Wisconsin.

Chen, Y., and Banin, A. (1975). Scanning electron microscope (SEM) observations of soil structure changes induced by sodium-calcium exchange in relation to hydraulic conductivity. *Soil Sci.* **120**, 428–436.

Chepil, W. S. (1958). Soil conditions that influence wind erosion. U.S. Dept. Agriculture Tech. Bull. 1185.

Chepil, W. S. (1962). A compact rotary sieve and the importance of dry sieving in physical soil analysis. *Soil Sci. Soc. Am. Proc.* **26**, 4–6.

Childs, E. C. (1940). The use of soil moisture characteristics in soil studies. *Soil Sci.* **50**, 239–252.

Childs, E. C. (1947). The water table equipotentials and streamlines in drained land. *Soil Sci.* **63**, 361–376.

Childs, E. C. (1969). "An Introduction to the Physical Basis of Soil Water Phenomena." Wiley, New York.

Childs, E. C., and Collis-George, N. (1950). The permeability of porous materials. *Proc. R. Soc. London Ser. A.* **201**, 392–405.

Childs, E. C., and Poulovassilis, A. (1962). The moisture profile above a moving water table. *J. Soil Sci.* **13**, 272–285.

Chow, V. T. (ed.) (1964). "Handbook of Applied Hydrology." McGraw-Hill, New York.

Chudnovskii, A. F. (1966). "Fundamentals of Agrophysics." Israel Program for Scientific Translations, Jerusalem.

Clarke, R. T., and Newson, M. D. (1978). Some detailed water balance studies of research catchments. *Proc. R. Soc. London Series A* **363**, 21–42.

Coelho, M. A. (1974). Spatial variability of water related soil physical properties. Ph.D. Dissertation, Univ. of Ariz., Tucson, Arizona (available as 75-11,061 from Xerox Univ. Microfilms, Ann Arbor, Michigan).

Cohron, G. T. (1971). Forces causing soil compaction. In "Compaction of Agricultural Soils," (K. K. Barnes, ed.), pp. 106–122. Monograph, Am. Soc. Agr. Eng., St. Joseph, Michigan.

Coleman, E. A., and Bodman, G. B. (1945). Moisture and energy conditions during downward entry of water into moist and layered soils. *Soil Sci. Soc. Am. Proc.* **9**, 3–11.

Collis-George, N., and Youngs, E. G. (1958). Some factors determining water table heights in drained homogeneous soils. *J. Soil Sci.* **9**, 332–338.

Colman, E. A., and Hendrix, T. M. (1949). Fiberglass electrical soil moisture instrument. *Soil Sci.* **67**, 425–438.

Cooper, A. W., and Trouse, A. C., and Dumas, W. T. (1969). Controlled traffic in row crop production. *Proc. Int. Cong. Agric. Eng. (CIGR), 7th* Baden-Baden, Germany. Section III, 1–6.

Corey, A. T. (1954). The interrelation between gas and oil relative permeabilities. Oil Producer's Monthly, Vol. XIX, No. 1. November.

Corey, A. T. (1977). "Mechanics of Heterogeneous Fluids in Porous Media." Water Resources Publications, Fort Collins, Colorado.

Corey, A. T. and Brooks, R. H., Drainage characteristics of soils. *Soil Sci. Soc. Am. Proc.* **39**, No. 2, March–April 1975, pp. 251–255.

Covey, W. (1963). Mathematical study of the first stage of drying of a moist soil. *Soil Sci. Soc. Am. Proc.* **27**, 130–134.

Cowan, I. R. (1965). Transport of water in the soil-plant atmosphere system. *J. Appl. Ecol.* **2**, 221–229.

Cowan, I. R., and Milthorpe, F. L. (1968). Plant factors influencing the water status of plant tissues. *In* "Water Deficits and Plant Growth" (T. T. Kozlowski, ed.), pp. 137–193. Academic Press, New York.

Crank, J. (1956). "The Mathamatics of Diffusion." Oxford Univ. Press, London and New York.

Currie, J. A. (1961). Gaseous diffusion in porous media Part 3—Wet granular material. *Brit. J. Appl. Phys.* **12**, 275–281.

Currie, J. A. (1975). Soil respiration. *In* "Soil Physical Conditions and Crop Production." Tech. Bull. 29, Min. of Agr., Fisheries and Food, HMSO, London.

Dalton, F. N., and Rawlins, S. L. (1968). Design criteria for Peltier effect thermocouple psychrometers. *Soil Sci.* **105**, 12–17.

Dane, J. H. (1978). Calculation of hydraulic conductivity decreases in the presence of mixed Na-Cl$_2$ solutions. *Can. J. Soil Sci.* **58**, 145–152.

Dane, J. H., and Klute, A. (1977). Salt effects on the hydraulic properties of a swelling soil. *Soil Sci. Soc. Am. J.* **41**, 1043–1049.

Darcy, H. (1856). "Les Fontaines Publique de la Ville de Dijon." Dalmont, Paris.

Dasberg, S., and Bakker, J. W. (1970). Characterizing soil aeration under changing soil moisture conditions for bean growth. *Agr. J.* **62**, 689–692.

Dasberg, S., Hillel, D., and Arnon, I. (1966). Response of grain sorghum to seedbed compaction. *Agron. J.* **58**, 199–201.

Davidson, D. T. (1965). Penetrometer measurements. *In* "Methods of Soil Analysis," Part I. Monograph 9, Am. Soc. Agron., Madison, Wisconsin.

Davidson, J. M., Nielsen, D. R., Biggar, J. W., and Cassel, D. K. (1966). Soil water diffusivity and water content distribution during outflow experiments. Water in the unsaturated zone. Int. Assoc. Sci. Hydrology. *Proc. Wageningen Symp.* 214–223.

Davidson, J. M., Stone, L. R., Nielsen, D. R., and La Rue, M. E. (1969). Field measurement and use of soil water properties. *Water Resources Res.* **5**, 1312–1321.

Davies, D. B., Finney, J. B., and Richardson, S. J. (1973). Relative effects of tractor weight and wheel slip in causing soil compaction. *J. Soil Sci.* **24**, 399–408.

Davies, J. A., and Allen, C. D. (1973). Equilibrium, potential and actual evaporation from cropped surfaces in S. Ontario. *J. Appl. Meteorol.* **12**, 649–657.

Day, O. R. (1965). Particle fractionation and particle size analysis. *In* "Methods of Soil Analysis," pp. 545–567. Monograph 9, Am. Soc. Agron., Madison, Wisconsin.

Day, P. R., and Luthin, J. (1956). A numerical solution of the differential equation of flow for a vertical drainage problem. *Soil Sci. Soc. Am. Proc.* **20**, 443–447.

Deacon, E. L., Priestley, C. H. B., and Swinbank, W. C. (1958). Evaporation and the water balance. Climatol-Rev. Res. Arid Zone Res. 9–34 (UNESCO).

de Boer, J. H. (1953). "The Dynamical Character of Adsorption." Oxford Univ. Press, London and New York.

DeBoodt, M. (1972a). Improvement of soil structure by chemical means. *In* "Optimizing the Soil Physical Environment Toward Greater Crop Yields" (D. Hillel, ed.), pp. 43–55. Academic Press, New York.

DeBoodt, M. (ed.) (1972b). *Proc. Symp. Fundamentals Soil Conditioning,* State Univ. of Ghent, Belgium.

DeBoodt, M., and DeLeenheer, L. (1958). Proposition pour l'evaluation de la stabilite des aggregates sur le terrain. *Proc. Int. Symp. Soil Structure, Ghent, Belgium,* pp. 234–241.

DeBoodt, M., DeLeenheer, L., and Kirkham, D. (1961). Soil aggregate stability indexes and crop yields. *Soil Sci.* **91,** 138–146.

Decker, W. L. (1959). Variations in the net exchange of radiation from vegetation of different heights. *J. Geophys. Res.* **64,** 1617–1619.

de Groot, S. R. (1963). "Thermodynamics of Irreversible Processes." North-Holland Publ., Amsterdam.

de Jong, E. (1968). Applications of thermodynamics to soil moisture. *Proc. Hydrol. Symp., 6th* pp. 25–48. National Research Council of Canada.

DeLeenheer, L., and DeBoodt, M. (1954). Discussion on aggregate analysis of soils by wet sieving. *Trans. Int. Congr. Soil Sci., 5th Leopoldville* **2,** 111–117.

Delhomme, J. P. (1976). Applications de la theorie des variables regionalises dans les sciences de l'eau. Thesis de Docteur-Ingenieur, Université Pierre et Marie Curie, Paris. 160 pp.

Denmead, O. T., and Shaw, R. H. (1962). Availability of soil water to plants as affected by soil moisture content and meteorological conditions. *Agron. J.* **54,** 385–390.

Deresiewicz, H. (1958). Mechanics of granular matter. *Adv. Appl. Mech.* **5,** 233–306.

Deryaguin, B. V., and Melnikova, M. K. (1958). Mechanism of moisture equilibrium and migration in soils. Water and its conduction in soils. *Int. Symp. Highway Res. Board* Spec. Rep. 40, pp. 43–54.

de Vries, D. A. (1975). The thermal conductivity of soil. Med. Landbouwhogeschool Wageningen.

de Vries, D. A. (1963). Thermal properties of soils. *In* "Physics of the Plant Environment" (van Wijk, W. R., ed). North-Holland, Amsterdam.

de Vries, D. A. (1975). Heat transfer in soils. *In* "Heat and Mass Transfer in the Biosphere." (D. A. de Vries and N. H. Afgan, eds.), pp. 5–28. Scripta Book Co., Washington, D.C.

de Vries, D. A., and Peck, A. J. (1958). On the cylindrical probe method of measuring thermal conductivity with special reference to soils. *Aust. J. Phys.* **11,** 255–271; 409–423.

De Wiest, R. J. M. (ed.) (1969). "Flow Through Porous Media," Academic Press, New York.

de Wit, C. T. (1958). Transpiration and plant yields. *Versl. Landbouwk. Onderz.* **646,** 59–84.

de Wit, C. T., and van Keulen, H. (1972). "Simulation of Transport Processes in Soils." PUDOC, Wageningen, Netherlands.

d'Hollander, E., and Impens, I. (1975). Hybrid simulation of a dynamic model for water movement in a soil-plant-atmosphere continuum. *In* "Computer Simulation of Water Resources Systems" (G. C. Vansteenkiste, ed.), pp. 349–360. North-Holland Publ., Amsterdam.

Diamond, S. (1970). Pore size distribution in clays. *Clays Clay Mineral.* **18,** 7–24.

Dick, D. A. T. (1966). "Cell Water." Butterworth, London.

Dirksen, C., and Miller, R. D. (1966). Closed-system freezing of unsaturated soil. *Soil Sci. Soc. Am. Proc.* **30,** 168–173.

Dixon, R., and Linden, M. (1972). Soil air pressure and water infiltration under border irrigation. *Soil Sci. Soc. Am. Proc.* **36,** 948–953.

Domenico, P. A. (1972). "Concepts and Models in Groundwater Hydrology." McGraw-Hill, New York.

Donahue, R. L., Shickluna, J. C., and Robertson, L. S. (1971). "Soils—An Introduction to Soils and Plant Growth," 3rd ed. Prentice-Hall, Englewood Cliffs, New Jersey.

Donnan, W. W. (1947). Model tests of a tile-spacing formula. *Soil Sci. Soc. Am. Proc.* **11,** 131–136.

Doorenbos, J., *et al.* Irrigation. FAO, Rome.

Dudal, R. (1968). Definitions of soil units for the soil map of the world. FAO, Rome.

Dumas, W. T., Trouse, A. C., Smith, L. A., Kummer, F. A., and Gill, W. R. (1973). Development and evaluation of tillage and other cultural practices in a controlled traffic system for cotton in the Southern Coastal Plain. *Trans. Am. Soc. Agr. Eng.* **16,** 872–875, 880.

Dumas, W. T., Trouse, A. C., Smith, L. A., Kummer, F. A., and Gill, W. R. (1975). Traffic control as a means of increasing cotton yields by reducing soil compaction. Paper No. 75–1050, 1975 presented at the Annual Meeting of Am. Soc. Agric. Eng., Davis, California.

Eagleson, P. S. (1970). "Dynamic Hydrology." McGraw-Hill, New York.

Edlefsen, N. E., and Anderson, A. B. C. (1943). Thermodynamics of soil moisture. *Hilgardia* **15,** 31–298.

Edwards, R. S. (1956). A mechanical sieve designed for experimental work on tilths. *Empire J. Exp. Agr.* **24,** 317–322.

Edwards, W. M., and Larson, W. E. (1970). Infiltration of water into soils as influenced by surface seal development. *Soil Sci. Soc. Am. Proc.* **34,** 101.

Eisenberg, D., and Kauzmann, W. (1969). "The Structure and Properties of Water." Oxford Univ. Press, London and New York.

Ekern, P. C. (1950). Raindrop impact as the force initiating soil erosion. *Soil Sci. Soc. Am. Proc.* **15,** 7–10.

Elrick, D. E., and Laryea, K. B. (1979). Sorption of water in soils: a comparison of techniques for solving the diffusion equation. *Soil Sci.* **128,** 210.

Elrick, D. E., Laryea, K. B., and Groenevelt, P. H. (1979). Hydrodynamic dispersion during infiltration of water into soil. *Soil Sci. Soc. Am. Proc.* **43,** 856–865.

Elrick, D. E., Scandrett, J. H., and Miller, E. E. (1959). Tests of capillary flow scaling. *Soil Sci. Soc. Am. Proc.* **23,** 329–332.

Emerson, E. W. (1959). The structure of soil crumbs. *J. Soil Sci.* **10,** 235.

Emerson, W. W., and Grundy, G. M. F. (1954). The effect of rate of wetting on water uptake and cohesion of soil crumbs. *J. Agr. Sci.* **44,** 249–253.

Emerson, W. W. (1959). The structure of soil crumbs. *J. Soil Sc.* **10,** 235.

Epstein, E. (1973). Roots. *Sci. Am.* **228,** 48–58.

Epstein, E. (1977). The role of roots in the chemical economy of life on earth. *Bioscience* **27,** 783–787.

Erickson, A. E. (1972). Improving the water properties of sand soil. *In* "Optimizing the Soil Physical Environment Toward Greater Crop Yield" (D. Hillel, ed.), pp. 35–42. Academic Press, New York.

Erickson, A. E., and van Doren, D. M. (1960). The relation of plant growth and yield to soil oxygen availability. *Trans. Int. Congr. Soil Sci., 7th, Madison, Wisconsin* **3,** 428–434.

Ernst, L. F. (1962). Groundwater flow in the saturated zone and its calculation when parallel horizontal open conduits are present. *Versl. Landb. Ond.* **67.15,** 189 p.

Evans, D. D. (1965). Gas movement. *In* "Methods of Soil Analysis," Part I, pp. 319–330. Monograph 9, Am. Soc. Agron., Madison, Wisconsin.

Evans, I., and Sherratt, G. G. (1948). A simple and convenient instrument for measuring the shearing resistance of clay soils. *J. Sci. Inst.* **25,** 411–414.

Evenari, M., Aharoni, Y., Shanan, L., and Tadmor, N. H. (1958). The ancient agriculture of the Negev. III. Early beginnings. *Israel Explor. J.* **8,** 231–268.

Evenari, M., and Koller, D. (1956). Masters of the desert. *Sci. Am.* **194,** 39–45.

Everett, D. H. (1961). The thermodynamics of frost damage to porous solids. *Trans. Faraday Soc.* **57,** 1541–1551.

Fairbourn, M. L., and Cluff, C. B. (1974). Use gravel mulch to save water for crops. *Crops and Soil Magazine,* April–May.

Fairbourn, M. L., and Kemper, W. D. (1971). Microwatersheds and ground color for sugarbeet production. *Agron. J.* **63,** 101–104.

Farrell, D. A., Greacen, E. L., and Gurr, C. G. (1966). Vapor transfer in soil due to air turbulence. *Soil Sci.* **102,** 305–313.

Feddes, R. A., Bresler, E., and Neuman, S. P. (1974). Field test of a modified numerical model for water uptake by root systems. *Water Resources Res.* **10,** 1199–1206.

Feddes, R. A., Neuman, S. P., and Bresler, E. (1976). Finite element analysis of two-dimensional flow in soils considering water uptake by roots: II. Field applications. *Soil Sci. Soc. Am. Proc.* **39,** 231–237.

Ferguson, H., and Gardner, W. H. (1962). Water content measurement in soil columns by gamma ray absorption. *Soil Sci. Soc. Am. Proc.* **26,** 11–14.

Ferraro, G. (1895). "Les Lois Psychologiques du Symbolisme." Publ., Paris.

Feyen, J., Belmans, C., and Hillel, D. (1980). Comparison between measured and simulated plant water potential during soil water extraction by potted ryegrass. *Soil Sci.* **129,** 180.

Fireman, M. (1957). Salinity and alkali problems in relation to high water tables in soils. *In* "Drainage of Agricultural Lands," pp. 505–513. Monograph 7, Am. Soc. Agron., Madison, Wisconsin.

Fisher, R. A. (1923). Some factors affecting the evaporation of water from soil. *J. Agr. Sci.* **13,** 121–143.

Fisher, R. A. (1928). Further note on the capillary forces in an ideal soil. *J. Agric. Sci.* **18,** 406–410.

Fleming, G. (1972). "Computer Simulation Techniques in Hydrology." American Elsevier, New York.

Fok, Y. S. (1970). One-dimensional infiltration into layered soils. *J. Irrig. Drainage Div. ASCE* **96,** 121–129.

Forchheimer, P. (1930). "Hydraulik," 3rd ed. Teubner, Leipzig and Berlin.

Forsgate, J. A., Hosegood, P. H., and McCullock, J. S. G. (1965). Design and installation of semienclosed hydraulic lysimeters. *Agr. Meteorol.* **2,** 43–52.

Foth, H. (1978). "Fundamental of Soil Science." Wiley, New York.

Fountaine, E. R., and Payne, P. C. J. (1951). The shear strength of top soils. Natl. Inst. Agr. Engr. Tech. Memo. 42.

Fountaine, E. R., and Payne, P. C. J. (1952). The effect of tractors on volume weight and other soil properties. Rept. 17, Nat'l Inst. Agr. Eng., Silsoe, England.

Frank, H. S., and Wen, W. (1957). Structural aspects of ion-solvent interaction in aqueous solutions: a suggested picture of water structure. *Dis. Faraday Soc.* **24,** 133–140.

Franzini, J. B. (1951). *Trans. Am. Geophys. Un.* **32,** 443.

Free, J. R., and Palmer, V. J. (1940). Relationship of infiltration, air movement, and pore size in graded silica sand. *Soil Sci. Soc. Am. Proc.* **5,** 390–398.

Freeze, R. A. (1969). The mechanism of natural groundwater recharge and discharge. *Water Resour. Res.* **5**, 153–171.

Freeze, R. A., and Cherry, J. A. (1979). "Groundwater." Prentice-Hall, Englewood Cliffs, New Jersey.

Frenkel, H., Goertzen, J. O., and Rhoades, J. D. (1978). Effects of clay type and content, exchangeable sodium percentage, and electrolyte concentration on clay dispersion and soil hydraulic conductivity. *Soil Sci. Soc. Am. Proc.* **42**, 32–39.

Fried, J. J. (1976). "Ground Water Pollution." Elsevier, Amsterdam.

Fritton, D. D., Kirkham, D., and Shaw, R. H. (1970). Soil water evaporation, isothermal diffusion, and heat and water transfer. *Soil Sci. Soc. Am. Proc.* **34**, 183–189.

Fuchs, M., and Tanner, C. B. (1967). Evaporation from a drying soil. *J. Appl. Meteorol* **6**, 852–857.

Fuchs, M., and Tanner, C. B. (1968). Calibration and field test of soil heat flux plates. *Soil Sci. Soc. Am. Proc.* **32**, 326–328.

Fuchs, M., Tanner, C. B., Thurtell, G. W., and Black, T. A. (1969). Evaporation from drying surfaces by the combination method. *Agron. J.* **61**, 22–26.

Gairon, S., and Swartzendruber, D. (1973). Streaming potential effects in saturated water flow through a sand-kaolinite mixture. *In* "Physical Aspects of Soil Water and Salts in Ecosystems" (A. Hadas *et al.*, eds.). Springer-Verlag, Berlin and New York.

Gardner, H. R., and Hanks, R. J. (1966). Evaluation of the evaporation in soil by measurement of heat flux. *Soil Sci. Soc. Am. Proc.* **30**, 425–428.

Gardner, W. (1920). The capillary potential and its relation to soil moisture constants. *Soil Sci.* **10**, 357–359.

Gardner, W. H. (1965). Water content. *In* "Methods of Soil Analysis," pp. 82–127. Monograph 9, Am. Soc. Agron., Madison, Wisconsin.

Gardner, W. H. (1972). Use of synthetic soil conditioners in the 1950's and some implications to their further development. *In Proc. Symp. Fundamentals Soil Conditioning, State Agr. Univ., Ghent, Belgium.* pp. 1046–1061.

Gardner, W. R. (1950). Some steady state solutions of the unsaturated moisture flow equation with application to evaporation from a water table. *Soil Sci.* **85**, 228–232.

Gardner, W. R. (1956a). Calculation of capillary conductivity from pressure plate outflow data. *Soil Sci. Soc. Am. Proc.* **20**, 317–320.

Gardner, W. R. (1956b). Representation of soil aggregate-size distribution by a logarithmic-normal distribution. *Soil Sci. Soc. Am. Proc.* **20**, 151–153.

Gardner, W. R. (1958). Some steady state solutions of the unsaturated moisture flow equation with application to evaporation from a water table. *Soil Sci.* **85**, (4).

Gardner, W. R. (1959). Solutions of the flow equation for the drying of soils and other porous media. *Soil Sci. Soc. Am. Proc.* **23**, 183–187.

Gardner, W. R. (1960a). Dynamic aspects of water availability to plants. *Soil Sci.* **89**, 63–73.

Gardner, W. R. (1960b). Soil water relations in arid and semi-arid conditions. *UNESCO* **15**, 37–61.

Gardner, W. R. (1962a). Approximate solution of a non-steady state drainage problem. *Soil Sci. Soc. Am. Proc.* **26**, 129–132.

Gardner, W. R. (1962b). Note on the separation and solution of diffusion type equations. *Soil Sci. Soc. Am. Proc.* **26**, 404.

Gardner, W. R. (1964). Relation of root distribution to water uptake and availability. *Agron. J.* **56**, 35–41.

Gardner, W. R. (1968). Availability and measurement of soil water. *In* "Water Deficits and Plant Growth," Vol. 1, pp. 107–135. Academic Press, New York.

Gardner, W. R. (1970). Field measurement of soil water diffusivity. *Soil Sci. Soc. Am. Proc.* **34,** 832.

Gardner W. R., and Brooks, R. H. (1956). A descriptive theory of leaching. *Soil Sci.* **83,** 295–304.

Gardner, W. R., and Fireman, M. (1958). Laboratory studies of evaporation from soil columns in the presence of a water table. *Soil Sci.* **85,** 244–249.

Gardner, W. R., and Hillel, D. (1962). The relation of external evaporative conditions to the drying of soils. *J. Geophys. Res.* **67,** 4319–4325.

Gardner, W. R., and Mayhugh, M. S. (1958). Solutions and tests of the diffusion equation for the movement of water in soil. *Soil Sci. Soc. Am. Proc.* **22,** 197–201.

Gardner, W. R., Hillel, D., and Benyamini, Y. (1970a). Post irrigation movement of soil water: I. Redistribution. *Water Resources Res.* **6,** 851–861.

Gardner, W. R., Hillel, D., and Benyamini, Y. (1970b). Post irrigation movement of soil water: II. Simultaneous redistribution and evaporation. *Water Resources Res.* **6,** 1148–1153.

Gardner, W., Israelsen, O. W., Edlefsen, N. W., and Clyde H. (1922). The Capillary Potential Function and its Relation to Irrigation Practice. Physical Review, second series, July–December, p. 196.

Gash, J. H. C., and Stewart, J. B. (1975). The average surface resistance of a pine forest. *Boundary-Layer Meteorol.* **8,** 453–464.

Gee, G. W., Stiver, J. F., and Borchert, H. R. (1976). Radiation hazard from Americium-Beryllium neutron moisture probes. *Soil Sci. Soc. Am. J.* **40,** 492–494.

Giesel, W., Lorch, S., and Tenger, M. (1970). Water flow calculation by means of gamma absorption and tensiometer field measurement in the unsaturated soil profile. *In* "Isotope Hydrology, 1970," Int. Atomic Energy Agency, Vienna.

Gill, W. R. (1959). The effect of drying on the mechanical strength of Lloyd clay. *Soil Sci. Soc. Am. Proc.* **23,** 253–257.

Gill, W. R. (1968). The influence of compaction hardening of soil on penetration resistance. *Trans. Am. Soc. Agr. Eng.* **11,** 741–745.

Gill, W. R. (1974). Tillage and soybean root growth. *Proc. Conf. on Soybean Production, Marketing and Use* (sponsored by TVA, Am. Soybean Assoc. and Natl. Soybean Crop Improvement Council), pp. 66–72. Chattanooga, Tennessee.

Gill, W. R. (1979). Tillage, *In* "The Encyclopedia of Soil Science," Part 1 (Fairbridge, R. W., and Finkl, C. W., Jr., eds.). pp. 566–571. Dowden, Hutchinson & Ross, Stroudsburg, Pennsylvania.

Gill, W. R., and Bolt, G. H. (1955). Pfeffer's studies of the root growth pressures exerted by plants. *Agron. J.* **47,** 166–168.

Gill, W. R., and Cooper, A. (1972). Soil compaction—its causes and remedies. *Proc. Western Cotton Production Conf.,* Bakersfield, California, pp. 11–13.

Gill, W. R., and McCreery, W. F. (1960). Relation of size of cut to tillage tool efficiency. *Agr. Eng.* **41,** 372–374, 381.

Gill, W. R., and Miller, R. D. (1956). A method for study of the influences of mechanical impedance and aeration on the growth of seedling roots. *Soil Sci. Soc. Am. Proc.* **20,** 154–157.

Gill, W. R., and Vanden Berg, G. E. (1967). "Soil Dynamics in Tillage and Traction." Handbook 316, Agr. Res. Service, U.S. Dept. Agriculture, Washington, D.C.

Gold, L. W. (1957). A possible force mechanism associated with the freezing of water in porous materials. *High. Res. Board, Bull.* **168,** 65–72.

Goldberg, D., and Shmueli, M. (1970). Drip irrigation—a method used under arid and desert conditions of high water and soil salinity. *Trans. ASAE* **13,** 38–41.

Goryachkin, W. P. (1968). "Collected Works," Vols. 1 and 2. Kolos, Moscow. (Translated from Russian by U.S. Department of Agriculture and National Science Foundation, TT-72-50023 and TT-71-50087.

Goss, M. J. (1970). Further studies on the effect of mechanical resistance on the growth of plant roots. Rept. Agr. Res. Coun. Letcombe Lab. ARCRL **20**, 43–45.

Gouy, M. (1910). Sur la constitution de la charge electrique a la surface d'un electrolyte. *Ann. Phys.* **9**, 457–468.

Grable, A. R. (1966). Soil aeration and plant growth. *Adv. Agron.* **18**, 57–106.

Grable, A. R., and Siemer, E. G. (1968). Effects of bulk density, aggregate size, and soil water suction on oxygen diffusion, redox potentials and elongation of corn roots. *Soil Sci. Soc. Am. Proc.* **32**, 180–186.

Grace, J. (1978). "Plant Response to Wind." Academic Press, New York.

Graham, W. G, and King, K. M. (1961). Fraction of net radiation utilized in evapotranspiration from a corn crop. *Soil Sci. Soc. Am. Proc.* **25**, 158–160.

Greacen, E. L., Barley, K. P., and Farrell, D. A. (1969). The mechanics of root growth in soils with particular reference to the implications for root distribution. *In* "Root Growth" (W. J. Whittington, ed.), pp. 256–269. Butterworths, London.

Greacen, E. L., Ponsana, P., and Barley, K. P. (1976). Resistance to water flow in the roots of cereals. *In* "Water and Plant Life, Ecological Studies 19," pp. 86–100. Springer-Verlag, Berlin and New York.

Greb, B. W. (1966). Effect of surface-applied wheat straw on soil water losses by solar distillation. *Soil Sci. Soc. Am. Proc.* **30**, 786–788.

Green, R. E., and Corey, J. C. (1971). Calculation of hydraulic conductivity: A further evaluation of predictive methods. *Soil Sci. Soc. Am. Proc.* **35**, 3–8.

Green, W. H., and Ampt, G. A. (1911). Studies on soil physics: I. Flow of air and water through soils. *J. Agr. Sci.* **4**, 1–24.

Greenland, D. J. (1965). Interaction between clays and organic compounds in soils. Part I. Mechanisms of interaction between clays and defined organic compounds. *Soil Fert.* **28**, 415–425.

Greenland, D. J., Lindstrom, G. R., and Quirk, J. P. (1962). Organic materials which stabilize natural soil aggregates. *Soil Sci. Am. Proc.* **26**, 236–371.

Greenwood, D. J. (1971). Soil aeration and plant growth. *Rep. Prog. Appl. Chem.* **55**, 423–431.

Grim, R. E. (1953). "Clay Mineralogy." McGraw-Hill, New York.

Groenevelt, P. H., and Bolt, G. H. (1969). Non-equilibrium thermodynamics of the soil-water system. *J. Hydrol.* **7**, 358–388.

Grover, B. L. (1956). Simplified air permeameter for soil in place. *Soil Sci. Soc. Am. Proc.* **19**, 414–418.

Guggenheim, E. A. (1959). "Thermodynamics." North-Holland Publ., Amsterdam.

Gumaa, G. S. (1978). Spatial variability of *in situ* available water. Ph.D. Dissertation. The Univ. of Ariz., Tucson, Arizona (available as 78-24365 from Xerox, University Microfilms, Ann Arbor, Michigan).

Gupta, S. C., and Larson, W. E. (1979). *Soil Sci. Soc. Am. J.* **43**, 758–764.

Gurr, C. G. (1962). Use of gamma rays in measuring water content and permeability in unsaturated columns of soil. *Soil Sci.* **94**, 224–449.

Gurr, C. G., Marshall, T. J., and Hutton, J. T. (1952). Movement of water in soil due to temperature gradients. *Soil Sci.* **74**, 333–345.

Hadas, A., and Fuchs, M. (1973). Prediction of the thermal regime of bare soils. *In* "Physical Aspects of Soil Water and Salts in Ecosystems" (A. Hadas, D. Swartzendruber, P. E. Rijtema, M. Fuchs, and B. Yaron, eds.). Springer-Verlag, Berlin and New York.

Hadas, A., and Hillel, D. (1968). An experimental study of evaporation from uniform columns in the presence of a water table. *Trans. Int. Congr. Soil Sci., 9th, Adelaide* **I,** 67–74.

Hagan, R. M. (1952). Soil temperature and plant growth. *In* "Soil Physical Conditions and Plant Growth" (B. T. Shaw, ed.), pp. 367–462. Academic Press, New York.

Haines, W. B. (1930). Studies in the physical properties of soils. V. The hysteresis effect in capillary properties and the modes of moisture distribution associated therewith. *J. Agr. Sci.* **20,** 97–116.

Halstead, M. H., and Covey, W. (1957). Some meteorological aspects of evapotranspiration. *Soil Sci. Soc. Am. Proc.* **21,** 461–464.

"Handbook of Chemistry and Physics." (1968). 49th ed. The Chemical Rubber Co., Cleveland, Ohio.

Hanks, R. J. (1980). Yield and water use relationships. *In* "Efficient Water Use in Crop Production." Am. Soc. Agron., Madison, Wisconsin (in press).

Hanks, R. J., and Bowers, S. A. (1960). Nonsteady-state moisture, temperature, and soil air pressure approximation with an electric simulator. *Soil Sci. Soc. Am. Proc.* **24,** 247–252.

Hanks, R. J., and Bowers, S. A. (1962). Numerical solution of the moisture flow equation for infiltration into layered soils. *Soil Sci. Soc. Am. Proc.* **26,** 530–534.

Hanks, R. J., and Gardner, H. R. (1965). Influence of different diffusivity-water content relations on evaporation of water from soils. *Soil Sci. Soc. Am. Proc.* **29,** 495–498.

Hanks, R. J., and Harkness, K. A. (1956). Soil penetrometer employs strain gages. *Agr. Eng.* **37,** 553–554.

Hanks, R. J., and Woodruff, N. P. (1958). Influence of wind on water vapor transfer through soil, gravel, and straw mulches. *Soil Sci.* **86,** 160–164.

Hanks, R. J., Bowers, S. B., and Boyd, L. D. (1961). Influence of soil surface conditions on net radiation, soil temperature, and evaporation. *Soil Sci.* **91,** 233–239.

Hanks, R. J., Gardner, H. R., and Fairbourn, M. L. (1967). Evaporation of water from soils as influenced by drying with wind or radiation. *Soil Sci. Soc. Am. Proc.* **31,** 593–598.

Hansen, G. K. (1975). A dynamic continuous simultation model of water state and transportation in the soil-plant-atmosphere system. *Acta Agr. Scand.* **25,** 129–149.

Hanway, J. J. (1963). Growth stages of corn (Zea mays, L.). *Agron. J.* **55,** 487–492.

Harlan, R. L. (1973). An analysis of coupled heat-fluid transport in partially frozen soil. *Water Resour. Res.* **9,** 1314–1325.

Harr, M. E. (1962). "Groundwater and Seepage." McGraw-Hill, New York.

Harris, R. F., Chester, G., and Allen, O. N. (1965). Dynamics of soil aggregation. *Adv. Agron.* **18,** 107–160.

Harris, W. L. (1960). Dynamic stress transducers and the use of continuum mechanics in the study of various soil stress-strain relationships. Ph.D. thesis, Michigan State Univ., East Lansing, Michigan.

Harris, W. L. (1971). The soil compaction process. *In* "Compaction of Agricultural Soils" (K. K. Barnes, ed.), pp. 9–44. Monograph, Am. Soc. Agr. Eng., St. Joseph, Michigan.

Hawkins, J. C. (1959). Cultivations. *Natl. Inst. Agric. Eng. Ann. Rep.,* 1–7, London.

Helling, C. S. *et al.* (1964). Contribution of organic matter and clay to soil cation exchange capacity as affected by the pH of the saturated solution. *Soil Sci. Soc. Am. Proc.* **28,** 517–520.

Herkelrath, W. N. (1975). Water uptake by plant roots. Ph.D. dissertation, Univ. of Wisconsin, Madison, Wisconsin.

Hesstvedt, E. (1964). The interfacial energy ice/water. Pub. #56, Norwegian Geotech. Inst.
Hettiaratchi, D. R. P., Whitney, B. D., and Reece, A. R. (1966). The calculation of passive pressure in two-dimensional soil failure. *J. Agr. Eng. Res.* **11**, 89–107.
Hide, J. C. (1954). Observation on factors influencing the evaporation of soil moisture. *Soil Sci. Soc. Am. Proc.* **18**, 234–239.
Hide, J. C. (1958). Soil moisture conservation in the Great Plains. *Adv. Agron.* **10**, 23–26.
Hill, E. D., and Parlange, J.-Y. Wetting front instability in layered soils. *Soil Sci. Soc. Am. Proc.* **36**, 697–702.
Hill, R. W., Hanks, R. J., Keller, J., and Rasmussen, V. P. (1974). Predicting corn growth as affected by water management: An example. CUSUSWASH 211(d)-6. Utah State Univ., Logan, Utah, p. 18.
Hillel, D. (1959). Studies of Loessial Crusts. Bull. 63, Israel Agr. Res. Inst., Beit Dagan, Israel.
Hillel, D. (1960). Crust formation in loessial soils. *Trans. Int. Soil Sci. Congr., 7th, Madison, Wisconsin* **I**, 330–339.
Hillel, D. (1964). Infiltration and rainfall runoff as affected by surface crusts. *Trans. Int. Soil Sci. Congr., 8th, Bucharest* **2**, 53–62.
Hillel, D. (1967). "Runoff Inducement in Arid Lands." Hebrew Univ. of Jerusalem, Israel.
Hillel, D. (1968). Soil Water Evaporation and Means of Minimizing It. Hebrew Univ. Faculty Agriculture Res. Rept., Rehovot, Israel.
Hillel, D. (1971). "Soil and Water: Physical Principles and Processes." Academic Press, New York.
Hillel, D., ed. (1972a). "Optimizing the Soil Physical Environment Toward Greater Crop Yields." Academic Press, New York.
Hillel, D. (1972b). Soil moisture and seed germination. *In* "Water Deficits and Plant Growth" (T. T. Kozlowski, ed.). pp. 65–89. Academic Press, New York.
Hillel, D. (1974). Methods of laboratory and field investigation of physical properties of soils. *Trans. Int. Soil Sci. Congr., 10th, Moscow* **I**, 301–308.
Hillel, D. (1975). Evaporation from bare soil under steady and diurnally fluctuating evaporativity. *Soil Sci.* **120**, 230–237.
Hillel, D. (1976a). On the role of soil moisture hysteresis in the suppression of evaporation from bare soil. *Soil Sci.* **122**, 309–314.
Hillel, D. (1976b). Soil management. *In* "McGraw-Hill Yearbook of Science and Technology." McGraw-Hill, New York.
Hillel, D. (1977). "Computer Simulation of Soil-Water Dynamics." Int. Dev. Res. Centre, Ottawa, Canada.
Hillel, D. (1980a). "Applications of Soil Physics." Academic Press, New York.
Hillel, D. (1980b). "Fundamentals of Soil Physics." Academic Press, New York.
Hillel, D., and Benyamini, Y. (1974). Experimental comparison of infiltration and drainage methods for determining unsaturated hydraulic conductivity of a soil profile in situ. *Proc. FAO/IAEA Symp. Isotopes and Radiation Techniques in Studies of Soil Physics, Vienna*, pp. 271–275.
Hillel, D. and Berliner, P. (1974). Water-proofing surface zone soil aggregates for water conservation. *Soil Sci.* **118**, 131–135.
Hillel, D., and Gardner, W. R. (1969). Steady infiltration into crust-topped profiles. *Soil Sci.* **108**, 137–142.
Hillel, D., and Gardner, W. R. (1970a). Measurement of unsaturated conductivity diffusivity by infiltration through an impeding layer. *Soil Sci.* **109**, 149.

Hillel, D., and Gardner, W. R. (1970b). Transient infiltration into crust-topped profiles. *Soil Sci.* **109**, 410–416.

Hillel, D., and Guron, Y. (1973). Relation between evapotranspiration rate and maize yield. *Water Resources Res.* **9**, 743–748.

Hillel, D., and Hornberger, G. M. (1979). Physical model of the hydrology of sloping heterogenous fields. *Soil Sci. Soc. Am. J.* **43**, 434–439.

Hillel, D., and Mottes, J. (1966). Effect of plate impedance, wetting method and aging on soil moisture retention. *Soil Sci.* **102**, 135–140.

Hillel, D., and Rawitz, E. (1972). Soil moisture conservation. *In* "Water Deficits and Plant Growth" (T. T. Kozlowski, ed.), pp. 307–337. Academic Press, New York.

Hillel, D., and Talpaz, H. (1976). Simulation of root growth and its effect on the pattern of soil water uptake by a nonuniform root system. *Soil Sci.* **121**, 307–312.

Hillel, D., and Talpaz, H. (1977). Simulation of soil water dynamics in layered soils. *Soil Sci.* **123**, 54–62.

Hillel, D., and van Bavel, C. H. M. (1976). Dependence of profile water storage on soil hydraulic properties: a simulation model. *Soil Sci. Soc. Am. J.* **40**, 807–815.

Hillel, D., Ariel, D., Orlowski, S., Stibbe, E., Wolf, D., and Yavnai, A. (1969). Soil-crop-tillage interactions in dryland and irrigated farming. Research report submitted to the U.S. Dept. Agric. by the Hebrew University of Jerusalem, Israel.

Hillel, D., Gairon, S., Falkenflug, V., and Rawitz, E. (1969). New design of a low-cost hydraulic lysimeter for field measurement of evapo-transpiration. *Israel J. Agr. Res. 3.* **19**, 57–63.

Hillel, D., Krentos, V. D., and Stylianou, Y. (1972). Procedure and test of an internal drainage method for measuring soil hydraulic characteristics *in situ*. *Soil Sci.* **114**, 395–400.

Hillel, D., Talpaz, H., and van Keulen, H. (1976). A macroscopic-scale model of water uptake by a nonuniform root system and of water and salt movement in the soil profile. *Soil Sci.* **121**, 242–255.

Hillel, D., van Beek, C., and Talpaz, H. (1975). A microscopic-scale model of soil water uptake and salt movement to plant roots. *Soil Sci.* **120**, 385–399.

Hoekstra, P., and Miller, R. D. (1967). On the mobility of water molecules in the transition layer between ice and a solid surface. *J. Colloid Interface Sci.* **25**, 166–173.

Hoekstra, P., Osterkamp, T. E., and Weeks, W. F. (1965). The migration of liquid inclusions in single ice crystals. *J. Geophys. Res.* **70**, 5035–5041.

Hoffman, O., and Sachs, G. (1953). "Introduction to the Theory of Plasticity." McGraw-Hill, New York.

Holmes, J. W. (1956). Calibration and field use of the neutron scattering method of measuring soil water content. *Aust. J. Appl. Sci.* **7**, 45–58.

Holmes, J. W., Greacen, E. L., and Gurr, C. G. (1960). The evaporation of water from bare soils with different titlths. *Trans. Int. Congr. Soil Sci. 7th, Madison, Wisconsin* **1**, 188–194.

Holtan, H. N. (1961). A concept for infiltration estimates in watershed engineering. U.S. Dept. Agr., Agr. Res. Service Publ. 41–51.

Hooghoudt, S. B. (1937). Bijdregen tot de kennis van eenige natuurkundige grootheden van de grond, 6. *Versl. Landb. Ond.* **43**, 461–676.

Hopkins, R. M., and Patrick, W. H. (1969). Combined effects of oxygen content and soil compaction on root penetration. *Soil Sci.* **108**, 408–413.

Horton, R. E. (1940). An approach toward a physical interpretation of infiltration-capacity. *Soil Sci. Soc. Am. Proc.* **5**, 399–417.

Hubbert, M. K. (1940). The theory of groundwater motion. *J. Geol.* **48**, 785–944.

Hubbert, M. K. (1956). Darcy's law and the field equations of the flow of underground fluids. *Am. Inst. Min. Met. Petl. Eng. Trans.* **207,** 222–239.

Huck, M. G., Klepper, B., and Taylor, H. M. (1970). Diurnal variation in root diameter. *Plant Physiol.* **45,** 529–530.

IBM. (1972). System/360 continuous system modeling program user's manual program number 360A-CS-16X, 5th ed. IBM Corp., Tech. Pub. Dept. White Plains, New York.

Isherwood, J. D. (1959). Water-table recession in tile-drained land. *J. Geophys. Res.* **64,** 795–804.

Israelson, O. W., and West, F. L. (1922). Water holding capacity of irrigated soils. Utah State Univ. Agr. Exp. Sta. Bull. 183.

Jackson, R. D. (1960). The Importance of Convection as a Heat Transfer Mechanism in Two-Phase Porous Materials. Unpublished Ph.D. dissertation, Colorado State Univ., Fort Collins, Colorado.

Jackson, R. D. (1964). Water vapor diffusion in relatively dry soil: I. Theoretical considerations and sorption experiments. *Soil Sci. Soc. Am. Proc.* **28,** 172–176.

Jackson, R. D. (1964a). Water vapor diffusion in relatively dry soil: I. Theoretical considerations and sorption experiments. *Soil Sci. Soc. Am. Proc.* **28,** 172–176.

Jackson. R. D. (1964b). Water vapor diffusion in relatively dry soil: II. Desorption experiment. *Soil Sci. Soc. Am. Proc.* **28,** 464–466.

Jackson, R. D. (1964c). Water vapor diffusion in relatively dry soil: III. Steady state experiments. *Soil Sci. Soc. Am. Proc.* **28,** 466–470.

Jackson, R. D. (1972). On the calculation of hydraulic conductivity. *Soil Sci. Soc. Am. Proc.* **36,** 380–383.

Jackson, R. D. (1973). Diurnal changes in soil water content during drying. *In* "Field Soil Water Regime," pp. 37–55. Soil Sci. Soc. Am., Madison, Wisconsin.

Jackson, R. D., and Taylor, S. A. (1965). Heat transfer. *In* "Methods of Soil Analysis," pp. 349–360. Monograph No. 9, Am. Soc. Agron., Madison, Wisconsin.

Jackson, R. D., Reginato, R. J., Kimball, B. A., and Nakayama, F. S. (1974). Diurnal soil-water evaporation: comparison of measured and calculated soil-water fluxes. *Soil Sci. Soc. Am. Proc.* **38,** 861–866.

Jackson, R. D., and Whisler, F. D. (1970). Approximate equations for vertical nonsteady-state drainage: I. Theoretical approach. *Soil Sci. Soc. Am. Proc.* **34,** 715–718.

Jackson, R. D., Kimball, B. A., Reginato, R. J., and Nakayama, S. F. (1973). Diurnal soil water evaporation: Time-depth-flux patterns. *Soil Sci. Soc. Am. Proc.* **37,** 505–509.

Jacob, C. E. (1946). Trans. Am. Geophys. Un. **27,** 198.

Jaeger, J. C. (1964). "Elasticity, Fracture, and Flow." Wiley, New York.

Jamil, A. (1976). Effect of Clay Swelling on the Hydraulic Parameters of Porous Sandstones. Ph.D. thesis submitted to Colorado State Univ., Fort Collins, Colorado.

Janert, H. (1934). The application of heat of wetting measurements to soil research problems. *J. Agr. Sci.* **24,** 136–145.

Jenny, H. F. (1941). "Factors of Soil Formation." McGraw-Hill, New York.

Jensen, M. E. (1973). (ed.) "Consumptive Use of Water and Irrigation Water Requirements." Amer. Soc. Civ. Eng., New York, 215 pp.

Jensen, M. E., and Hanks, R. J. (1967). Nonsteadystate drainage from porous media. *J. Irrig. Drain. Div. Am. Soc. Civil Eng.* **93,** IR3, 209–231.

Johnson, H. P., Frevert, R. K., and Evans, D. D. (1952). Simplified procedure for the measurement and computation of soil permeability below the water table. *Agr. Eng.* **33,** 283–289.

Johnston, J. R., and Hill, H. O. (1945). A study of the shrinkage and swelling of rendzina soils. *Soil Sci. Soc. Am. Proc.* **9,** 24–29.

Journel, A. G., and Huijbregts, Ch. J. (1978). "Mining Geostatistics." Academic Press, New York.

Jury, W. A. (1973). Simultaneous transport of heat and moisture through a medium sand. Ph.D. Thesis, Univ. of Wisconsin, Madison, Wisconsin.

Jury, W. A., and Bellantuoni, B. (1976). Heat and water movement under surface rocks in a field soil. *Soil Sci. Soc. Am. J.* **40,** 505–513.

Jury, W. A., and Tanner, C. B. (1975). Advective modifications of the Priestly and Taylor evapotranspiration formula. *Agron. J.* **67,** 840–842.

Kanemasu, E. T., Rasmussen, V. P., and Bagley, J. (1978). Estimating water requirements for corn with a "pocket calculator." *Agron. J.* **70,** 610.

Kanemasu, E. T., Stone, L. R., and Powers, W. L. (1976). Evapotranspiration model tested for soybean and sorghum. *Agron. J.* **68,** 569–572.

Katchalsy, A., and Curran, P. F. (1965). "Nonequilibrium Thermodynamics in Biophysics." Harvard Univ. Press, Cambridge, Massachusetts.

Kavanau, J. L. (1965). Water. *In* "Structure and Function in Biological Membranes," pp. 170–248. Holden-Day, San Francisco, California.

Kays, S. J., Nicklow, C. W., and Simons D. H. (1974). Ethylene in relation to the response of roots to mechanical impedance. *Plant Soil* **40,** 565–571.

Keen, B. A. (1931) "The Physical Properties of Soil." Longmans, Green, New York.

Keen, B. A., and Russell, E. W. (1937). Are cultivation standards wastefully high? *J. R. Agr. Soc.* **98,** 53–60.

Kemper, W. D. (1965). Aggregate stability. *In* "Methods of Soil Analysis." Am. Soc. Agron., Madison, Wisconsin.

Kemper, W. D., and Amemiya, M. (1957). Alfalfa growth as affected by aeration and soil moisture stress under flood irrigation. *Soil Sci. Soc. Am. Proc.* **21,** 657–660.

Kemper, W. D., and Chepil, W. S. (1965). Size distribution of aggregates. *In* "Methods of Soil Analysis." Am. Soc. Agron., Madison, Wisconsin.

Kemper, W. D., and Evans, N. A. (1963). Movement of water as affected by free energy and pressure gradients. II. Restriction of solutes by membranes. *Soil Sci. Soc. Am. Proc.* **27,** 485–490.

Kemper, W. D., and Maasland, D. E. L. (1964). Reduction in salt content of solution on passing through thin films adjacent to charged surfaces. *Soil Sci. Soc. Am. Proc.* **28,** 318–323.

Kemper, W. D., and Quirk, J. P. (1970). Graphic presentation of a mathematical solution for interacting diffuse layers. *Soil Sci. Soc. Am. Proc.* **34,** 347–351.

Kepner, R. A., Bainer, R., and Barger, E. L. (1972). "Principles of Farm Machinery." 2nd ed. AVI., Westport, Connecticut.

Kersten, M. S. (1949). Thermal Properties of Soils. Bull. 28, Univ. Minnesota Inst. Technol., St. Paul, Minnesota.

Kezdi, A. (1974). "Handbook of Soil Mechanics," Vol. I, Soil Physics. Elsevier, Amsterdam.

Kiesselbach, T. A. (1916). Transpiration as a factor in crop production. Neb. Bull. Agric. Exp. Stn. Res. **6.**

Kimball, B. A., and Lemon, E. R. (1971). Air turbulence effects upon soil gas exchange. *Soil Sci. Soc. Am. Proc.* **35,** 16–21.

Kimball, B. A., and Lemon, E. R. (1972). Theory of air movement due to pressure fluctuations. *Agr. Meteorol.* **9,** 163–181.

King, K. M., Tanner, C. B., and Suomi, V. E. (1956). A floating lysimeter and its evaporation recorder. *Trans. Am. Geophys. Un.* **37,** 738–742.

Kirkham, D. (1946). Field method for determination of air permeability of soil in its undisturbed state. *Soil Sci. Soc. Am. Proc.* **11,** 92–99.

Kirkham, D. (1949). Flow of ponded water into drain tubes in soil overlying an impervious layer. *Trans. Am. Geophys. Un.* **30,** 369–385.

Kirkham, D. (1957). The ponded water case. *In* "Drainage of Agricultural Lands," pp. 139–180. Monograph 7, Amer. Soc. Agron., Madison, Wisconsin.

Kirkham, D. (1958). Seepage of steady rainfall through soil into drains. *Trans. Am. Geophys. Un.* **32,** 892–908.

Kirkham, D. (1967). Explanation of paradoxes in Dupuit-Forchheimer seepage theory. *Water Resources Res.* **3,** 609–622.

Kirkham, D., DeBoodt, M. F., and DeLeenheer, L. (1959). Modulus of rupture determination on undisturbed soil core samples. *Soil Sci.* **87,** 141–144.

Kirkham, D., and Gaskell, R. E. (1951). The falling water table in tile and ditch drainage. *Soil Sci. Soc. Am. Proc.* **15,** 37–42.

Klotz, I. M. (1974). Water. *In* "Horizons in Biochemistry" (M. Kasha and B. Pullman, eds.). pp. 253–550. Academic Press, New York.

Klute, A. (1965a). Laboratory measurement of hydraulic conductivity of saturated soil. *In* "Methods of Soil Analysis," pp. 210–221. Monograph 9. Am. Soc. Agron., Madison, Wisconsin.

Klute, A. (1965b). Laboratory measurement of hydraulic conductivity of unsaturated soil. *In* "Methods of Soil Analysis," pp. 253–261. Monograph No. 9, Am. Soc. Agron., Madison, Wisconsin.

Klute, A., and Peters, D. B. (1969). Water uptake and root growth. *In* "Root Growth" (W. J. Whittington, ed.), pp. 105–133. Buttersworths, London.

Klute, A., Whisler, F. D., and Scott, E. H. (1965). Numerical solution of the flow equation for water in a horizontal finite soil column. *Soil Sci. Soc. Am. Proc.* **29,** 353–358.

Kohnke, H. (1968). "Soil Physics." McGraw-Hill, New York.

Kondner, R. L. (1958). A non-dimensional approach to vibratory cutting, compaction and penetration of soils. Tech. Rep. 8 by Johns Hopkins Univ. to U.S. Army Corps of Eng., Waterways Exp. Sta., Vicksburg, Mississippi.

Koopmans, R. W. R., and Miller, R. D. (1966). Soil freezing and soil water characteristic curves. *Soil. Sci. Soc. Am. Proc.* **30,** 680–685.

Kostiakov, A. N. (1932). On the dynamics of the coefficient of water-percolation in soils and on the necessity of studying it from a dynamic point of view for purposes of amelioration. *Trans. Com. Int. Soc. Soil Sci., 6th, Moscow* Part A, 17–21.

Kramer, P. J. (1956). Physical and physiological aspects of water absorption. *In* "Handbuch der Pflanzenphysiologie, Vol. III, Pflanze and Wasser, pp. 124–159. Springer-Verlag, Berlin and New York.

Kramer, P. J., and Coile, J. S. (1940). An estimation of the volume of water made available by root extension. *Plant Physiol.* **15,** 743–747.

Kravtchenko, J., and Sirieys, P. M. (eds.) (1966). "Rheology and Soil Mechanics." Springer-Verlag, Berlin and New York.

Kristenson, K. J., and Lemon, E. R. (1964). Soil aeration and plant root relations: III. Physical aspects of oxygen diffusion in the liquid phase of the soil. *Agron. J.* **56,** 295–301.

Kunze, R. J., and Kirkham, D. (1962). Simplified accounting for membrane impedance in capillary conductivity determinations. *Soil Sci. Soc. Am. Proc.* **26,** 421–426.

Kunze, R. J., Uehara, G., and Graham, K. (1968). Factors important in the calculation of hydraulic conductivity. *Soil Sci. Soc. Am. Proc.* **32,** 760–765.

Lagerwerff, J. V., Nakayama, F. S., and Frere, M. H. (1969). Hydraulic conductivity related to porosity and swelling of soil. *Soil Sci. Soc. Am. Proc.* **33,** 3–11.

Laliberte, G. E. (1969). A mathematical function for describing capillary pressure-desaturation data. *Bull. Int. Ass. Sci. Hydrol.* **14**(2), 131–149.

Lambe, T. W., and Whitman, R. V. (1969). "Soil Mechanics." Wiley, New York.

Lambert, J. R., and Penning de Vries, F. W. T. (1973). Dynamics of water in the soil plant atmosphere system: A model named Troika. *In* "Physical Aspects of Soil, Water and Salts in Ecosystems." Springer-Verlag, Berlin and New York.

Langmuir, I. (1918). The adsorption of gases on plane surfaces of glass, mica, and platinum. *J. Am. Chem. Soc.* **40,** 1361–1402.

Langmuir, I. (1938). Distribution of cations between two charged plates. *Science* **88,** 430–433.

Larson, W. E., and Gill, W. R. (1973). Soil physical parameters for designing new tillage systems. Proc. Natl. Conf. Conserv. Tillage (sponsored by Soil Conserv. Soc. Am., Am. Soc. Agron., and Am. Soc. Agr. Eng.). pp. 13–21, Des Moines, Iowa.

Law, J. P. (1964). Effect of fatty alcohol and a nonionic surfactant on soil moisture evaporation in a controlled environment. *Soil Sci. Sci. Am. Proc.* **28,** 695–699.

LeFur, B. (1962). Influences de la capillarité et de la gravité sur le déplacement nonmiscible unidimensionel dans un milieu poreux. *J. Mécan.* **1,** 59.

Lemon, E. R. (1956). The potentialities for decreasing soil moisture evaporation loss. *Soil Sci. Soc. Am. Proc.* **20,** 120–125.

Lemon, E. R. (1960). Photosynthesis under field conditions. II. An aerodynamic method for determining the turbulent carbon dioxide exchange between the atmosphere and a corn field. *Agron. J.* **52,** 697–703.

Lemon, E. R. (1962). Soil aeration and plant root relations. I. Theory. *Agron. J.* **54,** 167–170.

Lemon, E. R., and Erickson, A. E. (1952). The measurement of oxygen diffusion in the soil with a platinum microelectrode. *Soil Sci. Soc. Am. Proc.* **16,** 160–163.

Leopold, L. B. (1974). "Water: A Primer." Freeman, San Francisco, California.

Letey, J. (1968). Movement of water through soil as influenced by osmotic pressure and temperature gradients. *Hilgardia* **39,** 405–418.

Letey, J., and Stolzy, L. H. (1964). Measurement of oxygen diffusion rates with the platinum microelectrode: I. Theory and equipment. *Hilgardia* **55,** 545–554.

Letey, J., Stolzy, L. H., and Kemper, W. D. (1967). Soil aeration. *In* "Irrigation of Agricultural Lands," pp. 943–948. Amer. Soc. Agron., Madison, Wisconsin.

Lettau, H. H. (1962). A theoretical model of thermal diffusion in nonhomogeneous conductors. *Gerlands. Beitr. Geophys.* **71,** 257–271.

Levy, R., and Hillel, D. (1968). Thermodynamic equilibrium constants of Na-Ca exchange in some Israeli soils. *Soil Sci.* **106,** 393–398.

Lewis, W. M. (1973). No-till systems. *In* "Conservation Tillage," pp. 182–187. Soil Conserv. Soc. Am., Ames, Iowa.

Leyser, J. P., and Loch, J. P. G. (1972). Effect of xylem resistance on the water relations of plant and soil. Dept. Theoretical Production Ecology. Agr. Univ., Wageningen, Netherlands.

Lin, C. C. (1955). "The Theory of Hydrodynamic Stability." Cambridge Univ. Press, London and New York.

Linell, K. A., and Kaplar, C. W. (1959). The factor of soil and material type in frost action. *Highw. Res. Board, Bull.* **225,** 81–128.

Low, P. F. (1961). Physical chemistry of clay-water interactions. *Adv. Agron.* **13,** 269–327.

Low, P. F. (1965). The effect of osmotic pressure on the diffusion rate of water. *Soil Sci.* **80**, 95–100.

Lowdermilk, W. C. (1954). The use of flood water by the Nabataeans and the Byzantines. *Isr. Explora. J.* **4**, 50–51.

Luth, H. A., and Wismer, R. D. (1971). Performance of plane soil cutting blade in sand. *Trans. ASAE* **14**, 225–259, 262.

Luthin, J. N., (ed.) (1957). "Drainage of Agricultural Lands." Monograph 7; Am. Soc. Agron., Madison, Wisconsin.

Luthin, J. N. (1966). "Drainage Engineering." Wiley, New York.

Luthin, J. N. (1974). Drainage analogues. *In* "Drainage for Agriculture." Monograph 17, Am. Soc. Agron., Madison, Wisconsin.

Lutz, J. F. (1952). Mechanical impedance and plant growth. *In* "Soil Physical Conditions and Plant Growth" (B. T. Shaw, ed.), pp. 43–71. Academic Press, New York.

Lutz, J. F., and Leamer, R. W. (1939). Pore-size distribution as related to the permeability of soils. *Soil Sci. Soc. Am. Proc.* **4**, 28–31.

Lyles, L., and Woodruff, U. P. (1961). Surface soil cloddiness in relation to soil density at the time of tillage. *Soil Sci.* **91**, 178–182.

Maasland, M. (1957). Soil anisotropy and land drainage. *In* "Drainage of Agricultural Lands" (J. N. Luthin, ed.). Am. Soc. Agron., Madison, Wisconsin.

Maasland, M., and Kirkham, D. (1955). Theory and measurement of anisotropic air permeability in soil. *Soil Sci. Soc. Am. Proc.* **19**, 395–400.

McCalla, T. M. (1944). Water-drop method of determining stability of soil structure. *Soil Sci.* **58**, 117–121.

McClelland, J. H. (1956). Instrument for measuring soil condition. *Agr. Eng.* **37**, 480–481.

McGregor, K. C., Greer, J. D., and Gurley, G. E. (1975). Erosion control with no-till cropping practices. *Trans. Am. Soc. Agr. Eng.* **18**, 918–920.

McIlroy, I. C., and Angus, D. E. (1963). The Aspendale multiple weighted lysimeter installation. CSIRO Div. Meteorol. Phys. Tech. Paper No. 14. Melbourne, Australia.

McIntyre, D. S. (1958). Soil splash and the formation of surface crusts by raindrop impact. *Soil Sci.* **85**, 261–266.

McIntyre, D. S. (1970). The platinum microelectrode method for soil aeration measurement. *In* "Advances in Agronomy," Vol. 22, pp. 235–283. Academic Press, New York.

McIntyre, D. S., and Philip, J. R. (1964). A field method for measurement of gas diffusion into soils. *Aust. J. Soil Res.* **2**, 133–145.

Mackay, J. R. (1978). Sub-pingo water lenses, Tuktoyaktuk Peninsula, Northwest Territories. *Can. J. Earth Sci.* **15**, 1219–1227.

McMurdie, J. L. (1963). Some characteristics of the soil deformation process. *Soil Sci. Soc. Am. Proc.* **27**, 251–254

McNeal, B. L. (1974). Soil salts and their effects on water movement. *In* "Drainage for Agriculture" (J. van Schilfgaarde, ed.). Monograph 17, Am. Soc. Agron., Madison, Wisconsin.

McNeal, B. L., and Coleman, N. T. (1966). Effect of solution composition on soil hydraulic conductivity. *Soil Sci. Soc. Am. Proc.* **30**, 308–312.

McNeal, B. L., Layfield, D. A., Norvell, W. A., and Rhoades, J. D. (1968). Factors influencing hydraulic conductivity of soils in the presence of mixed-salt solutions. *Soil Sci. Soc. Am. Proc.* **32**, 187–190.

Manning, R. (1891). On the flow of water in open channels and pipes. *Trans. Inst. Civil Eng. Ireland* **20**, 161–207.

Marshall, C. E. (1964). "The Physical Chemistry and Mineralogy of Soils." Wiley, New York.

Marshall, T. J. (1958). A relation between permeability and size distribution of pores. *J. Soil Sci.* **9**, 1–8.

Marshall, T. J. (1959). The diffusion of gases through porous media. *J. Soil Sci.* **10**, 79–82.

Marshall, T. J., and Gurr, C. G. (1966). Movement of water and chlorides in relatively dry soil. *Soil Sci.* **77**, 147–152.

Matano, C. (1933). On the relation between the diffusion coefficients and concentrations of solid materials (the nickel-copper system). *Jpn. J. Phys.* **8**, 108–113.

Matheron, G. (1963). Principles of geostatistics. *Econ. Geol.* **58**, 1246–1266.

Matheron, G. (1971). The theory of regionalized variables and its applications. Ecole des Mines, Fountainebleu, France.

Mayanskas, I. S. (1959). Investigation of the pressure distribution on the surface of a plow share in work. *J. Agr. Eng. Res.* **4**, 186–190.

Mazurak, A. P. (1950). Effect of gaseous phase on water-stable synthetic aggregates. *Soil Sci.* **69**, 135–148.

Meidner, H., and Sheriff, D. W. (1976). "Water and Plants." Halsted Press, New York and Wiley, New York.

Mein, R. G., and Larson, C. L. (1973). Modeling infiltration during a steady rain. *Water Resources Res.* **9**, 384–394.

Miller, E. E. (1975). Physics of swelling and cracking soils. *J. Colloid Interface Sci.* **52**, 434–443.

Miller, E. E., and Elrick, D. E. (1958). Dynamic determination of capillary conductivity extended for non-negligible membrane impedence. *Soil Sci. Soc. Am. Proc.* **22**, 483–486.

Miller, E. E., and Klute, A. (1967). Dynamics of soil water. Part I—Mechanical forces. *In* "Irrigation of Agricultural Lands," pp. 209–244. Monograph 11, Am. Soc. Agron., Madison, Wisconsin.

Miller, E. E., and Miller, R. D. (1955a). Theory of capillary flow: I. Practical implications. *Soil Sci. Soc. Am. Proc.* **19**, 267–271.

Miller, E. E., and Miller, R. D. (1955b). Theory of capillary flow: II. Experimental information. *Soil Sci. Soc. Am. Proc.* **19**, 271–275.

Miller, E. E., and Miller, R. D. (1956). Physical theory for capillary flow phenomena. *J. Appl. Phys.* **27**, 324–332.

Miller, R. J., and Low, P. F. (1963). Threshold gradient for water flow in clay systems. *Soil Sci. Soc. Am. Proc.* **27**, 605–609.

Miller, R. D. (1970). Ice sandwich: Functional semipermeable membrane. *Science* **169**, 584–585.

Miller, R. D. (1973a). Soil freezing in relation to pore water pressure and temperature. *Int. Conf. Permafrost, 2nd, Natl. Acad. Sci.*, **344**–352.

Miller, R. D. (1973b). The porous phase barrier and crystallization. *Sep. Sci.* **8**, 521–535.

Miller, R. D. (1978). Frost heaving in non-colloidal soils. *Int. Conf. on Permafrost 3rd*, **1**, 708–720.

Miller, D. E., and Gardner, W. H. (1962). Water infiltration into stratified soil. *Soil Sci. Soc. Am. Proc.* **26**, 115–118.

Miller, R. D., Baker, J. H., and Kolaian, J. H. (1960). Particle size, overburden pressure, pore water pressure and freezing temperature of ice lenses in soil. *Trans. Int. Congress Soil Sci., 7th*, **1**, 122–129.

Miller, R. D., Loch, J. P. G., and Bresler, E. (1975). Transport of water and heat in a frozen permeameter. *Soil Sci. Soc. Am. Proc.* **39**, 1029–1036.

Millington, R. J. (1959). Gas diffusion in porous media. *Science* **130**, 100–102.

Millington, R. J., and Quirk, J. P. (1959). Permeability of porous media. *Nature (London)* **183**, 387–388.

Molz, F. J. (1975). Potential distribution in the soil-root system. *Agron. J.* **67,** 726–729.

Molz, F. J. (1976). Water transport in the soil-root system: Transient analysis. *Water Resour. Res.* **12,** 805–807.

Molz, F. J., and Ikenberry, E. (1974). Water transport through plant cells and cell walls: Theoretical development. *Soil Sci. Soc. Am. Proc.* **38,** 699–704.

Molz, F. J., and Remson, I. (1970). Extraction term models of soil moisture use by transpiring plants. *Water Resources Res.* **6,** 1346–1356.

Molz, F. J., and Remson, I. (1971). Application of an extraction term model to the study of moisture flow to plant roots. *Agron. J.* **63,** 72–77.

Monteith, J. L. (1965). Evaporation and environment. *Symp. Soc. Exp. Biol., 19th,* 205–234.

Monteith, J. L. (1973). "Principles of Environmental Physics." American Elsevier, New York.

Monteith, J. L. (1978). Models and measurement in crop climatology. *Proc. Int. Soil Sci. Congress, 11th* Edmonton.

Monteith, J. L., Szeicz, G., and Waggoner, P. E. (1965). The measurement and control of stomatal resistance in the field. *J. Appl. Ecol.* **2,** 345–357.

Moore, R. E. (1939). Water conduction from shallow water tables. *Hilgardia* **12,** 383–426.

Morel-Seytoux, H. J., and Noblanc, A. (1972). Infiltration predictions by a moving strained coordinates method. *In* "Physical Aspects of Soil, Water and Salts in Ecosystems" (A. Hadas *et al.,* eds). Springer-Verlag, Berlin and New York.

Morgan, J., and Warren, B. E. (1938). X-ray analysis of the structure of water. *J. Chem. Phys.* **6,** 666–673.

Morin, J., Goldberg, D., and Seginer, I. (1967). A rainfall simulator with a rotating disk. *ASAE Trans.* **10,** 74–77.

Mortland, M. M., and Kemper, W. D. (1965). Specific surface. *In* "Methods of Soil Analysis," pp. 532–544. Monograph 9, Am. Soc. Agron., Madison, Wisconsin.

Morton, C. T., and Buchele, W. F. (1960). Emergence energy of plant seedlings. *Agr. Eng.* **41,** 428–431.

Mualem, Y. (1976). A new model for predicting the hydraulic conductivity of unsaturated porous media. *Water Resour. Res.* **12,** 513–522.

Muskat, M. (1946). "The Flow of Homogeneous Fluids Through Porous Media." Edwards, Ann Arbor, Michigan.

Myers, L. E. (1961). Water proofing soil to collect precipitation. *J. Soil Water Conserv.* **16,** 281–282.

Myers, L. E. (1963). Water harvesting. Special publication of the U.S. Water Conservation Laboratory, Tempe, Arizona.

Narten, A. H., and Levy, H. A. (1969). Observed diffraction pattern and proposed model of liquid water. *Science* **165,** 447–454.

Nemethy, G., and Scheraga, H. A. (1962). Structure of water and hydraulic bonding in proteins. I. A model for the thermodynamic properties of liquid water. *J. Chem. Phys.* **36,** 3382–3400.

Nerpin, S., Pashkina, S., and Bondarenko, N. (1966). The evaporation from bare soil and the way of its reduction. *Symp. Water Unsaturated Zone, Wageningen.*

Newman, E. I. (1969). Resistance to water flow in soil and plant: 1. Soil resistance in relation to amount of roots: Theoretical estimates. *J. Appl. Ecol.* **6,** 1–12.

Newman, E. I. (1974). Root and soil water relations. *In* "The Plant Root and Its Environment" (E. W. Carson, ed.), pp. 363–440. Univ. Press of Virginia, Charlottesville, Virginia.

Nichols, M. L. (1929). Methods of research in soil dynamics as applied to implement design. *Ala. Agric. Exp. Stn. Bull. 229, Agric. Eng.* (1932), **13,** 279–285.

Nichols, M. L., and Randolph, J. W. (1925). A method of studying soil stresses. *Agr. Eng.* **6,** 134–135.

Nichols, M. L., and Reed, I. F. (1934). Soil dynamics: VI. Physical reactions of soils to moldboard surfaces. Agr. Eng. **15,** 187–190.

Nielsen, D. R., and Biggar, J. W. (1961). Miscible displacement in soils: I. Experimental information. *Soil Sci. Soc. Am. Proc.* **25,** 1–5.

Nielsen, D. R., and Biggar, J. W. (1962). Miscible displacement: III. Theoretical consideration. *Soil Sci. Soc. Am. Proc.* **26,** 216–221.

Nielsen, D. R., Jackson, R. D., Cary, J. W., and Evans, D. D. (eds.) (1972). "Soil Water." Am. Soc. Agron., Madison, Wisconsin.

Nielsen, D. R., Biggar, J. W., and Erh, K. T. (1973). Spatial variability of field-measured soil-water properties. *Hilgardia* **42,** 215–259.

Nimah, M. N., and Hanks, R. J. (1973). Model for estimating soil, water, plant, and atmosphere interactions: 1. Description and sensitivity. *Soil Sci. Soc. Am. Proc.* **37,** 522–527.

Nixon, P. R., and Lawless, G. P. (1960). Translocation of moisture with time in unsaturated soil profiles. *J. Geophys. Res.* **65,** 655–661.

Nobel, P. S. (1970). "Plant Cell Physiology: A Physiochemical Approach." Freeman, San Francisco, California.

Nobel, P. S. (1974). "Introduction to Biophysical Plant Physiology." Freeman, San Francisco, California.

Nutting, P. G. (1943). Some standard thermal dehydration curves of minerals. U.S.G.S. Professional paper 197-E.

Ogata, G., and Richards, L. A. (1957). Water content changes following irrigation of bare field soil that is protected from evaporation. *Soil Sci. Soc. Am. Proc.* **21,** 355–356.

Ogata, G., Richards, L. A., and Gardner, W. R. (1960). Transpiration of alfalfa determined from soil water content changes. *Soil Sci.* **89,** 179–182.

Olsen, H. W. (1965). Deviations from Darcy's law in saturated clays. *Soil Sci. Soc. Am. Proc.* **29,** 135–140.

Onafenko, O., and Reece, A. R. (1967). Soil stresses and deformations beneath rigid wheels. *J. Terramech.* **4,** 59–80.

Oster, J. D., and Schroer, F. W. (1979). Infiltration as influenced by irrigation water quality. *Soil Sci. Soc. Am. J.* (in press).

Oster, J. D., and Willardson, L. S. (1971). Reliability of salinity sensors for the management of soil salinity. *Agron. J.* **63,** 695–698.

Overbeek, J. T. G. (1952). *In* "Colloid Science" (H. R. Kruyt, ed.), Chapters 4–6. Elsevier, Amsterdam.

Page, J. B. (1948). Advantages of the pressure pycnometer for measuring the pore space in soils. *Soil Sci. Soc. Am. Proc.* **12,** 81–84.

Papendick, R. I. and Runkles, J. R. (1965). Transient-state oxygen diffusion in soil. I. The case when rate of oxygen consumption is constant. *Soil Sci.* **100,** 251–261.

Parker, J. J., and Taylor, H. M. (1965). Soil strength and seedling emergence relations. *Agron. J.* **57,** 289–291.

Parlange, J.-Y. (1971). Theory of water movement in soils. I. One dimensional absorption. *Soil Sci.* **111,** 134–137.

Parlange, J.-Y. and Hill, D. E. (1976). Theoretical analysis of wetting front instability in soils. *Soil Sci.* **122,** 236–239.

Payne, P. C. J. (1956). The relationship between the mechanical properties of soil and the performance of simple cultivation implements. *J. Agr. Eng. Res.* **1,** 23–28.

Peacemen, D. W., and Rachford, H. H. Jr. (1955). The numerical solution of parabolic and elliptic differential equations, *J. Soc. Ind. Appl. Math.* **3,** 28–41.

Pearse, J. F., Oliver, T. R., and Newitt, D. M. (1949). The mechanism of the drying of solids: Part I. The forces giving rise to movement of water in granular beds during drying. *Trans. Inst. Chem. Eng. (London)* **27**, 1–8.

Peck, A. J. (1965). Moisture profile development and air compression during water uptake by bounded porous bodies: 3. Vertical columns. *Soil Sci.* **100**, 44–51.

Peck, A. J. (1969). Entrapment, stability, and persistence of air bubbles in soil water. *Aust. J. Soil Res.* **7**, 79–90.

Peck, A. J. (1970). Redistribution of soil water after infiltration. *Aust. J. Soil Res.* **7**.

Pelton, W. L. (1961). The use of lysimetric methods to measure evapotranspiration. *Proc. Hydrol. Symp.* **2**, 106–134 (Queen's Printer, Ottawa, Canada. Cat. No. R32–361/2).

Penman, H. L. (1940). Gas and vapor movements in the soil: I. The diffusion of vapors through porous solids. *J. Agr. Sci.* **30**, 437–461.

Penman, H. L. (1948). Natural evaporation from open water, bare soil and grass. *Proc. R. Soc. London Ser. A* **193**, 120–146.

Penman, H. L. (1949). The dependence of transpiration on weather and soil conditions. *J. Soil Sci.* **1**, 74–89.

Penman, H. L. (1953). The physical basis of irrigation control. *Rep. Int. Hort. Congress, 13th,* **2**, 913–924.

Penman, H. L. (1955). The evaporation calculations. *In* "Lake Eyre, The Great Flooding of 1949–1950." Royal Geographical Society of Australia.

Penman, H. L. (1956). Evaporation: an introductory survey. *Neth. J. Agr. Sci.* **4**, 9–29.

Penner, E. (1959). The mechanism of frost heaving in soils. *Highw. Res. Board. Bull.* **225**, 1–13.

Penner, E., and Ueda, T. (1978). A soil frost-susceptibility test and a basis for interpreting heaving rates. *Int. Conf. Permafrost, 3rd,* **1**, 721–727.

Pereira, H. C., and Jones, P. A. (1954). A tillage study in Kenya coffee: Part II. The effect of tillage practices on the structure of the soil. *Emp. J. Expt. Agric.* **22**, 323–327.

Peters, D. B. (1965). Water availability. *In* "Methods of Soil Analysis" (C. A. Black, ed), pp. 279–285. Monograph 9, Am. Soc. Agron., Madison, Wisconsin.

Philip, J. R. (1955a). Numerical solution of equations of the diffusion type with diffusivity concentration dependent. *Trans. Faraday Soc.* **51**, 885–892.

Philip, J. R. (1955b). The concept of diffusion applied to soil water. *Proc. Nat. Acad. Sci. India* **24A**, 93–104.

Philip, J. R. (1957). Numerical solution of equations of the diffusion type with diffusivity concentration-dependent II. *Aust. J. Phys.* **10**, 29–42.

Philip, J. R. (1957a). The theory of infiltration: 2. The profile at infinity. *Soil Sci.* **83**, 435–448.

Philip, J. R. (1957b). The theory of infiltration: 3. Moisture profiles and relation to experiment. *Soil Sci.* **84**, 163–178.

Philip, J. R. (1957c). The theory of infiltration: 4. Sorptivity and algebraic infiltration equations. *Soil Sci.* **84**, 257–264.

Philip, J. R. (1957d). Evaporation, moisture and heat fields in the soil. *J. Meteorol.* **14**, 354–366.

Philip, J. R. (1960). Absolute thermodynamic functions in soil-water studies. *Soil Sci.* **89**, 111.

Philip, J. R. (1964). Similarity hypothesis for capillary hysteresis in porous materials. *J. Geophys. Res.* **69**, 1553–1562.

Philip, J. R. (1966a). Absorption and infiltration in two- and three-dimensional systems. *In* "Water in the Unsaturated Zone" (R. E. Rijtema and H. Wassink, eds.). Vol. 2, pp. 503–525. IASH/UNESCO Symp., Wageningen.

Philip, J. R. (1966b). Plant water relations: some physical aspects. *Ann. Rev. Plant Physiol.* **17**, 245–268.

Philip, J. R. (1969a). Theory of infiltration. *Adv. Hydrosci.* **5**, 215–290.

Philip, J. R. (1969b). Hydrostatics and hydrodynamics in swelling soils. *Water Resources Res.* **5**, 1070–1077.

Philip, J. R. (1972). Hydrology of swelling soils. *In* "Salinity and Water Use" (a national symposium on hydrology sponsored by the Australian Academy of Science). Macmillan, New York.

Philip, J. R. (1975a). Water movement in soil. *In* "Heat and Mass Transfer in the Biosphere" (D. A. de Vries and N. H. Afgan, eds.), pp. 29–47. Halsted Press-Wiley, New York.

Philip, J. R. (1975b). Stability analysis of infiltration. *Soil Sci. Soc. Am. Proc.* **39**, 1042–1049.

Philip, J. R., and deVries, D. A. (1957). Moisture movement in porous materials under temperature gradients. *Trans. Am. Geophys. Un.* **38**, 222–228.

Phillips, S. H., and Young, H. M. (1973). "No-tillage farming." Reiman, Milwaukee, Wisconsin.

Plummer, F. L., and Dore, S. M. (1940). "Soil Mechanics and Foundations." Pitman, New York.

Poulovassilis, A. (1962). Hysteresis of pore water, an application of the concept of independent domains. *Soil Sci.* **93**, 405–412.

Prigogine, I. (1961). "Introduction to Thermodynamics of Irreversible Processes." Wiley, New York.

Pringle, J. (1975). The assessment and significance of aggregate stability in soil. *In* "Soil Physical Conditions and Crop Productions," pp. 249–260. Tech. Bull. 29, Min. Agr., Fisheries and Food, London.

Proctor, R. R. (1948). Laboratory compaction methods, penetration resistance measurements, and indicated saturation penetration resistance. *Proc. Int. Conf. Soil Mech. Found. 2nd* **5**, 242–245.

Pruitt, W. O., and Angus, D. E. (1960). Large weighing Lysimeter for measuring evapotranspiration. *Trans. Am. Soc. Agr. Eng.* **3**, 3–15, 18.

Puri, A. N., and Puri, B. R. (1939). Physical characteristics of soils: II. Expressing mechanical analysis and state of aggregation of soils by single values. *Soil Sci.* **47**, 77–86.

Quastel, J. H. (1954). Soil conditioners. *Res. Plant Physiol.* **5**, 75–92.

Quirk, J. P., and Schofield, R. K. (1955). The effect of electrolyte concentration on soil permeability. *J. Soil Sci.* **6**, 163–178.

Raats, P. A. C. (1973). Unstable wetting fronts in uniform and nonuniform soils. *Soil Sci. Soc. Am. Proc.* **37**, 681–685.

Radhakrishnan, S. (1958). "Indian Philosophy." Macmillan, New York.

Raney, W. A. (1950). Field measurement of oxygen diffusion through soil. *Soil Sci. Soc. Am. Proc.* **14**, 61–65.

Rasmussen, V. P., and Hanks, R. J. (1978). Model for predicting spring wheat yields with limited climatological and soil data. *Agron. J.* **70**, 940–944.

Ravina, I., and Low, P. F. (1972). Relation between swelling, water properties and b-dimension in montmorillonite-water systems. *Clay Miner.* **20**, 109–123.

Rawitz, E. (1969). The dependence of growth rate and transpiration on plant and soil physical parameters under controlled conditions. *Soil Sci.* **110**, 172–182.

Rawitz, E., and Hillel, D. (1974). Progress and problems of drip irrigation in Israel. *Proc. Int. Conf. Drip Irrig., San Diego, California.*

Rawitz, E., Margolin, M., and Hillel, D. (1972). An improved variable-intensity sprinkling infiltrometer. *Soil Sci. Soc. Am. Proc.* **36**, 533–535.

Rawlins, S. L., and Raats, P. A. C. (1975). Prospects for high frequency irrigation. *Science* **188**, 604–610.

Reaves, C. A., and Cooper, A. W. (1960). Stress distribution in soils under tractor loads. *Agr. Eng.* **41**, 20–21, 31.

Reeve, R. C. (1965a). Air-to-water permeability ratio. *In* "Methods of Soil Analysis." Am. Soc. Agron., Madison, Wisconsin.

Reeve, R. C. (1965b). Modulus of rupture. *In* "Methods of Soil Analysis," Pt. I. Monograph 9, Am. Soc. Agron., Madison, Wisconsin.

Reeve, R. C., and Bower, C. A. (1960). Use of high salt water as a flocculant and source of divalent cations for reclaiming sodic soils. *Soil Sci.* **90**, 139–144.

Reichardt, K., Nielsen, D. R., and Biggar, J. W. (1972). Scaling of horizontal infiltration into homogeneous soils. *Soil Sci. Am. Proc.* **36**, 241–245.

Reicosky, D. C., Cassel, D. K., Blevius, R. L., Gill, W. R., and Naderman, G. C. (1977). Conservation tillage in the Southeast. *J. Soil Water Conserv.* **32**, 13–19.

Reicosky, D. C., and Ritchie, J. T. (1976). Relative importance of soil resistance and plant resistance in root water absorption. *Soil Sci. Soc. Am. J.* **40**, 293–297.

Reid, C. E. (1960). "Principles of Chemical Thermodynamics." Van Nostrand-Reinhold, Princeton, New Jersey.

Reitemeyer, R. F., and Richards, L. A. (1944). Reliability of the pressure-membrane method for extraction of soil solution. *Soil Sci.* **57**, 119–135.

Remson, I., Drake, R. L., McNeary, S. S., and Walls, E. M. (1965). Vertical drainage of an unsaturated soil. *Am. Soc. Civil Eng. Proc. J. Hyd. Div.* **9**, 55–74.

Remson, I., Fungaroli, A. A., and Hornberger, G. M. (1967). Numerical analysis of soil moisture systems. *Am. Soc. Civil Eng. Proc. J. Irrig. Drain. Div.* **3**, 153–166.

Remson, I., Hornberger, G. M., and Molz, F. (1971). "Numerical Methods in Subsurface Hydrology." Wiley (Interscience), New York.

Retta, A., and Hanks, R. J. (1979). Corn and alfalfa production as influenced by limited irrigation. *Irrig. Sci.* (in press).

Retzer, J. L., and Russell, M. B. (1941). Differences in the aggregation of a Prairie and a Gray Brown Podzolic soil. *Soil Sci.* **52**, 47–58.

Rhoades, J. D. (1974). Drainage for salinity control. *In* "Drainage for Agriculture" (J. van Schilfgaarde, ed.). Monograph 17, Am. Soc. Agron., Madison, Wisconsin.

Richards, L. A. (1928). The usefulness of capillary potential to soil-moisture and plant investigators. *J. Agri. Res.* **37**, 719–742.

Richards, L. A. (1931). Capillary conduction of liquids in porous mediums. *Physics* **1**, 318–333.

Richards, L. A. (1953). Modulus of rupture of soils as an index of crusting of soil. *Soil Sci. Soc. Am. Proc.* **17**, 321–323.

Richards, L. A. (ed.) (1954). "Diagnosis and Improvement of Saline and Alkali Soils." U.S. Dept. Agr. Handbook 60.

Richards, L. A. (1952). Report of the subcommittee on permeability and infiltration, Committee on Terminology, Soil Science Society of America. *Soil Sci. Soc. Am. Proc.* **16**, 85–88.

Richards, L. A. (1960). Advances in soil physics. *Trans. Int. Congr. Soil Sci., 7th, Madison* **I**, 67–69.

Richards, L. A. (1965). Physical condition of water in soil. *In* "Methods of Soil Analysis," pp. 128–152. Monograph 9, Am. Soc. Agron., Madison, Wisconsin.

Richards, L. A., and Moore, D. C. (1952). Influence of capillary conductivity and depth of wetting on moisture retention in soil. *Trans. Am. Geophys. Un.* **33**, 4.

Richards, L. A., and Wadleigh, C. H. (1952). Soil water and plant growth. *In* "Soil Physical Conditions and Plant Growth," p. 13. Am. Soc. Agron. Monograph 2.

Richards, L. A., and Weaver, L. R. (1944). Fifteen atmosphere percentage as related to the permanent wilting percentage. *Soil Sci.* **56,** 331–339.

Richards, L. A., Gardner, W. R., and Ogata, G. (1956). Physical processes determining water loss from soil. *Soil Sci. Am. Proc.* **20,** 310–314.

Richards, S. J. (1965). Soil suction measurements with tensiometers. *In* "Methods of Soil Analysis," pp. 153–163. *Am. Soc. Agron.,* Monograph 9.

Ripple, C. D., Rubin, J., and van Hylkama, T. E. A. (1972). Estimating steady-state evaporation rates from bare soils under conditions of high water table. U.S. Geol. Survey, Water Supp. Pap. 2019-A.

Ritchie, J. T., and Adams, J. E. (1974). Field measurement of evaporation from soil shrinkage cracks. *Soil Sci. Am. Proc.* **38,** 131–134.

Robins, J. S., Pruitt, W. O., and Gardner, W. H. (1954). Unsaturated flow of water in field soils and its effect on soil moisture investigations. *Soil Sci. Soc. Am. Proc.* **18,** 344–348.

Robinson, F. E. (1964). A diffusion chamber for studying soil atmosphere. *Soil Sci.* **83,** 465–469.

Römkens, M. J. M., and Miller, R. D. (1973). Migration of mineral particles in ice with temperature gradient. *J. Colloid Interface Sci.* **42,** 103–111.

Rose, C. W. (1966). "Agricultural Physics." Pergamon, Oxford.

Rose, C. W. (1968). Evaporation from bare soil under high radiation conditions. *Trans. Int. Congr. Soil Sci., 9th, Adelaide* **I,** 57–66.

Rose, C. W., and Stern, W. R. (1967). The drainage component of the water balance equation. *Aust. J. Soil Res.* **3,** 95–100.

Rose, C. W., and Stern, W. R. (1967a). Determination of withdrawal of water from soil by crop roots as a function of depth and time. *Aust. J. Soil Res.* **5,** 11–19.

Rose, C. W., Stern, W. R., and Drummond, J. E. (1965). Determination of hydraulic conductivity as a function of depth and water content for soil *in situ. Aust. J. Soil Res.* **3,** 1–9.

Rose, C. W., Byrne, G. F., and Begg, J. E. (1966). An accurate hydraulic lysimeter with remote weight recording. CSIRO Div. Land Res. and Reg. Survey, Tech. Paper 27. Canberra, Australia.

Rose, C. W., Byrne, G. F., and Hansen, G. K. (1976). Water transport from soil through plant to atmosphere: A lumped-parameter model. *Agric. Meteorol.* **16,** 171–184.

Rosenberg, N. J. (1974). "Microclimate: The Biological Environment." Wiley, New York.

Rubin, J. (1966). Theory of rainfall uptake by soils initially drier than their field capacity and its applications. *Water Resour. Res.* **2,** 739–749.

Rubin, J. (1967). Numerical method for analyzing hysteresis-affected, post-infiltration redistribution of soil moisture. *Soil Sci. Soc. Am. Proc.* **31,** 13–20.

Rubin, J., and Steinhardt, R. (1963). Soil water relations during rain infiltration: I. Theory. *Soil Sci. Soc. Am. Proc.* **27,** 246–251.

Rubin, J., and Steinhardt, R. (1964). Soil water relations during rain infiltration: III. Water uptake at incipient ponding. *Soil Sci. Soc. Am. Proc.* **28,** 614–619.

Rubin, J., Steinhardt, R., and Reiniger, P. (1964). Soil water relations during rain infiltration: II. Moisture content profiles during rains of low intensities. *Soil Sci. Soc. Am. Proc.* **28,** 1–5.

Russell, G. (1980). Crop evaporation, surface resistance, and soil water status. *Agric. Meteorol.* **21,** 213–226.

Russel, E. J. (1912). "Soil Conditions and Plant Growth." Longman, London.

Russell, E. W. (1973). "Soil Conditions and Plant Growth," 10th ed., Longman, London.

Russell, M. B. (1941). Pore-size distribution as a measure of soil structure. *Soil Sci. Soc. Am. Proc.* **6,** 108–112.

Russell, M. B.(1949). Methods of measuring soil structure and aeration. *Soil Sci.* **68,** 25–35.

Russell, M. B. (1952). Soil aeration and plant growth. *In* "Soil Physical Conditions and Plant Growth"(B. T. Shaw, ed.), pp. 253–301. Academic Press, New York.

Russell, M. B., and Feng, C. L. (1947). Characterization of the stability of soil aggregates. *Soil Sci.* **63,** 299–304.

Russell, R. S., and Goss, M. J. (1974). Physical aspects of soil fertility—the response of roots to mechanical impedance. *Neth. J. Agr. Sci.* **22,** 305–318.

Russo, D. and Bresler, E. (1977). Analysis of saturated-unsaturated hydraulic conductivity in a mixed sodium-calcium soil system. *Soil Sci. Soc. Am. J.* **41,** 706–710.

Rutter, A. J. (1967). An analysis of evaporation from a stand of Scots pine. *In* Forest Hydrology (W. E. Sopper and H. W. Lull, eds), pp. 403–417. Pergamon Press, Oxford.

Saffman, P. G., and Taylor, G. I. (1958). The penetration of a fluid into a porous medium or Hele-Shaw cell containing a more viscous liquid. *Proc. R. Soc. London Ser. A.* **245,** 312–331.

Sahin, T. (1973). Transport of water in frozen soil. M.S. Thesis, Cornell Univ., Ithaca, New York.

Salberg, J. R. (1965). Shear strength. *In* "Methods of Soil Analysis," Pt. I. Monograph 9, Am. Soc. Agron., Madison, Wisconsin.

Schamp, N. (1971). Soil conditioning by means of organic polymers. *Pedologie* **21,** 100.

Scheidegger, A. E. (1957). "The Physics of Flow through Porous Media." Macmillan, New York.

Schofield, C. S. (1940). Salt balance in irrigated ares. *J. Agr. Res.* **61,** 17–39.

Schofield, R. K. (1935). The pF of the water in soil. *Trans. Int. Cong. Soil Sci., 3rd.* **2,** 37–48.

Schofield, R. K. (1946). Ionic forces in thick films between charged surfaces. *Trans. Faraday Soc.* **42B,** 219–225.

Scotter, D. R., and Raats, P. A. C. (1968). Dispersion in porous mediums due to oscillating flow. *Water Resources Res.* **4,** 1201–1206.

Seaton, K. A., Landsberg, J. J., and Sedgley, R. H. (1977). Transpiration and leaf water potentials of wheat in relation to changing soil water potential. *Aust. J. Agr. Res.* **28,** 355–367.

Seed, H. B., Woodward, R. J., and Lundgren, R. (1964). Fundamental aspects of the Atterbert limits. *Proc. Am. Soc. Civil Eng. J. Soil Mech. Found. Div.* **90,** No. SM6.

Selim, H. M., and Kirkham, D. (1970). Soil temperature and water content changes during drying as influenced by cracks: A laboratory experiment. *Soil Sci. Soc. Am. Proc.* **34,** 565–569.

Sellers, W. D. (1965). "Physical Climatology." Univ. of Chicago Press, Chicago, Illinois.

Sellin, R. H. J. (1969). "Flow in Channels." Macmillan, New York.

Shainberg, I. (1973). Ion exchange properties in irrigated soils. *In* "Arid Zone Irrigation" (B. Yaron, *et al.,* eds.). Springer-Verlag, Berlin and New York.

Shalhevet, J., Mantell, A., Bielorai, A., and Shimshi, D. (1976). Irrigation of field and orchard crops under semi-arid conditions. Int. Irrig. Inf. Ctr. Publ. No. 1.

Shanan, L., Tadmor, N. H., Evenari, M., and Reiniger, P. (1970). Runoff farming in the desert. III. Microcatchments for improvement of desert range. *Agron. J.* **62,** 445–449.

Shantz, H. L., and Piemeisel, L. N. (1927). The water requirement of plants at Akron, Colo. *J. Agric. Res.* **34,** 1093–1190.

Shaw, R. H., and Buchele, W. F. (1957). The effect of the shape of the soil surface profile on soil temperature and moisture. *Iowa State College J. Sci.* **32,** 95–104.

Sheppard, M. I., Kay, B. D., and Loch, J. P. G. (1978). Development and testing of a computer model for heat and mass flow in freezing soils. *Soil Sci. Soc. Am. J.* **42**, 38.

Shuttleworth, W. J. (1976). A one-dimensional theoretical description of the vegetation-atmosphere interaction. *Boundary Layer Meteorol.* **10**, 273–302.

Simmons, C. S., Nielsen, D. R., and Biggar, J. W. (1979). Scaling of field-measured soil water properties. *Hilgardia* (in press).

Skaggs, R. W., Huggins, L. F., Monke, E. J., and Foster, G. R. (1969). Experimental evaluation of infiltration equations. *Trans. Am. Soc. Agr. Eng.* **12**, 822–828.

Skapski, A., Billups, R. and Rooney, A. (1957). The capillary cone method for determination of surface tension of solids. *J. Chem. Phys.* **26**, 1350.

Slater, P. J., and Williams, J. B. (1965). The influence of texture on the moisture characteristics of soils. I. A critical comparison of techniques for determining the available water capacity and moisture characteristic curve of a soil. *J. Soil Sci.* **16**, 1–12.

Slatyer, R. O. (1967). "Plant Water Relationships." Academic Press, New York.

Slatyer, R. O., and McIlroy, I. C. (1961). "Practical Microclimatology." CSIRO, Australia.

Slichter, C. S. (1899). U.S. Geol. Sur. Ann. Rep. 19-II, pp. 295–384.

Smiles, D. E. (1974). Infiltration into swelling material. *Soil Sci.* **117**, 110–116.

Smiles, D. E. (1976). On the validity of the theory of flow in saturated swelling materials. *Aust. J. Soil Res.* **14**, 389–395.

Smiles, D. E., and Rosenthal, M. J. (1968). The movement of water in swelling materials. *Aust. J. Soil Res.* **6**, 237–248.

Smiles, D. E., Philip, J. R., Knight, J. H., and Elrick, D. E. (1978). Hydrodynamic dispersion during sorption of water by soil. *Soil Sci. Soc. Am. J.* **42**, 229–234.

Smith, A. (1932). Seasonal subsoil temperature variations. *J. Agr. Res.* **44**, 421–428.

Smith, G. D. (1965). "Numerical Solution of Partial Differential Equations." Oxford Univ. Press, London and New York.

Smith, G. D., Newhall, F., Robinson, L. H., and Swanson, D. (1964). "Soil Temperature Regimes, Their Characteristics and Predictability." U.S. Dept. Agr. Soil Conservation Service SCS-TP-144, Washington, D.C.

Smith, R. E., and Woolhiser, D. A. (1971). Overland flow on an infiltrating surface. *Water Resources Res.* **5**, 114–152.

Smythe, W. R. (1950). "Static and Dynamic Electricity." McGraw-Hill, New York.

So, H. B., Aylmore, L. A. C., and Quirk, J. P. (1976). The resistance of intact maize roots to water flow. *Soil Sci. Soc. Am. J.* **40**, 222–225.

Soane, B. D. (1970). The effect of traffic and implements on soil compaction. *Agr. Eng.* **25**, 115–128.

Soane, B. D. (1975). Studies on some soil physical properties in relation to cultivations and traffic. *In* "Soil Physical Conditions and Crop Production," pp. 249–260. Tech. Bull. 29, Min. Agr., Fisheries and Food, London.

Söhne, W. H. (1953). Pressure distribution in the soil and soil deformation under tractor tires. *Grundl. d. Landtech.* **5**, 49–63.

Söhne, W. H. (1956). Some basic considerations of soil mechanics applied to agricultural engineering. *Grundl. d. Landtech.* **7**, 11–27.

Söhne, W. H. (1958). Fundamentals of pressure distribution and soil compaction under tractor tires. *Agr. Eng.* **39**, 276–281, 290.

Söhne, W. H. (1966). Characterization of tillage tools. Särtryck ur Grundförbättring **1**, 31–48.

Soil. (1957). Yearbook of Agriculture, U.S. Dept. of Agriculture.

Soil Survey Manual. (1951). U.S. Dept. of Agriculture Handbook No. 18.

Sor, K., and Kemper, W. D. (1959). Estimation of surface area of soils and clays from the amount of adsorption and retention of ethylene glycol. *Soil Sci. Soc. Am. Proc.* **23,** 105–110.

Sorensen, V. M., Hanks, R. J., and Cartee, R. L. (1979). Cultivation during early season and irrigation influences on corn production. *Agron. J.* (in press).

Sourisseau, J. H. (1935). Determination and study of physico-mechanical properties of soil. *Organ. Raps. II Congr. Int. Genie Rural, Madrid,* pp. 159–194.

Southwell, R. V. (1946). "Relaxation Methods in Engineering Science." Oxford Univ. Press, London and New York.

Sowers, G. F. (1965). Consistency. *In* "Methods of Soil Analysis," pp. 391–399. Monograph No. 9., Am. Soc. Agron., Madison, Wisconsin.

Speckhart, F. H., and Green, W. L. (1976). "A Guide to Using CSMP." Prentice-Hall, Englewood Cliffs, New Jersey.

Stanhill, G. (1965). Observation on the reduction of soil temperature. *Agr. Met.* **2,** 197–203.

Staple, W. J. (1969). Comparison of computed and measured moisture redistribution following infiltration. *Soil Sci. Soc. Am. Proc.* **33,** 206.

Steinhardt, R., and Hillel, D. (1966). A rainfall simulator for laboratory and field use. *Soil Sci. Soc. Am. Proc.* **30,** 680–682.

Stewart, R. B., and Rouse, W. R. (1977). Substantiation of the Priestley-Taylor parameter for potential evaporation in high latitudes. *J. Appl. Meteorol.* **16,** 649–650.

Stewart, J. I., Danielson, R. E., Hanks, R. J., Jackson, E. B., Hagan, R. M., Pruitt, W. O., Franklin, W. T., and Riley, J. P. (1977). Optimizing crop production through control of water and salinity levels in the soil. Utah Water Lab. PRWG 151–1. p. 191. Logan, Utah.

Stolzy, L. H., and Barley, K. P. (1968). Mechanical resistance encountered by roots entering compact soil. *Soil Sci.* **105,** 297–301.

Stolzy, L. H., and Letey, J. (1964). Characterizing soil oxygen conditions with a platinum microelectrode. *Agr. J.* **16,** 249–279.

Stotzky, G. (1965). Microbial respiration. *In* "Methods of Soil Analysis," pp. 1550–1569. Monograph 9, Am. Soc. Agron., Madison, Wisconsin.

Su, C., and Brooks, R. H. (1975). Soil hydraulic properties from infiltration tests. Watershed Management Proceedings, Irrigation and Drainage Division, ASCE. Logan, Utah. August 11–13, pp. 516–542.

Su, C., and Brooks, R. H. (1979). Measurement of water retention data for soils. Submitted for publication in ASCE.

Suklje, L. (1969). "Rheological Aspects of Soil Mechanics." Wiley (Interscience), New York.

Sutcliffe, J. (1968). "Plants and Water," Edward Arnold, London.

Swain, R. W. (1975). Subsoiling. *In* "Soil Physical Conditions and Crop Production," Tech. Bull. 29, 189–204. Min. Agr. Fish Food. HMSO, London.

Swarzendruber, D. (1962). Non Darcy behavior in liquid saturated porous media. *J. Geophys. Res.* **67,** 5205–5213.

Swartzendruber, D. (1969). The flow of water in unsaturated soils. *In* "Flow Through Porous Media" (R. J. M. DeWiest, ed.), Chapter 6, pp. 215–292. Academic Press, New York.

Swartzendruber, D., and Hillel, D. (1973). The physics of infiltration. *In* "Physics of Soil, Water and Salts in Ecosystems" (A. Hadas *et al.,* eds.). Springer-Verlag, Berlin and New York.

Swartzendruber, D., and Hillel, D. (1975). Infiltration and runoff for small field plots under constant intensity rainfall. *Water Resources Res.* **11,** 445–451.

Szeicz, G., van Bavel, C. H. M., and Takami, S. (1973). Stomatal factor in water use and dry matter production by sorghum. *Agr. Meteorol.* **12,** 361–389.

Taber, S. (1930). The mechanics of frost heaving. *J. Geol.* **38,** 303–317.

Tackett, J. L., and Pearson, R. W. (1965). Some characteristics of soil crusts formed by simulated rainfall. *Soil Sci.* **99,** 407–413.

Tadmor, N. H., Evenari, M., Shanan, L., and Hillel, D. (1957). The ancient desert agriculture of the Negev. *Isr. J. Agric. Res.* **8,** 127–151.

Takagi, S. (1960). Analysis of the vertical downward flow of water through a two-layered soil. *Soil Sci.* **90,** 98–103.

Talsma, T. (1963). The control of saline ground water. *Med. Landb. Wageningen* **63** (10), 1–68.

Talsma, T. (1960). Comparison of field methods of measuring hydraulic conductivity. *Trans. Congr. Irrigat. Drainage* **4,** C145–C156.

Tanner, C. B. (1957). Factors affecting evaporation from plants and soils. *J. Soil Water Conserv.* **12,** 221–227.

Tanner, C. B. (1960). Energy balance approach to evapotranspiration from crops. *Soil Sci. Soc. Am. Proc.* **24,** 1–9.

Tanner, C. B. (1968). Evaporation of water from plants and soil. *In* "Water Deficits and Plant Growth." Academic Press, New York.

Tanner, C. B., and Elrick, D. E. (1958). Volumetric porous (pressure) plate apparatus for moisture hysteresis measurements. *Soil Sci. Soc. Am. Proc.* **22,** 575–576.

Tanner, C. B., and Lemon, E. R. (1962). Radiant energy utilized in evaporation. *Agron. J.* **54,** 207–212.

Tanner, C. B., and Mamaril, C. P. (1959). Pasture soil compaction by animal traffic. *Agron. J.* **51,** 329–331.

Tanner, C. B., and Pelton, W. L. (1960). Potential evapotranspiration estimates by the approximate energy balance method of Penman. *J. Geophys. Res.* **65,** 3391–3413.

Taylor, D. W. (1948). "Fundamentals of Soil Mechanics." Wiley, New York.

Taylor, G. I. (1950). The instability of liquid surfaces when accelerated in a direction perpendicular to their planes. *Proc. R. Soc. London Ser. A* **201,** 192–196.

Taylor, H. M., and Bruce, R. R. (1968). Effect of soil strength on root growth and crop yield

Taylor, H. M., and Gardner, H. R. (1963). Penetration of cotton seedling tap roots as influenced by bulk density, moisture content, and strength of soil. *Soil Sci.* **96,** 153–156.

Taylor, H. M., and Klepper, B. (1976). Water uptake by cotton root systems: An examination of assumptions in the single root model. *Soil Sci.* **120,** 57–67.

Taylor, H. M., and Ratliff, L. F. (1969). Root elongation rates of cotton and peanuts as a function of soil strength and water content. *Soil Sci.* **108,** 113–119.

Taylor, H. M., Robertson, G. M., and Parker, J. J. (1966). Soil strength-root penetration relations for medium- to coarse-textured soil materials. *Soil Sci.* **102,** 18–22.

Taylor, J. H., and Vanden Berg, G. E. (1966). Role of displacement in a simple traction system. *Trans. Am. Soc. Agr. Eng.* **9,** 10–13.

Taylor, S. A., and Cary, J. W. (1960). Analysis of the simultaneous flow of water and heat or electricity with the thermodynamics of irreversible processes. *Int. Congr. Soil Sci. Trans., 7th* **1,** 80–90.

Taylor, S. A., and Cary, J. W. (1964). Linear equations for the simultaneous flow of matter and energy in a continuous soil system. *Soil Sci. Soc. Am. Proc.* **28,** 167–172.

Taylor, S. A., and Jackson, R. D. (1965). Soil Temperature. *In* "Methods of Soil Analysis," pp. 331–344. Monograph 9, Am. Soc. Agron., Madison, Wisconsin.

Terry, C. W., and Wilson, H. M. (1953). The soil penetrometer in soil compaction studies. *Agr. Eng.* **34,** 831–834.

Terzaghi, K. (1953). "Theoretical Soil Mechanics." Wiley, New York.

Terzaghi, K., and Peck, R. B. (1948). "Soil Mechanics in Engineering Practice." New York.

Thom, A. S. (1975). Momentum, mass and heat exchange of plant communities. *In* "Vegetation and the Atmosphere" (J. L. Monteith, ed.). Academic Press, New York.

Thom, A. S., and Oliver, H. R. (1977). On Penman's equation. *Q. J. R. Meteorol. Soc.* **103,** 345–357.

Thorthwaite, C. W. (1948). An approach toward a rational classification of climate. *Geograph. Rev.* **38,** 55–94.

Timoshensko, S., and Goodier, J. N. (1951). "Theory of Elasticity." McGraw-Hill, New York.

Tiulin, A. F. (1928). Questions on soil structure: II. Aggregate analysis as a method for determining soil structure. *Perm. Agr. Exp. Sta. Div. Agr. Chem. Rep.* **2,** 77–122.

Todd, D. K. (1967). "Ground Water Hydrology," 6th printing. Wiley, New York.

Topp, G. C. (1969). Soil water hysteresis measured in a sandy loam and compared with the hysteresis domain model. *Soil Sci. Soc. Am. Proc.* **33,** 645–651.

Topp, G. C. and Miller, E. E. (1966). Hysteresis moisture characteristics and hydraulic conductivities for glass-bead media. *Soil Sci. Soc. Am. Proc.* **30,** 156–162.

Tovey, R. 1963. Consumptive use and yield of alfalfa in the presence of static water tables. Tech. Bull. 232. *Nev. Agric. Exp. Stn.*

Trouse, A. C. (1971). Soil Conditions as they affect plant establishment, development, and yield. *In* "Compaction of Agriculture Soils." Am. Soc. Agr. Eng., St. Joseph, Michigan.

Trouse, A. C. (1978). Tillage and traffic effects on soil. Paper presented at 33rd Annu. Meet. Soil Conserv. Soc. Am., Denver, Colorado.

Tschebotarioff, G. P. (1973). "Foundations, Retaining and Earth Structures." McGraw-Hill, New York.

van Bavel, C. H. M. (1949). Mean weight diameter of soil aggregates as a statistical index of aggregation. *Soil Sci. Soc. Am. Proc.* **14,** 20–23.

van Bavel, C. H. M. (1952). Gaseous diffusion and porosity in porous media. *Soil Sci.* **73,** 91–104.

van Bavel, C. H. M. (1963). Neutron scattering measurement of soil moisture: Development and current status. *Proc. Int. Symp. Humidity Moisture,* pp. 171–184, Washington, D. C.

van Bavel, C. H. M. (1966). Potential evaporation: The combination concept and its experimental verification. *Water Resources Res.* **2,** 455–467.

van Bavel, C. H. M. (1972). Soil temperature and crop growth. *In* "Optimizing the Soil Physical Environment Toward Greater Crop Yields" (D. Hillel, ed.), pp. 23–33. Academic Press, New York.

van Bavel, C. H. M., and Ahmed, J. (1976). Dynamic simulation of water depletion in the root zone. *Ecol. Modelling* **2,** 189–212.

van Bavel, C. H. M., and Hillel, D. (1975). A simulation study of soil heat and moisture dynamics as affected by a dry mulch. Proc. Summer Simulat. Conf., San Francisco, California. Simulation Councils, La Jolla, California.

van Bavel, C. H. M., and Hillel, D. (1976). Calculating potential and actual evaporation from a bare soil surface by simulation of concurrent flow of water and heat. *Agr. Meteorol.* **17,** 453–476.

van Bavel, C. H. M., and Myers, L. E. (1962). An automatic weighing lysimeter. *Agr. Eng.* **43,** 580–583.

van Bavel, C. H. M., Stirk, G. B., and Brust, K. J. (1968a). Hydraulic properties of a clay loam soil and the field measurement of water uptake by roots: I. Interpretation of water content and pressure profiles. *Soil Sci. Soc. Am. Proc.* **23,** 310–317.

van Bavel, C. H. M., Brust, K. J., and Stirk, G. B. (1968b). Hydraulic properties of a clay loam soil and the field measurement of water uptake by roots: II. The water balance of the root zone. *Soil Sci. Soc. Am. Proc.* **23,** 317–321.

Vanden Berg, G. E. (1958). Application of Continuum Mechanics to Compaction in tillable soils. Ph.D. thesis, Michigan State Univ., East Lansing, Michigan.

Vanden Berg, G. E. (1962). Requirements for a soil mechanic. *Trans. Am. Soc. Agr. Eng.* **4,** 234–238.

Vanden Berg, G. E., and Gill, W. R. (1962). Pressure distribution between a smooth tire and soil. *Trans. Am. Soc. Agr. Eng.* **5,** 105–107.

Van De Pol, R. M., Wierenga, P. J., and Nielsen, D. R. (1977). Solute movement in a field soil. *Soil Sci. Soc. Am. J.* **41,** 10–13.

van der Molen, W. H., (1956). Desalinization of saline soils as a column process. *Soil Sci.* **81,** 19–27.

van Duin, R. H. A. (1956). "On the Influence of Tillage on Conduction of Heat, Diffusion of Air, and Infiltration of Water in Soil," p. 62. Versl. Landbouwk, Onderz.

van Genuchten, M. Th. (1978a). Calculating the unsaturated hydraulic conductivity with a new closed-form analytical model. Publication of the Water Resour. Prog. Dept. Civ. Eng., September, 1978, Princeton Univ., Princeton, New Jersey.

van Genuchten, M. Th. (1978b). Numerical solutions of the one-dimensional saturated-unsaturated flow equation. Res. Rep. 78-WR-9, Water Resour. Prog., Dept. Civil Eng., Princeton Univ., Princeton, New Jersey.

van Genuchten, M. Th., and Wierenga, P. J. (1974). Simulation of one-dimensional solute transfer in porous media. *Agric. Exp. Stn. Bull.* 628, New Mexico State Univ., Las Cruces, New Mexico.

van Keulen, H., and Hillel, D. (1974). A simulation study of the drying-front phenomenon. *Soil Sci.* **118,** 270–273.

Van Olphen, H. (1963). "An Introduction to Clay Colloid Chemistry." Wiley (Interscience), New York.

van Rooyen, M., and Winterkorn, H. F. (1959). Structural and textural influences on thermal conductivity of soils. *Highway Res. Bd. Proc.* **38,** 576–621.

van Schilfgaarde, J. (1957). Approximate solutions to drainage flow problems. *In* "Drainage of Agricultural Lands," pp. 79–112. *Am. Soc. Agron.,* Monograph 7.

van Schilfgaarde, J. (ed.) (1974). "Drainage for Agriculture." Monograph 17, Am. Soc. Agron., Madison, Wisconsin.

van Wijk, W. R., and de Vries, D. A. (1963). Periodic temperature variations in homogeneous soil. *In* "Physics of Plant Environment" (W. R. van Wijk, ed.). Noth-Holland Publ., Amsterdam.

Varga, R. S. (1962). "Matrix Iterative Analysis." Prentice Hall, Englewood Cliffs, New Jersey.

Veihmeyer, F. J., and Hendrickson, A. J. (1927). Soil moisture conditions in relation to plant growth. *Plant Physiol.* **2,** 71–78.

Veihmeyer, F. J., and Hendrickson, A. H. (1931). The moisture equivalent as a measure of the field capacity of soils. *Soil Sci.* **32,** 181–193.

Veihmeyer, F. J., and Hendrickson, A. H. (1949). Methods of measuring field capacity and wilting percentages of soils. *Soil Sci.* **68,** 75–94.

Veihmeyer, F. J., and Hendrickson, A. H. (1950). Soil moisture in relation to plant growth. *Ann. Rev. Plant Physiol.* **1,** 285–304.

Veihmeyer, F. J., and Hendrickson, A. H. (1955). Does transpiration decrease as the soil moisture decreases? *Trans. Am. Geophys. Un.* **36,** 425–448.

Vennard, J. K. (1961). "Elementary Fluid Mechanics," 4th ed. Wiley, New York.

Verwey, E. J. W., and Overbeek, J. Th. G. (1948). "Theory of the Stability of Lyophobic Colloids." Elsevier, New York.

Viets, F. G., Jr. (1962). Fertilizers and the efficient use of water. *Adv. Agron.* **14,** 228–261.

Viets, F. G., Jr. (1966). Increasing water use efficiency by soil management. *In* "Plant Environment and Efficient Water Use" (W. H. Pierre, D. Kirkham, J. Pesek, and R. Shaw, eds.), pp. 259–274. Am. Soc. Agron. and Soil Sci. Soc. of Am., Madison, Wisconsin.

Vilain, M. (1963). L'aeration du sol. Mise au point bibliographique. *Ann. Agron.* **14,** 967–998.

Visser, W. C. (1959). Crop Growth and Availability of Moisture. Inst. of Land and Water Management, Wageningen, Netherlands, Tech. Bull. No. 6.

Visser, W. C. (1966). Progress in the knowledge about the effect of soil moisture content on plant production. Inst. Land Water Management, Wageningen, Netherlands, Tech. Bull. 45.

Vomocil, J. A. (1954). In situ measurement of soil bulk density. *Agr. Eng.* **35,** 651–654.

Vomocil, J. A. (1965). Porosity. *In* "Methods of Soil Analysis," pp. 299–314. Am. Soc. Agron., Madison, Wisconsin.

Vomocil, J. A., and Flocker, W. J. (1961). Effect of soil compaction on storage and movement of soil air and water. *Trans. Am. Soc. Agr. Eng.* **4,** 242–246.

Vomocil, J. A., Waldron, L. J., and Chancellor, W. J. (1961). Soil tensile strength by centrifugation. *Soil Sci. Soc. Am. Proc.* **25,** 176–180.

Voorhees, W. B., and Hendrick, J. G. (1977). Compaction—good and bad effects on energy needs. *Crops and Soils Magazine* **29,** 11–13.

Vries, D. A. de. (1963). Thermal properties of soils. *In* "Physics of Plant Environment." North-Holland, Amsterdam.

Wadleigh, C. H. (1946). The integrated soil moisture stress upon a root system in a large container of saline soil. *Soil Sci.* **64,** 225–238.

Wadleigh, C. H., and Ayers, A. D. (1945). Growth and biochemical composition of bean plants as conditioned by soil moisture tension and salt concentration. *Plant Physiol.* **20,** 106–132.

Wagenet, R. J., and Jurinak, J. J. (1978). Spatial variability of soluble salt content in a Mancos shale watershed. *Soil Sci.* **126,** 342–349.

Waggoner, P. E., Miller, P. M., and De Roo, H. C. (1960). "Plastic Mulching—Principles and Benefits." Bull. No. 634, Connecticut Agr. Exp. Sta., New Haven, Connecticut.

Wang, F. C., and Lakshminarayana, V. (1968). Mathematical simulation of water movement through unsaturated nonhomogeneous soil. *Soil Sci. Soc. Am. Proc.* **32,** 329–334.

Warrick, A. W., and Amoozegar-Fard, A. (1979). Infiltration and drainage calculations using spatially scaled hydraulic properties. *Water Resour. Res.* **15,** 1116–1120.

Warrick, A. W., Mullen, G. J., and Nielsen, D. R. (1977). Scaling field measured soil hydraulic properties using a similar media concept. *Water Resour. Res.* **13,** 355–362.

Watson, K. K. (1966). An instaneous profile method for dertermining the hydraulic conductivity of unsaturated porous materials. *Water Resources Res.* **2,** 709–715.

Wesseling, J. (1962). Some solutions of the steady-state diffusion of carbon dioxide through soils. *Neth. J. Agr. Sci.* **10,** 109–117.

Wesseling, J. (1974). Crop growth and wet soils. *In* "Drainage for Agriculture" (J. van Schilfgaarde, ed.), pp. 7–38. Monograph 17, Am. Soc. Agron., Madison, Wisconsin.

Whisler, F. D., and Millington, R. J. (1968). Analysis of steady-state evapotranspiration from a soil column. *Soil Sci. Soc. Am. Proc.* **32,** 167–174.

Whisler, F. D., Klute, A., and Millington, R. J. (1968). Analysis of steady-state evapotranspiration from a soil column. *Soil Sci. Soc. Am. Proc.* **32,** 167–174.

White, N. F., Duke, H. R., Sunada, D. K., and Corey, A. T. (1970). Physics of desaturation in porous materials. *J. IR Div.,* ASCE Proc. **IR-2,** 165–191.

White, N. F., Duke, H. R., Sunada, D. K., and Corey, A. T. (1972). Boundary effects in the desaturation of porous media. *Soil Sci.* **113,** 7–12.

Whittig, L. D. (1965). X-ray diffraction techniques for mineral identification and mineralogical composition. *In* "Methods of Soil Analysis, Part I," Chap. 49, pp. 671–697. ASA, Madison, Wisconsin.

Wiegand, C. L., and Taylor, S. A. (1961). Evaporative Drying of Porous Media, Spec. Rep. 15, Agr. Exp. Sta., Utah State Univ., Logan, Utah.

Wierenga, P. J., and de Wit, C. T. (1970). Simulation of heat transfer in soils. *Soil Sci. Soc. Am. Proc.* **32,** 326–328.

Wiersum, L. K. (1957). The relationship of the size and structural rigidity of pores to their penetration by roots. *Pl. Soil* **9,** 75–85.

Wilkinson, G. E., and Klute, A. (1959). Some tests of the similar media concept of capillary flow: II. Flow systems data. *Soil Sci. Soc. Am. Proc.* **22,** 432–437.

Willardson, L. S., and Hurst, R. L. (1965). Sample size estimates in permeability studies. *J. Irr. Am. Soc. Civil Engr.* **91,** 1–9.

Willey, C. R., and Tanner, C. B. (1963). Membrane-covered electrode for measurement of oxygen concentration in soil. *Soil Sci. Soc. Am. Proc.* **27,** 511–515.

Williams, P. J. (1966). Pore pressures at a penetrating frost line and their prediction. *Geotechique* **16,** 187–208.

Williams, P. J. (1972). Use of the ice-water surface tension concept in engineering practice. *Highw. Res. Rec.* **393,** 19–29.

Williams, P. J., and Burt, T. P. (1974). Measurement of hydraulic conductivity of frozen soils. *Can. Geotech. J.* **11,** 647–650.

Willis, W. O. (1960). Evaporation from layered soils in the presence of a water table. *Soil Sci. Soc. Am. Proc.* **24,** 239–242.

Wind, G. P. (1955). Flow of water through plant roots. *Neth. J. Agric. Sci.* **3,** 259–264.

Wind, G. P. (1959). A field experiment concerning capillary rise of moisture in a heavy clay soil. *Neth. J. Agric. Sci.* **3,** 60–69.

Winger, R. J. (1960). In-place permeability tests and their use in subsurface drainage. Off. of Drainage and Ground Water Eng., Bur. of Reclamation, Denver, Colorado.

Winterkorn, H. F. (1936). Surface chemical factors influencing the engineering properties of soil. *Proc. Nat. Res. Council Highway Res. Board Ann. Meeting 16th, Washington, D. C.,* pp. 293–301.

Wise, M. E. (1952). Dense random packing of unequal spheres. *Philips Res. Rep.* **7,** 321–343.

Wittmus, H., Olson, L., and Delbert, L. (1975). Energy requirements for conventional versus minimum tillage. J. Soil Water Conserv. **30,** 72–75.

Wittmus, H. D., Triplett, G. B., Jr., and Greb, B. W. (1973). Concepts of conservation tillage using surface mulches. *In* "Conservation Tillage," pp. 5–12. Soil Conserv. Soc. Am., Ames, Iowa.

Wolf, J. M. (1968). The Role of Root Growth in Supplying Moisture to Plants. Unpublished Doctoral Dissertation, Univ. of Rochester, Rochester, New York.

Wong, J. Y. (1967). Behavior of soil beneath rigid wheels. *J. Agr. Eng. Res.* **12,** 257–269.

Wooding, R. A. (1965). A hydraulic model for the catchment-stream problem. 1. Kinematic wave theory. *J. Hydrol.* **3,** 254–267.

Wooding, R. A. (1969). Growth of fingers at an unstable diffusing interface in a porous medium or Hele-Shaw cell. *J. Fluid Mech.* **39,** 477–495.

Woodside, W. (1958). Probe for thermal conductivity measurement of dry and moist materials. *Am. Soc. Heating and Air-Conditioning Eng. J. Sect., Heating, Piping, and Air Conditioning,* 163–170.

Yamaguchi, M., Howard, F. D., Hughes, D. L., and Flocker, W. J. (1962). *Soil Sci. Soc. Am. Proc.* **26,** 512–513.

Yang, S. J., and DeJong, E. (1971). Effect of soil water potential and bulk density on water uptake patterns and resistance to flow of water in wheat plants. *Can. J. Soil Sci.* **51,** 211–220.

Yatsuk, E. P. (1971). "Rotary Soil Working Machines" (Rotatsionnye Pochvoobrabatyvayushchie Mashiny). Machine Construction, Moscow.

Yavorsky, B., and Detlaf, A. (1972). "Handbook of Physics." Mir, Moscow.

Yoder, R. E. (1936). A direct method of aggregate analysis and a study of the physical nature of erosion losses. *J. Am. Soc. Agron.* **28,** 337–351.

Yong, R. N., and Osler, J. C. (1966). On the analysis of soil deformation under a moving rigid wheel. *Proc. Int. Conf. Soc. Terrain-Vehicle Sys.* p. 341.

Yong, R. N., and Warkentin, B. P. (1975). "Soil Properties and Behaviour." Elsevier, Amsterdam.

Youker, R. E., and McGuinness, J. L. (1956). A short method of obtaining mean weight-diameter values of aggregate analyses of soils. *Soil Sci.* **83,** 291–294.

Youngs, E. G. (1964). An infiltration method of measuring the hydraulic conductivity of unsaturated porous materials. *Soil Sci.* **109,** 307–311.

Youngs, E. G. (1958a). Redistribution of moisture in porous materials after infiltration. *Soil Sci.* **86,** 117–125.

Youngs, E. G. (1958b). Redistribution of moisture in porous materials after infiltration. *Soil Sci.* **86,** 202–207.

Youngs, E. G. (1960a). The drainage of liquids from porous materials. *J. Geophys. Res.* **65,** 4025–4030.

Youngs, E. G. (1960b). The hysteresis effect in soil moisture studies. *Trans. Int. Soil Sci. Congr. 7th, Madison* **1,** 107–113.

Youngs, E. G., and Towner, G. E. (1970). Comment on paper by Philip (1969). *Water Resources Res.* **6,** 1246.

Zelenin, A. N., Balovnev, V. I., and Kerov, I. P. (1975). "Machines for Earthmoving Work" (Mashiny dlya Zemlyanteykh Rabot). Machine Construction, Moscow. 423 p.

# Index

## A

## B

## C

Colloids and colloidal systems, 7, 21, 25
Compaction of soil, 4, 10, 77, 136, 176–
191, 201, 257
control of, 189–191
density moisture curves, 179, 180
occurrence in agriculture fields, 180–181
proctor test, 180
in relation to wetness, 177–180
by vibration, 43, 177
Conduction of heat in soil, 158–160
Conservation of energy, 304
Conservation of mass, 114, 240, 304
Consolidation (soil), 191–196
coefficient of, 195
preconsolidation, 195
theory of (Terzaqhi), 194
Contact angle (water on solids), 79
Continuity equation, 114, 141, 158
Core sampler, 46
Cracks and cracking, 4, 10, 224, 270
Crust and crusting (of soil), 51–52, 53,
180, 201, 213, 215, 216, 220, 224
infiltration through, 226–228
Crust test (hydraulic conductivity), 120–
122
Cultivation (tillage), 44, 53, 188
primary and secondary, 201, 204

**D**

Damping depth, temperature wave, 168
Darcy's law, 92–95, 98, 115, 140, 219, 237,
262, 276
Deep percolation, 237
Degree of saturation, 11, 12, 16, 17
Density of solids, 9
Diffusion equation, 124
*see also* "Fick's law"
Diffusion of gases, 135, 142–144
coefficient of, 142–143, 151
Diffusivity, hydraulic, 114–116, 220, 221,
281
relation to wetness, 116, 117
weighted mean, 253, 280, 285
Disk plow, 205
Disperse systems, 7
Dispersing agent, 31
Double layer, electrostatic, 24, 25, 27, 28
Drainable porosity, 252, 254
Drainage
surface, 256
groundwater, 256–265
Drainage front, 253

Drains, depths and spacings, 264
Drawdown of water-table, 251
Drip irrigation, 104
Drying front, 277
Drying of bare soils, 275–278
analysis of, 278–281
stage of, 275–276
Dryland farming, 245
Dupuit–Forchheimer assumptions, 256

**E**

Effective stress, 193
Electrical resistance blocks, 60–61
Eluviation and illuviation, 13
Emerson's model of soil aggregates, 45–
46
Emissivity coefficient, 156
Energy balance, 311, 313
Energy conservation law, 158
Energy state (potential) of soil water, 64–
67, 74
capillary potential, 69
gravitational potential, 67–68
matric potential, 69–75
osmotic potential, 71–73
pneumatic potential, 70
pressure potential, 68–70, 73
submergence potential, 68
total potential, 66
units and dimensions, 73–75
Energy transfer, modes of, 156–157
Erosion, 3, 41, 48, 49, 53, 200
Evaporation from bare soils, 268–282
physical conditions, 269–271
in presence of shallow water-table, 271,
274
reduction of, 281–282
Evaporative demand, 269, 288
Evaporativity, atmospheric, 269, 273, 275,
279, 280, 286, 296
Evapotranspiration, 248, 262, 268, 301
actual, 307
potential, 247, 316–319
Exchangeable sodium percentage (ESP),
52
Exchangeable ions, 24, 25, 28

**F**

Fick's law (diffusion), 95, 115, 116, 142,
151
Field air capacity, 137

*I do not know what I may appear to the world; but to myself I seem to have been only like a boy playing on the seashore, diverting myself in now and then finding a smoother pebble or a prettier shell than ordinary, whilst the great ocean of truth lay all undiscovered before me.*

*Sir Isaac Newton*
*1642 – 1727*